ADVANCES IN CHEMICAL PHYSICS

VOLUME XIX

Advances in

CHEMICAL PHYSICS

EDITED BY

I. PRIGOGINE

University of Brussels, Brussels, Belgium

AND

STUART A. RICE

Department of Chemistry
and
The James Franck Institute
The University of Chicago
Chicago, Illinois

VOLUME XIX

WILEY—INTERSCIENCE

A DIVISION OF JOHN WILEY AND SONS
NEW YORK · LONDON · SYDNEY · TORONTO

Library of Congress Catalog Card Number: 58-9935

ISBN O 471 69924 1

Printed in the United States of America

10 9 8 7 6 5 4 3 2 1

INTRODUCTION

In the last decades, chemical physics has attracted an ever-increasing amount of interest. The variety of problems, such as those of chemical kinetics, molecular physics, molecular spectroscopy, transport processes, thermodynamics, the study of the state of matter, and the variety of experimental methods used, makes the great development of this field understandable. But the consequence of this breadth of subject matter has been the scattering of the relevant literature in a great number of publications.

Despite this variety and the implicit difficulty of exactly defining the topic of chemical physics, there are a certain number of basic problems that concern the properties of individual molecules and atoms as well as the behavior of statistical ensembles of molecules and atoms. This new series is devoted to this group of problems which are characteristic of modern chemical physics.

As a consequence of the enormous growth in the amount of information to be transmitted, the original papers, as published in the leading scientific journals, have of necessity been made as short as is compatible with a minimum of scientific clarity. They have, therefore, become increasingly difficult to follow for anyone who is not an expert in this specific field. In order to alleviate this situation, numerous publications have recently appeared which are devoted to review articles and which contain a more or less critical survey of the literature in a specific field.

An alternative way to improve the situation, however, is to ask an expert to write a comprehensive article in which he explains his view on a subject freely and without limitation of space. The emphasis in this case would be on the personal ideas of the author. This is the approach that has been attempted in this new series. We hope that as a consequence of this approach, the series may become especially stimulating for new research.

Finally, we hope that the style of this series will develop into something more personal and less academic than what has become the standard scientific style. Such a hope, however, is not likely to be completely realized until a certain degree of maturity has been attained—a process which normally requires a few years.

At present, we intend to publish one volume a year, and occasionally several volumes, but this schedule may be revised in the future.

In order to proceed to a more effective coverage of the different aspects of chemical physics, it has seemed appropriate to form an editorial board. I want to express to them my thanks for their cooperation.

I. PRIGOGINE

CONTRIBUTORS TO VOLUME XIX

RODERICK K. CLAYTON, Division of Biological Sciences and Department of Applied Physics, Cornell University, Ithaca, New York

LEWIS FRIEDMAN, Department of Chemistry, Brookhaven National Laboratory, Upton, New York

GEORGE A. GRAY, Oregon Graduate Center for Study and Research, Portland, Oregon

EDWARD H. KERNER, Physics Department, University of Delaware, Newark, Delaware

JOHN C. LIGHT, James Franck Institute and the Department of Chemistry, The University of Chicago, Chicago, Illinois

G. NICOLIS, Faculté des Sciences de l'Université Libre de Bruxelles, Bruxelles, Belgium

BRYAN G. REUBEN, Department of Chemistry, Brookhaven National Laboratory, Upton, New York

CONTENTS

ADVANCES IN CHEMICAL PHYSICS

VOLUME XIX

QUANTUM THEORIES OF CHEMICAL KINETICS

JOHN C. LIGHT

James Franck Institute and the Department of Chemistry
The University of Chicago, Chicago, Illinois

CONTENTS

I. INTRODUCTION

Studies of simple gas phase reactions between neutral and ionized species are now being carried out at many centers with increasingly sophisticated experimental and theoretical techniques. The goal is, of course, to obtain sufficient knowledge such that reasonably detailed information about gas phase reactive collisions is easily available for use in many areas, ranging from chemical lasers and the study of upper atmospheric phenomena to cracking of petroleum distillates. One type of work being avidly pursued in the academic community is directed toward the classification and detailed understanding of very simple gas phase reactions, usually those involving only a few atoms or ions (usually three). The obvious reason for this is scientific rather than applied; such systems provide the simplest reactions which are interesting and for which we, in the present state of the art, feel that real progress toward understanding the dynamics of such collision processes can be made.

The information desired about simple reactive systems varies, but the maximum information obtainable theoretically or experimentally would consist of the (angular) differential cross sections for scattering from each

1

internal state i of channel α to every state f of channel β at any relative energy E. Here α and β refer to the chemical species involved as reactants and products respectively. From such differential cross sections, one can obtain all other quantities desired by summation or integration: the total cross section from α, i to β, f is obtained by integration over angles; the total reactive cross section at energy E is then given by summing over final states and averaging over initial states; and the rate constant is found by first performing the velocity average of $v\sigma$ for each pair of internal states, then averaging and summing over initial and final states respectively. Thus the fundamental quantities which we desire to calculate are the reactive cross sections for any reaction at any energy.

The status of experimental studies for simple gas phase systems has been reviewed elsewhere[1-3] and will not be discussed here. What we shall be concerned with is some of the theoretical work directed toward yielding the above information. The problems are not very tractable mathematically. So although the fundamental theory of reactive collision processes is well understood, both classically and quantum mechanically,[4-6] many simplifications, approximations, and models have been introduced in order to obtain results. These procedures vary greatly in complexity and validity, and it is only because no single approach has been preeminently successful that an article such as this is justifiable. In this paper we attempt to point out those quantum approaches that, in the biased eye of the author, seem most promising at the moment.

The question that naturally arises at this point is why someone is not just doing everything correctly starting from the Schrödinger equation. The answer seems to be twofold. First, at the moment we do not know how, even though attempts (some of which are described later) are being made in this direction. Second, and more important, is a question of relative values. Kinetics today is still only a semiquantitative science. Absolute cross sections known to within 50% are rare. The interaction potentials between the atoms involved in chemical reactions are not reliably known for short ranges. Activation energies are still largely derived from thermal rate constant measurements, and the actual Born–Oppenheimer potential barriers are uncertain to a kilocalorie or two in the best of cases. Thus the systems we must deal with have Hamiltonians which are uncertain within ranges which may strongly affect the kinetics. Any dynamical calculations done on such systems must reflect these uncertainties, and thus the value of very accurate dynamical calculations is greatly reduced. Such calculations cannot be viewed as yielding numbers of great significance in themselves; they are, at present, merely tools, probing potential energy surfaces for features of significance, and correlating scattering results with particular types of surfaces. In view of this, very extensive (and expensive) calculations must be used sparingly, and much room is still available for simple models which predict reactive behavior well. In addition, the cost of accurate direct quantum calculations goes up as approximately the third power of the number of states involved and thus will be limited, in the next few years at least, to relatively light simple systems at low energies. There is always the possibility that an advance in the formal theory will be made which will greatly reduce these computational difficulties, but such avenues are relegated to speculation at the end of this paper.

Of the four general approaches to the general problem now in use, transition state theory, statistical and dynamic models; classical trajectory calculations; quantum calculations with coupled equations; and quantum approximations, we shall discuss in detail only the last two. The other approaches have been studied in more detail and have already been reviewed.[7,8] Although to date the classical trajectory calculations have yielded by far the most accurate and pertinent results for cross sections and angular and energy distributions,[9,10,11] there are problems which these studies leave unanswered. Only quantum methods can be used to give, for instance, the effects of tunneling or of collision processes involving electronically excited species in which transitions between electronic energy surfaces may occur. Both the classical and accurate quantum calculations suffer, however, from excessive computational requirements. They both may use many hours of computer time to yield detailed numerical results. This severely limits their present utility as predictive tools, and the search for simpler, more approximate methods which yield the detailed information desired (this usually excludes the transition state theory) should not be abandoned.

Since Levine[4] has reviewed very well and very extensively the status of quantum approaches to chemical dynamics as well as many of the model theories which were published previous to 1968, we shall be concerned in this article largely with work published in the last two years. This encompasses approximately twenty-five articles devoted to a wide variety of approaches, but we shall also refer to some of the more germane work published earlier.

As in many fields, the work can readily be divided into approximate and model approaches, and attempts at more realistic numerical calculations. In Section II we discuss the perturbation calculations and model calculations of reactive systems on highly simplified potential energy surfaces. These calculations are useful because they exhibit, in relatively simple fashion, the most important characteristics to be expected from more realistic approaches.

In Section III some of the more detailed calculations on more realistic potential energy surfaces are discussed. In particular, the use of reaction coordinates in such calculations is investigated, and some of the more important results, particularly for the problem of tunneling, are reviewed.

In Section IV an attempt is made to define some of the more fundamental problems which must be resolved before we can really apply quantum mechanics to the more realistic problems of interest. A few of the more promising approaches, none of which has yet been tried, are presented with the hope that they may be useful in the future.

II. MODEL POTENTIALS AND PERTURBATION THEORY

Although it has been about forty years since Michael Polanyi's experiments on alkali halide diffusion flames were performed and published, many of his suggestions[12] about the relationships between potential energy surfaces and product energy distributions have only recently been put on a reasonably quantitative basis.[9] The characteristics of the collinear potential energy surface which give rise to vibrational excitation of products in bimolecular exchange reactions have now been studied rather extensively by classical mechanics, but the purely quantum effects of tunneling and the discrete vibrational energy levels have only very recently been studied. Since the use of quantum scattering theory for more or less realistic interactions leading to bimolecular exchange reactions is in its infancy, it is presumptuous to write a review article about the subject. There are, however, some aspects which both (a) have been studied and (b) must probably be taken into account in future similar work, which should make this section useful.

In considering bimolecular exchange reactions, one of the fundamental problems in quantum approaches is the choice of basis sets. The Hamiltonian for a reaction of the type

$$A + BC \longrightarrow AB + C$$
$$(\alpha) \qquad\qquad (\beta)$$

can be written most simply in two asymptotic forms

$$H = H_{0\alpha} + T_\alpha + V_\alpha \qquad\qquad (1a)$$

$$H = H_{0\beta} + T_\beta + V_\beta \qquad\qquad (1b)$$

where $H_{0\alpha}$ is the Hamiltonian for the internal motions of A and BC, $H_{0\beta}$ is the corresponding internal Hamiltonian for AB and C, T_α and T_β are the kinetic energy operators for the relative motion in the respective systems, and the V's are the appropriate interaction potentials. In the limit, as the particles separate and the potentials go to zero, the most appropriate basis sets for the wave functions of reactants and products are

$$\Psi_{\alpha i} = \phi_{\alpha i}(r_\alpha)\psi_{\alpha i}(R_\alpha) \tag{2}$$

and

$$\Psi_{\beta i} = \phi_{\beta i}(r_\beta)\psi_{\beta i}(R_\beta)$$

where

$$H_{0\alpha}\phi_{\alpha i}(r_\alpha) = \varepsilon_{\alpha i}\phi_{\alpha i}(r_\alpha)$$

and $\psi_{\alpha i}(R_\alpha)$ is the appropriate free particle wave function for the configuration α. In the standard integral representation from formal scattering theory,[4-6] it is permissible to use either basis, and one normally writes

$$\frac{d\sigma_{\beta\alpha}}{d\Omega} = \frac{\mu_\beta \mu_\alpha k_\beta}{(2\pi\hbar)^2 k_\alpha} |\langle \Psi_\beta | V_\beta | \Psi_\alpha^+ \rangle|^2 \tag{3}$$

where Ψ_α^+ is the solution of the integral equation

$$\Psi_\alpha^+ = \Psi_\alpha^{(0)} + [E_\alpha - H_\alpha + i\eta]^{-1} V_\alpha \Psi_\alpha^+ \tag{4}$$

and $\Psi_\alpha^{(0)}$ is the initial state incoming wave and $[E_\alpha - H_\alpha + i\eta]^{-1}$ represents the integral of the general Green's function operation for outgoing waves in all states of the channel α.

Although these equations are exact, there are several problems which appear when an accurate solution is attempted. First, the number of open (positive kinetic energy) states which may be involved can be large, and, perhaps more disturbing, quite accurate solutions of equation (4) are required in regions in which the overlap integrals of equation (3) are most important. These regions are likely to be those in which the internal α states are strongly perturbed, that is, regions in which both V_α is large and in which Ψ_α^+ can only be represented by a *large* linear combination of open and closed states of the α channel.

It is for such reasons that an accurate numerical solution of the equations in this form has never been attempted for realistic potentials, even for collinear collisions. It seems likely, however, that the recent work of Kouri et al.[13,14] in which a noniterative method of solution of the integral equations of scattering was developed, will soon be applied to simple reactive scattering problems. Hopefully, the problems mentioned above will not prove too serious.

During the past two years, a number of papers devoted to the problem at hand have been published. These vary, naturally, both in the approach to the problem and in the types of systems treated, but they can be divided roughly into four types: perturbation calculations; model soluble potentials; collinear calculations on "realistic" potential surfaces; and analysis of the problem and numerical methods not yet applied. The first two will be discussed in this section.

A. Perturbation Theory

The exact T matrix element in equation (3) between a reactant (α) and product state (β) is given by

$$T_{\beta\alpha} = \langle \Psi_\beta | V_\beta | \Psi_\alpha^+ \rangle \tag{5}$$

where Ψ_β is the solution of the noninteracting Hamiltonian of the final state, and Ψ_α^+ is the solution of the Lippmann–Schwinger equation (4), and V_β is defined in equation (1). In a perturbation approach to this problem, Ψ_α^+ is approximated by some known function which, it is assumed, is a good approximation in the region in which the integrand of equation (5) is large. If perturbation theory is to be valid, the approximation to Ψ_α^+ must consist basically of the product of the distorted initial translational wave and a wave function of the initial internal state (which may be distorted also). When such a function is used for Ψ_α^+, the approximation is called the distorted wave (perturbed stationary state) Born approximation.

Several such calculations have been done with varying degrees of sophistication concerning the distortion of the waves. The two most realistic calculations were performed by Micha[15] and by Karplus and Tang,[16] both for the system $H + H_2$ which is particularly favorable at low energies since only the ground vibrational state is accessible energetically.

Micha used a very simplified form of the interaction potential and simple forms for the distorted waves in the interaction region in order to obtain analytic results. He found quite reasonable threshold behavior of the cross section, with the total energy at threshold somewhat larger than the minimum energy path over the barrier. Above this threshold, however, the calculated cross section shows wide oscillations with energy, which are probably artifacts of the approximations used.

Karplus and Tang[16] did a much more sophisticated calculation, comparing the results obtained with unperturbed and perturbed internal (vibrational) states as well as obtaining the differential cross sections as a function of the energy and rotational states. Some qualitative features seem to be correct: strong backward peaking of the differential cross section, reasonably smooth (two peaks) behavior of the total cross section with energy, and rotational excitation of the product. However, the total cross section ($J = 0$ to $J' = 0$) changes by a factor of 20 when the distortion of the internal state is introduced, even though the energy was well above threshold. This raises, of course, serious questions as to the validity of such perturbation methods since no reliable means are available (within perturbation theory) to check on the accuracy.

Another perturbation approach which would seem to offer the possibility of greater accuracy is to use the adiabatic basis set defined in terms of reaction coordinates. In this way the distortion of the internal states is probably included rather accurately, but unless the system is, in fact, rather adiabatic vibrationally, first order perturbation theory will still fail.

B. Model Potentials

It is possible to construct a potential which is simple enough for "analytic" solutions of Schrödinger's equation to be found for collinear systems. These calculations were first attempted by Hulbert and Hirschfelder[17] in 1943, but a discrepancy was recently found by Tang, Kleinman, and Karplus.[18] Even more recently, the original method of Hulbert and Hirschfelder was used with a complete basis set[19] to solve the problem,[20] now in agreement with Tang et al.

The potential used for three atoms on a line is the most complex which allows "exact" solution. (See Figure 1.) Even in this very simple case the

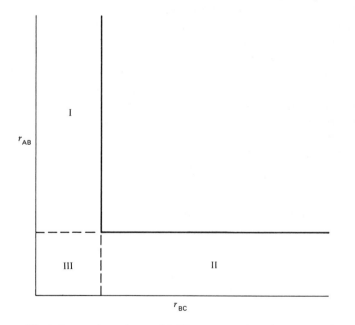

Fig. 1. Square channel potential. V is a constant in regions I, II, and III, and infinite elsewhere. The apparent origin is at the hard sphere diameters.

solution is nontrivial, requiring the solution of a large number of algebraic equations to obtain the coefficients of the wave function in regions
I and III. Because of the discontinuous nature of the potential and the
well known diffraction effects which are produced in quantum wave
functions by this discontinuity, this model may be expected to show strong
quantum effects. In fact, they are found in great abundance, with sharp
resonance spikes in the cross sections (transition probabilities) being very
evident. In the calculations of Tang et al.,[18] the problem was solved for a
variety of possible potentials and mass ratios over an energy range in which
one or two bound vibrational states were open.

Two interesting facts emerge from these studies. First, the highest
transmission coefficients to the ground states of the products occur at
energies just below the threshold for production of the first excited vibrational state of the product. This resonance behavior is suggestive of the
experimental results obtained for electron scattering by atoms, but has
never been observed in molecular beam reactive scattering. The second
interesting point is the amount of tunneling in comparison with classical
calculations of Jepsen and Hirschfelder.[21] The problem of tunneling in
two dimensional (collinear) systems will be discussed more fully later, but
in the present studies it was also found to be significant as compared with
classical calculations, and for energies at which the local translational
energy becomes negative even though the total energy (translational plus
vibrational) remained positive.

Although this type of potential allows computations to be done accurately with reasonable ease, there is some question as to whether the qualitative features (such as resonances) will remain so well defined when smoother,
more "realistic" potentials are used for the collinear reactive scattering
problem. In view of the results to be presented in the next section, it seems
likely that many of these features will disappear and that the transmission
coefficients as a function of energy will be rather smooth.

III. NUMERICAL CALCULATIONS FOR COLLINEAR COLLISIONS

This section is differentiated from the preceding for two reasons. First,
the method of solution of the Schrödinger equation is significantly different when smooth potentials are used, and second, the potentials used are
considered to be much more realistic. In these calculations, in fact, it is
only the dimensionality of the problem which prevents the results from
being compared directly with experiment. This is, of course, a strong con-

straint, but at present it seems to be necessary to obtain solutions. Within the next few years the problem will undoubtedly be done in more generality. The numerical methods of solving the Schrödinger equation for three atoms constrained to lie on a line can be divided into two broad categories. The problem has two independent degrees of freedom and thus the Schrödinger equation is a partial differential equation in two variables. The first method consists of writing the Schrödinger equation as a set of finite difference equations on a two dimensional grid of points and solving these, subject to the appropriate asymptotic boundary conditions, by some standard technique. The second method involves an expansion of the wave function in terms of a complete basis set for one of the variables, then solving the resultant coupled differential equations by appropriate numerical techniques.

A. Finite Difference Methods

The first attempt to solve the time-independent Schrödinger equation accurately by numerical techniques for a collinear reactive system was made by Mortensen and Pitzer[22] for the H and H_2 system. They used the finite difference method of solving the Schrödinger equation (in the form of a two-dimensional partial differential equation) in the interaction region, and then iterated the solution until the boundary values matched the incoming and outgoing WKB solutions. Since only a single vibrational state was open at the energies considered, the boundary value problem was not too difficult, and they were able to obtain reasonably accurate solutions. The potential surface used was quite realistic, being basically a Sato[23] potential.

More recently Mortensen[24] studied the various isotopic $H + H_2$ reactions by the same method to obtain more information on the permeabilities (transmission coefficients) as a function of energy. Only the ground vibrational state was needed over the range of energies considered for most of the calculations, but in the $D + D_2$ case, the first excited vibrational state was also included. Although the energy range was from 9 to 18 kcal/mole, no evidence of quantum oscillations in the probability of reaction versus energy was observed. The reaction probabilities followed a smooth sigmoidal curve from zero near 9 kcal to near unity at about 14 kcal, and then started to fall slowly. The effect of quantum mechanical tunneling (described in more detail later) was, of course, found to be important particularly for the rate constant at lower temperatures ($<500°K$).

In both this calculation and the earlier one by Mortensen and Pitzer, a "bending correction" was made to account in some way for the extra degrees of freedom not explicitly considered. This correction essentially adds the zero point energy of the bending modes (as a function of the nuclear separations) to the linear potential energy surface, thus contributing slightly to the effective potential. In these calculations the role of symmetry with respect to exchange of identical nuclei was not considered, a common procedure which has never been fully justified.

An alternative, but related method of solution was recently proposed by Diestler and McKoy.[25] In this method, the finite difference equations corresponding to the two-dimensional Schrödinger equation are solved several times for different arbitrary sets of boundary conditions. Each solution is then projected onto the asymptotic wave functions at the boundary of the interaction region, and appropriate linear combinations of these solutions are taken such that the true boundary conditions are satisfied. The method of solution is again by a two-dimensional finite difference equation, but in this method the very large set of algebraic simultaneous equations derived from the finite difference equations is solved. Diestler and McKoy applied their method to both the infinite wall, flat channel potential mentioned in the last section and to a "harmonic channel" problem again with no barrier or exothermicity.

On balance, these methods seem to have disadvantages which make them somewhat less useful than the methods involving reaction coordinates described below. The disadvantage stems from the fact that the values of the wave function are determined over a two-dimensional mesh of points. For high accuracy, this mesh must be fine, and large numbers of linear equations must then be solved (up to $10^3 \times 10^3$ matrices). The number of times this must be done is proportional, essentially, to the number of open channels of reactants and products. In fact, however, the only requisite information is contained in the projections of the wave function at the appropriate boundaries onto the asymptotic states. Thus, much "extra" information is calculated, leading, most likely, to larger computer time requirements.

These methods do, on the other hand, have some advantages, the first and foremost being that only asymptotic forms of the wave function need to be known analytically. In addition, no matrix elements of the potential need to be taken between bound states, since only the numerical values of the potential in the interaction regions are needed.

More recently, the practitioners of these finite difference methods have adopted alternative methods in their attacks on this problem. Diestler[26] has again studied the infinite wall, flat channel problem, using as a basis set the "internal" states of *each* channel. These solutions are integrated numerically for arbitrary boundary conditions, matched on a surface separating the channels with the other basis set, and then the solutions are combined to satisfy the appropriate boundary conditions. This method worked satisfactorily as did that of Tang et al.,[18] but has not yet been extended to more realistic surfaces. In his method, problems may arise[26] when the mass of the central atom is not infinite, since in this case the bounding surface between channels will not, in general, be perpendicular to the *coordinates* used in the close coupled equations. This leads to difficulties in the projection procedures at the boundary between reactants and products.

Mortensen and Gucwa,[27] on the other hand, have recently used variational techniques to solve the H + H$_2$ problem on the Sato surface[23] used earlier.[22] This is the first application of these techniques, developed formally by Kohn[28] to chemical reactions, and this will be discussed more fully later.

B. Reaction Coordinates and Coupled Channel Calculations

As mentioned earlier, the two problems of many open channels and different natural basis sets (or coordinates) plague the numerical approaches to the computation of chemical reaction cross sections. There is, apparently, no good way of avoiding the problem of many open channels (this problem is discussed in Section IV), but the problem of the different natural basis sets for reactants and products can, in large measure, be resolved. Marcus[29,30] recently introduced in a mathematical fashion reaction coordinates, which have been basic to the conceptual framework of chemical kinetics since the work of Eyring et al.[31] in the 1930s. In this section we shall generalize them very slightly to the form we found most useful in dealing with collinear reactive collisions. We shall then discuss briefly the coupled equations for reactive scattering and some of the results obtained for simple systems.

For a three-body collinear reactive system, the potential energy surface can be plotted as a contour map in the two interatomic distances, r_{AB} and r_{BC} as in Figure 2. In terms of these coordinates, however, the kinetic energy operator is not diagonal, and it is more convenient to start with the coordinates r and R, representing, for example, r_{AB}, and the distance from the center of mass of the AB pair to the particle C, respectively. In terms of these coordinates, the valley corresponding to the bound state of the products is not, asymptotically, at right angles to the entrance valley but rather is skewed by an angle θ which depends on the relative masses of the particles as

$$\tan \theta = \left[(M_A + M_B + M_C) \frac{M_B}{M_A M_C} \right]^{-1/2} \qquad (6)$$

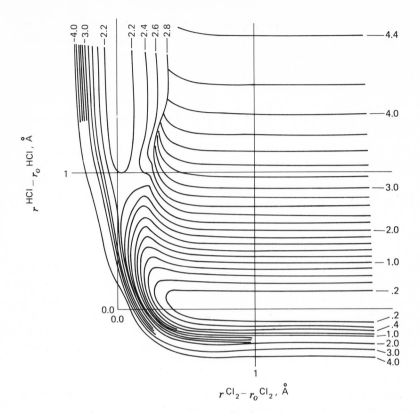

Fig. 2. "Realistic" potential energy surface for $H + Cl_2$. Energy in eV.

such that asymptotically, the product "valley" lies along the line (see Figures 3 and 4):

$$R \propto \cot \theta \qquad (7)$$

The reaction *path* may be defined as the minimum energy path (perpendicular to the energy contours) leading from reactants to the products. Marcus[29] used this path to define a new coordinate system. We have found it preferable, however, to distinguish between the reaction path and the reaction coordinate system. We define a reaction coordinate system by any smooth curve (with curvature of one sign only) leading asymptotically from the reactant valley to the product valley. We may, for instance use the curve

$$R_c = \frac{\gamma}{r_c} + \alpha r_c \qquad (8)$$

$$\alpha = \cot \theta$$

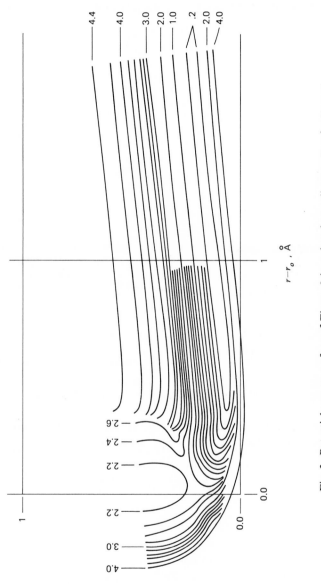

Fig. 3. Potential energy surface of Figure 2 in reduced coordinates, R, r for $H + Cl_2$.

to define the reaction coordinates while the reaction path does not lie on the curve as shown in Figure 4. In terms of this curve, a set of reaction coordinates may be defined as, for instance, shown in Figure 5, with

$$u = u_0 - \frac{\gamma}{r_c} + \beta r_c \qquad (9)$$

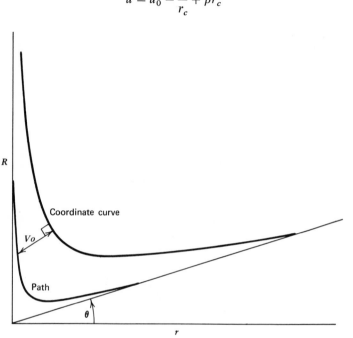

Fig. 4. Reaction path and coordinate curve. $v_0(u)$ is displacement.

and v a perpendicular distance from the curve. The transformation is, with the definitions above, completed by the equation

$$r = r_c(u) + v \cos \phi(u)$$
$$R = R_c(u) + v \sin \phi(u) \qquad (10)$$

with

$$\frac{dR_c}{dr_c} = \tan \phi$$

There are three advantages in defining the reaction coordinates independently of the path. First, one can choose simple analytic forms such as those above for which the curvature and other factors are elementary functions; second, if the curvature of the coordinates is kept small, the kinetic energy operator may be simplified; and third, the triple-valued region of the coordinate system (shown in Figure 6) can, if the curvature is reasonably small, be kept in regions of high potential in which the wave function is exponentially small. In addition, the potential may now be changed at will without affecting the kinetic energy operator.

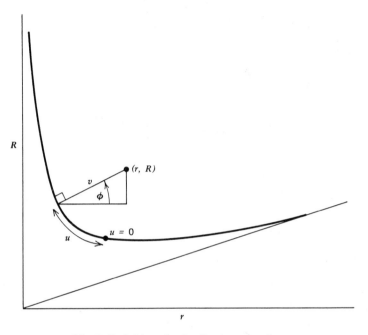

Fig. 5. Definition of point R, r in terms of u, v

The kinetic energy operator in terms of these coordinates is easily found to be

$$T = -\frac{\hbar^2}{2\mu_{AB}}\left\{\frac{1}{\eta}\frac{\partial}{\partial u}\frac{1}{\eta}\frac{\partial}{\partial u} + \frac{1}{\eta}\frac{\partial}{\partial v}\eta\frac{\partial}{\partial v}\right\} \tag{11}$$

$$\eta \equiv [1 - K(u)v]\left(\frac{ds}{du}\right)$$

where s is the arc length and, for the coordinates defined by equations (8) and (9), the curvature K is given by

$$K = \frac{2\gamma}{r_c^3}\left[1 + \left(\alpha - \frac{\gamma}{r_c^2}\right)^2\right]^{-3/2} \tag{12}$$

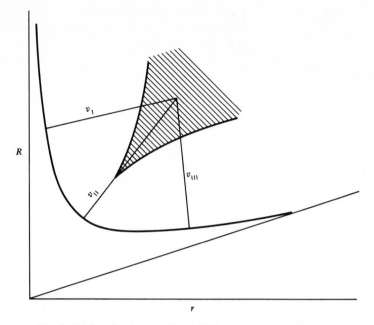

Fig. 6. Triple valued region (hatched) for reaction coordinates.

and the Jacobian, η, is explicitly

$$\eta = [1 - Kv][1 + (\alpha - \gamma/r_c^2)^2]^{1/2}[\beta + \gamma/r_c^2] \tag{13}$$

Since equation (9) is easily solved for r_c as a function of u, the terms η and K are determined analytically as is the kinetic energy operator.

Given the reaction coordinates of this type which go over asymptotically to the internal and translational coordinates of reactants and products, respectively, one is now free to define the potential energy surface. Since v is always a vibration type coordinate, quite a flexible and accurate surface can be constructed from the functional form

$$V(u, v) = \frac{2\mu_{AB}}{\hbar^2} [V_1(u) + V_2(u, v - v_0(u))] \tag{14}$$

where V_2 is a Morse type potential displaced from the reaction coordinate by an amount $v_0(u)$ (see Figure 4). The parameters of the Morse function, $\alpha(u)$ and $D(u)$ are dependent on the path coordinate, u. $V_1(u)$ is the potential along the reaction path.

Using this Hamiltonian, the problem of reactive scattering of these collinear atoms is now easily reduced to a set of coupled second order differential equations in u, similar, in fact, to the equations for inelastic scattering. This procedure was followed (with somewhat different coordinates) in recent work[32,33] and was shown to be a practical method, at least for the $H + Cl_2$ system.

Specifically the Hamiltonian and wave functions are written $\left(\text{after multiplying through by } \frac{2\mu}{\hbar^2}\right)$ as

$$H = -\frac{1}{\eta}\frac{\partial}{\partial u}\frac{1}{\eta}\frac{\partial}{\partial u} - \frac{1}{\eta}\frac{\partial}{\partial v}\eta\frac{\partial}{\partial v} + V(u, v) \tag{15a}$$

$$\psi(u, v) = \frac{X(u, v)}{\eta^{1/2}} \tag{15b}$$

$$H\psi = E\psi = \frac{1}{\eta^{1/2}}\left\{-\frac{\partial}{\partial u}\frac{1}{\eta^2}\frac{\partial}{\partial u} - \frac{\partial^2}{\partial v^2} + V(u, v) + Q(u, v)\right\}X(u, v) \tag{15c}$$

where $Q(u, v)$ is an effective potential term introduced by the transformation:

$$Q(u, v) = \frac{\eta''}{2}\left(\frac{1}{\eta} + \frac{1}{\eta^3}\right) - \frac{1}{4}\left(\frac{\eta'}{\eta}\right)^2\left(1 + \frac{5}{\eta^2}\right) \tag{16}$$

$$\eta' = \frac{d\eta}{du}$$

It is now possible to find a set of basis functions for the vibration which depend parametrically on u in such a fashion that for $u \to \pm\infty$, they become the reactant (or product) vibrational wave functions. We may, for example, take an harmonic oscillator basis set:

$$H_0\,\phi_{i,\,u}(v) = \varepsilon_i(u)\phi_{i,\,u}(v)$$

$$H_0 = -\frac{\partial^2}{\partial v^2} + \frac{\mu\kappa(u)}{\hbar^2}(v - v_0(u))^2 \equiv -\frac{\partial^2}{\partial v^2} + V_0(u, v)\frac{2\mu}{\hbar^2} \tag{17}$$

The wave function, X, is then expanded in terms of this basis set, and the matrix equations are then set up:

$$X(u, v) = \sum_i f_i(u)\phi_{i,\,u}(v) \tag{18}$$

substituting this into equation (15a), multiplying by $\eta^{1/2}\phi_{i,u}(v)$ and integrating over v, we obtain the desired coupled equations (after multiplying again by η^2)

$$\frac{d^2}{du^2}f_n(u) + \sum_m\left[\hat{V}_{nm}\frac{df_m}{du} + V_{nm}f_m\right] = 0 \tag{19}$$

where

$$\hat{V}_{nm} = 2\left\langle n \left| \frac{\partial}{\partial u} - \frac{1}{\eta}\frac{\partial \eta}{\partial u} \right| m \right\rangle \tag{20a}$$

$$V_{nm} = \left\langle n \left| \frac{\partial^2}{\partial u^2} - \frac{2}{\eta}\frac{\partial \eta}{\partial u}\frac{\partial}{\partial u} + \frac{2\eta^2\mu}{\hbar^2}[E - \varepsilon_m(u) - V_2(u, v) + V_0(u, v)] \right| m \right\rangle \tag{20b}$$

$$\langle n| g(u, v) |m \rangle \equiv \int_{-\infty}^{\infty} dv \phi_{n,u}^*(v) g(u, v) \phi_{m,u}(v) \tag{20c}$$

There are several points about these equations worth mentioning. First, we have multiplied by η^2 before integration over v in order that the second order differential operator on f_n be diagonal [see equation (15c) and remember that η is a function of both u and v]. Second, in equations (20a) and (20b), the operators $\frac{\partial}{\partial u}$ are operating on the functions $\phi_{m,u}(v)$, that is, on the parametric dependence of these on u. It is largely through this operation that the transitions between different vibrational states are introduced since the v dependence of the potentials $V_2(u, v)$ and $V_0(u, v)$ largely cancels in equation (20b). This means that the basis set is similar to the *adiabatic* basis discussed by Marcus. This also means, however, that $\mathbf{V}$ and $\hat{\mathbf{V}}$ are not symmetric matrices.

One large advantage of the use of the reaction coordinates as distinct from the reaction path is that many of the off-diagonal terms involving η are now very small. Going back to the definitions of η in equation (11), we can write it as

$$\eta = [1 - K(u)v_0(u) - K(u)(v - v_0(u))]\frac{ds}{du} \tag{12}$$

The last term in brackets, which will be proportional to the vibrational coordinate, and causes transitions, is quite small now if the radius of curvature of the reaction coordinate is large with respect to the vibrational amplitude. Thus, several of the terms involving η are almost diagonal, representing, in effect, only modifications of the potential.

One of the interesting features of this set of equations is the fact that they are in a representation in which the S matrix, calculated for a *truncated* basis set, will not be exactly unitary. This is easily seen by truncation at one term. We then would have

$$\frac{d^2}{du^2}f_0 + \hat{V}_{00}\frac{df_0}{du} + V_{00} f_0 = 0 \tag{13}$$

This operator is not Hermitian, in general, with respect to the weight function unity. However, the unitarity of the S matrix is restored exactly by using a complete basis set. In practice, however, the deviation from unitarity even with a small basis set is very slight provided the numerical integrations are done properly.

With respect to the numerical integration of equation (19), we will only outline the method briefly here. The boundary conditions are rather different from those normally employed in inelastic close coupling solutions in that one has outgoing waves in all open channels in the limits as $(u \to \pm \infty)$, and the number of open channels on each side may not be the same. We have found it most efficacious to solve the set of first order equations (of twice the dimensions) corresponding to equation (19) by a modified exponential method,[34] employing some of the techniques recently devised by Gordon.[35]

Briefly, we consider the vector made up from the f_n's and their first derivatives,

$$
\mathscr{F}(u) \equiv
\begin{pmatrix}
f_1(u) \\
\vdots \\
f_n(u) \\
\dfrac{df_1}{du} \\
\vdots \\
\dfrac{df_n}{du}
\end{pmatrix}
$$

This vector satisfies the matrix differential equation:

$$
\frac{d\mathscr{F}}{du} = \begin{pmatrix} 0 & \mathbf{1} \\ -\mathbf{V} & -\hat{\mathbf{V}} \end{pmatrix} \mathscr{F} \tag{18}
$$

Writing $\mathscr{F}(u)$ in terms of a translation matrix, $\mathbf{U}$, we define $\mathbf{U}$ by

$$
\mathscr{F}(u) = \mathbf{U}(u, -\infty)\mathscr{F}(-\infty) \tag{19}
$$

and find that $\mathbf{U}$ satisfies the equation similar to equation (18)

$$
\frac{d\mathbf{U}}{du}(u, u_0) = \begin{pmatrix} 0 & \mathbf{1} \\ -\mathbf{V} & -\hat{\mathbf{V}} \end{pmatrix} \mathbf{U}(u, u_0) \tag{20}
$$

subject to the boundary condition

$$
\mathbf{U}(u_0, u_0) = \mathbf{1} \tag{21}
$$

U is then approximated by the product of an exponential matrix in each interval:

$$\mathbf{U}(u_n, u_0) = \prod_{i=0}^{n-1} \mathbf{U}(u_{i+1}, u_i) = \mathbf{U}(u_n, u_{n-1}) \cdots \mathbf{U}(u_1, u_0) \qquad (22)$$

where the order of the product is taken with larger values of u_i to the left. Each individual $\mathbf{U}(u_{i+1}, u_i)$ is approximated by the exponential form

$$\mathbf{U}(u_{i+1}, u_i) = \exp\left\{\begin{matrix} 0 & (u_{i+1} - u_i)\mathbf{1} \\ -\mathbf{V}(\bar{u}_i)(u_{i+1} - u_i) & -\hat{\mathbf{V}}(\bar{u}_i)(u_{i+1} - u_i) \end{matrix}\right\} \qquad (23)$$

where $\bar{u}_i = \dfrac{(u_{i+1} + u_i)}{2}$. This form is correct to order $(u_{i+1} - u_i)^3$ times derivatives of the potential at $\bar{u}_i$, and was found to be quite accurate. Higher accuracy can be obtained by using higher order terms in the Magnus series.[36] The exponential is evaluated most easily by diagonalization of the dominant block matrix, $\mathbf{V}$.

Once the matrix $\mathbf{U}$ is determined over the range of the variable, u, for which the potential is acting, the S matrix is easily obtained by matrix algebra. We can write the asymptotic forms of the F vectors as follows for each open channel if the incoming particles were in the j state:

$$\lim_{u \to -\infty} \mathscr{F}^{(j)}(u) = \begin{pmatrix} 0 \\ \vdots \\ \exp(ik_j^- u) \\ 0 \\ \vdots \\ ik_j^- \exp(ik_j^- u) \\ \vdots \\ 0 \end{pmatrix} + \begin{pmatrix} g_{j1}^- \exp(-ik_1^- u) \\ \vdots \\ g_{jn}^- \exp(-ik_n^- u) \\ -ik_1^- g_{j1}^- \exp(-ik_1^- u) \\ \vdots \\ -ik_n^- g_{jn}^- \exp(-ik_n^- u) \end{pmatrix}$$

$$= \mathscr{I}\mathscr{N} + \mathscr{R} \qquad (24)$$

$$\lim_{u \to +\infty} \mathscr{F}^{(j)}(u) = \begin{pmatrix} g_{j1}^+ \exp(ik_1^+ u) \\ \vdots \\ g_{jn}^+ \exp(ik_n^+ u) \\ ik_1^+ g_{j1}^+ \exp(ik_1^+ u) \\ \vdots \\ ik_n^+ g_{jn}^+ \exp(ik_n^+ u) \end{pmatrix} = \mathscr{P}$$

If we write the matrix $\mathbf{U}(\infty, -\infty)$ in terms of its blocks,

$$\mathbf{U} = \begin{pmatrix} \mathbf{U1} & \mathbf{U2} \\ \mathbf{U3} & \mathbf{U4} \end{pmatrix}$$

using equations (19) and (24), we can write, after some algebraic manipulation, the vector of reflected waves, $\mathscr{R}$, in terms of the incoming vector, $\mathscr{I}\mathscr{N}$:

$$\mathscr{R} = [1 + \mathbf{Q}^{-1}\mathbf{L}]^{-1}(1 - \mathbf{Q}^{-1}\mathbf{L})\mathscr{I}\mathscr{N} = \mathbf{S}^{-}\mathscr{I}\mathscr{N} \tag{25}$$

where

$$\mathbf{Q} = \mathbf{U1} + i(\mathbf{k}^{+})^{-1}\mathbf{U3}$$

$$\mathbf{L} = i\mathbf{U2}(\mathbf{k}^{-}) + \mathbf{k}^{+}\mathbf{U4}\mathbf{k}^{-}$$

and $\mathbf{k}^{+}$ and $\mathbf{k}^{-}$ are diagonal matrices of the k^{+}s and k^{-}s. The magnitudes of the elements of $\mathbf{S}^{-}$ are the magnitudes of the S matrix elements for inelastic (nonreactive) scattering. Using these derived values, the product vector $\mathbf{P}$ is easily found:

$$\mathbf{P} = (\mathbf{U1} - i\mathbf{U2}\mathbf{k}^{-})\mathscr{R} + (\mathbf{U1} + i\mathbf{U2}\mathbf{k}^{-})\mathscr{I}\mathscr{N} \tag{26}$$

The magnitudes of the S matrix elements for reactive scattering are then given by

$$\mathbf{S}^{+}(\text{mag}) = [(\mathbf{U1} - i\mathbf{U2}\mathbf{k}^{-})\mathbf{S}^{-} + (\mathbf{U1} + i\mathbf{U2}\mathbf{k}^{-})]_{(\text{mag})} \tag{27}$$

In the case that some states are closed initially for reactants, but open for the products, the same type of equations result, but with the modifications that (a) the ks of closed channels are imaginary, with the sign chosen to represent solutions which grow exponentially as u is moved toward the classically allowed region, and (b) the actual S matrix elements are determined only for open channels.

In practice this method of solution, which not trivial to program, is found to be both reasonably fast (2.5 minutes for 6 open channels at one energy on an IBM 7094), and reasonably accurate.

We shall now complete this section with a brief review of the results obtained by the method above in comparison with other quantum approaches to the problem of chemical reactions.

The methods above have, with some variation, been applied by Wyatt[37] and by Rankin et al.[32,33] Wyatt studied the $H + H_2$ reaction for collinear collisions with one of the more accurate potential energy surfaces.[38] He also included the effect of rotation of the entire linear system which adds an effective potential. For the energies studied, only the ground vibrational state was open, and Wyatt studied the effects of several approximations (perturbed stationary state, etc.) in some detail. With only one vibrational state included, the nonadiabatic effects must be neglected, and the exact definition of the vibrational wave function along the reaction path becomes very important. Wyatt found significant differences in the transition (reaction) probabilities with different approximations to the vibrational wave function. The curvature was also shown to have a significant effect on tunneling, and this will be discussed later.

Rankin et al.[32,33] used moderately realistic potential energy surfaces for the H + Cl$_2$ reaction and studied the effects of variation of the energy and the surface itself on both the probability of reaction and the vibrational excitation of the product, HCl. In particular the parameters of the potential energy surface which governed the position of the col, the steepness of the potential along the reaction path, and the curvature of the reaction path were varied in order to determine the effects of the gross features of the potential energy surface. It was found, perhaps surprisingly, that the steepness of the reaction path had little effect either on the reaction probability or on the vibrational excitation. However, the vibrational excitation of the products was increased by a combination of (a) moving the col toward the reactant side and (b) making the reaction path very sharply curved. These results are a confirmation of some of the qualitative features found in classical trajectory calculations.[9] In these calculations, however, the reaction coordinates were set along the reaction path, and large scale variations of the potential energy surface were hampered by the necessity of keeping the potential high in the region in which the coordinates were triple-valued.

The calculations are now being extended using the coordinate systems above, and it appears[39] that surfaces corresponding to Polanyi's[9] characterization of "mixed energy release" can be constructed and used. These combine sharp curvature of the reaction path and col on the reactant side with lowered vibrational force constants in the strongly interacting region and lead to much higher excitation of vibration in the products. They correspond to very nonadiabatic (vibrational) potential energy surfaces or, correspondingly, to a "loose" complex in the interaction region.

These calculations, although they represent the most general and realistic to date, leave much room for improvement. They are still much more difficult than inelastic scattering calculations because the basis set expansion, being a function of both coordinates, makes the interaction matrix non-Hermitian, there are different numbers of open channels initially and finally which means that accurate treatment of several closed channels over some regions is necessary, and finally, the Hamiltonian is more complex due to the use of reaction coordinates. However, with care, all of these problems can be handled, and it seems likely that such calculations will, in the near future, provide us with a much more accurate picture of the quantum effects to be expected in simple chemical reactions.

C. Quantum Tunneling in Reactive Collisions

One of the more significant and interesting questions for which answers are now becoming available is the question of quantum tunneling; its definition, dependence on potential parameters, and importance to accurate cross sections and rate constants. Although tunneling has an obvious definition in one-dimensional quantum problems, it is much more complicated by dynamical effects in two or more dimensions.

In one dimension, tunneling occurs when the particle penetrates a region of space in which the local kinetic energy, $E - V$, becomes negative. This corresponds exactly to passage through a region which is classically inaccessible at that energy. In considering motion on a two-dimensional surface such as shown in Figures 1 and 2, however, there is good reason to question whether these criteria are the most relevant. There exists, of course, a classical trajectory which goes over the energy barrier at the col with infinitesimal translational energy. This trajectory, however, will in general not be one in which the asymptotic vibrational energy corresponds to a bound vibrational state of the system. If, on the other hand, one requires all classical trajectories to start with a "proper" quantized amount of vibrational energy, then one normally will find that none of these trajectories can cross the col with zero translational energy. The dynamics apparently do not normally permit it.

Probably the most reasonable definition of tunneling in such a system, in which one coordinate is translational in character and the other corresponds to bound oscillatory behavior, is that tunneling occurs when the system crosses a region in which the local translational energy is negative. This depends, of course, on the basis set used to describe the oscillatory motion near the col, and the most obvious choice is to use the vibrationally adiabatic basis defined in reaction coordinates.

This procedure differs from that of Johnston and Rapp,[40] who took a weighted average of the one-dimensional tunneling probabilities calculated for the series of potentials obtained by taking cuts of the potential energy surface parallel to the reaction path at the col. Such a method neglects many specific two-dimensional effects such as the dynamic effect of the curvature of the reaction path, and relies instead on a relatively ad hoc weighting of the different portions of the surface.

Recently Wyatt,[37] Rankin et al.[33] and Russell[41] have used reaction coordinates to study this question in more detail than was done by Marcus originally.[29] Wyatt studied the $H + H_2$ system for which the three atoms are constrained to be collinear but are allowed to rotate as a whole. Thus the effective potential contains an angular momentum term. The major effects of the two-dimensional character of the problem arise from two sources. First, the adiabatic vibrational state depends parametrically on the position along the reaction path. This has two effects, in that the zero point vibrational energy varies along the path (as in transition state theory) and second, the translational kinetic energy operator operates on these functions producing some effective potential terms. The other major effect arises from the curvature of the reaction path and the consequent centrifugal type effective potentials which occur. The curvature affects both the effective vibrational potential[37] and the resultant effective translational potential.[29]

If the modified coordinates above are used, it is easily shown that the most important terms can be reduced to effective potentials for translation along the reactive path. These important correction terms are:[41]

$$\langle m \mid [E - E_m(u) - V_2(u, v) + V_0(u, v)][K(u)v_0(u)]^2 \mid m \rangle$$

where v_0 (u) is shown in Figure 4 and the curvature, K, is that of the defining curve for the reaction path.

The effects of these correction terms are quite important for tunneling corrections at low energies (corresponding to temperatures up to $1000°K$). It is well known that the one-dimensional tunneling corrections increase the transmission probabilities at low energies. These calculations, however, seriously overestimate the amount of tunneling. When the more correct two-dimensional calculations are done, the tunneling is significantly reduced because the effective barrier is both raised and broadened by the two-dimensional corrections. Even so, tunneling is an important factor in low energy reactive collisions. This is shown clearly in the comparison of quantum[33] and classical[42] calculations on the same potential energy surface for the $H + Cl_2$ reaction (see Figure 7). In this case the calculated effective activation energy at $450°K$ is raised by approximately 1.5 kcal/mole in the classical calculations which ignore tunneling.

One further point should be mentioned with respect to the effects of the two-dimensional nature of the surface on isotopic reactions. The change of mass not only affects the reduced mass used in the calculation (as in the one-dimensional case) but also affects the curvature of the reaction path. (The asymptotic skewing of the R, r coordinate system in Figure 3 depends on the mass ratios.) Thus in considering the reaction rates of isotopic pairs of reactions both these effects must be taken into account. It is clear that although we now have a much better understanding of the important parameters involved in tunneling in chemically reactive systems, considerable work is necessary before these can be developed into simply applicable correction factors which have been the goal of transition state theorists for many years.[43]

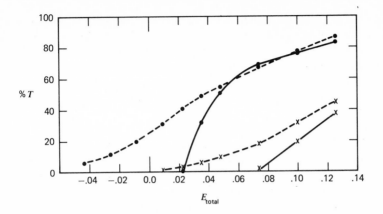

Fig. 7. Comparison of reaction probability for H + Cl₂ with total energy for quantum (- - - -) and classical (———) calculations [33,42] for the surface of Figure 2. Cl₂ ($v = 0$) results shown with dots, $v = 1$ shown with xs.

IV. ALTERNATIVE APPROACHES

Although the calculation of chemical reaction cross sections (or transition probabilities) by means of the close coupled sets of ordinary differential equations, using reaction coordinates, appears to be the most accurate and feasible method at present, there are some very significant disadvantages and difficulties. The extension to three dimensions (3 internal coordinates and 3 Euler angles) brings in serious difficulties in that reaction coordinates[30] which retain significant physical meaning (that is, correspond to bending or stretching vibrations in the interaction region and to vibrations and rotations of reactants and products asymptotically) have very complicated forms for the quantum kinetic energy operator. Conversely, reaction coordinates which possess simpler kinetic energy operators seem less physical and thus it is more difficult to determine the proper potential energy surface using them, and the choice of a nearly adiabatic basis set for the internal degrees of freedom becomes more difficult. For instance, one may find that there is no coordinate which defines the symmetric stretch well in the interaction region.

A second, more fundamental, difficulty is the number of accessible states which may be closely coupled, and the number of total angular momentum values for which the cross sections may be large. It appears that even for such a simple exothermic reaction as H + Cl₂ → HCl + Cl, there are approximately 10^3 final states open, and perhaps total angular momenta up to $J = 15$ will contribute. Since the fastest current methods[34,35] of solving even simpler coupled equations for inelastic processes take at least $10^{-4}N^3$ minutes (when N is the number of states) on a computer per calculation, it appears unprofitable to consider the direct application of these methods to any but the simplest of reactions.

Since quantum effects are important, particularly with respect to tunneling and quantized energy distributions, there is a strong stimulus to develop new approaches which perhaps can give us, with lesser accuracy, the cross sections of interest.

One approach, perturbation theory, has already been discussed. In this section we shall briefly indicate three other possible approaches; variational procedures, semiclassical techniques, and equivalent (or optical) potentials.

Mortensen,[27] as mentioned briefly earlier, has used the Kohn[28] variational technique to compute the transmission coefficients for the collinear $H + H_2$ reactions. The method is based on the fact that the functional

$$I_t = \langle \psi_{f,t}^- | E - H | \psi_{i,t}^+ \rangle$$

is stationary with respect to variation of the trial wave functions $\psi_{i,t}^+$ and $\psi_{f,t}^-$ which satisfy the appropriate boundary conditions when $\psi_{i,t}^+$ is the exact wave function, ψ_i^+. The application of this principle is difficult in the case of interest here since the trial wave functions are functions (in the 3-d case) of six variables. As Mortensen used it, the trial wave function was expanded in terms of the vibrational eigenfunctions of the reactants and products plus a combination of products of vibrational functions in the interaction region. Although the calculation actually involved integrating numerically to find the "translational" portions of the wave function, these are uncoupled ordinary differential equations. For energies low enough so that only one vibrational channel was open, Mortensen[27] found that 16 to 25 variational central functions gave adequate accuracy (about 2 decimal places).

Although Mortensen does not discuss the application of this method to problems in which more channels are open, or to problems of higher dimensionality, it appears that the difficulties will be formidable. It has been shown[44,4] that the Kohn variation is satisfied on a truncated basis set if that basis set is a solution of the close coupled equations. This means that a given number of basis functions in a variational calculation cannot, in general, produce more accurate results than the solutions of the same number of coupled equations. It is thus not readily apparent that the variational methods will be more successful than the close coupled equations. In addition, the variational principle may suffer from instabilities due to the poles of the inverse operator $(E - H)^{-1}$. These were first pointed out by Schwartz.[45] However, Miller[46] recently pointed out that the most serious singularities due to continuum contributions can be avoided and essentially exact results can, in principle, be obtained from a finite square integrable basis set. Miller also demonstrates in the same paper that in the application to rearrangement process of the variational method, the coupled equations for the energetically accessible states of the system should be solved as a first step; a confirmation of the conclusions above.

A fundamentally different approach is to attempt to derive equations for the dynamics of the system which yield not the full S matrix for the scattering problem, but only those transition probabilities or cross sections of interest. This is, of course, the result of using first order perturbation theory, but in this case the results have only limited applicability. A more general formulation of the problem can be given in terms of an equivalent (or optical) potential[44,4] in which the *exact* result is given formally in terms of the solution of a reduced set of equations.

Optical potentials have been used for several years now by chemists to parameterize the elastic scattering behavior of chemically reactive systems.[47-50] In these applications, the equation for elastic scattering is solved with an effective (or optical) component in the potential which is complex, representing the disappearance of amplitude from the elastic scattering channel into inelastic channels not explicitly considered. It has proved possible[47] to calculate the optical potential explicitly (numerically) for some very simple cases by adiabatic perturbation theory.

Of more interest here is the potential use of equivalent potentials to calculate inelastic or reactive cross sections without recourse to large basis set expansions. Mittleman and Pu[44] gave the original derivation of the method in 1962, but it has not yet been applied or investigated in great detail. It is, I believe, of sufficient potential value in treating these complicated, strongly coupled systems to be mentioned here.

Consider the projector, P, which projects out of an arbitrary wave function, $\psi(\{x\}, R)$ those eigenfunctions of the internal coordinates, $\{x\}$, of interest.

$$P\psi = \sum_{i \in P} f_i(R)\phi_i(\{x\}) \tag{28}$$

Let Q be the complement of P, where P and Q are standard projection operators:

$$P^2 = P; \qquad P + Q = 1; \qquad PQ = QP = 0$$

Operating on the Schrödinger equations with P or Q, we find

$$P(E - H)\psi = 0 = P(E - H)(P + Q)\psi$$
$$Q(E - H)\psi = 0 \tag{29}$$

If the initial state is in the P manifold, then the Lippman–Schwinger equation for $Q\psi^+$ can be written

$$Q\psi^+ = (E^+ - QHQ)^{-1}QHP\psi^+ \tag{30}$$

Thus the integro-differential equation for the components of ψ of interest is

$$(E^+ - PHP)P\psi^+ = PH(E^+ - QHQ)^{-1}QHP\psi^+ \tag{31}$$

subject to the appropriate boundary conditions. A physical interpretation of this equation is possible if one remembers that it is only the interaction portion of H which does not commute with P and Q. The left-hand side is equivalent to the set of coupled differential equations among the states in P. The right-hand side is the contribution to these states from second (and higher) order scattering processes involving the Q states. Reading the right-hand side from right to left, the first $QHP\psi^+$ factor represents the scattering from the P manifold to the Q manifold at some point r' in space, the integral operator $(E^+ - QHQ)^{-1}$ represents further scattering of this within the Q manifold, and finally, the PH factor yields that portion scattered back into the P manifold at r which contributes to the wave functions there.

The case of most interest would be when P contains only two states, the initial and final. In this case, Q represents all intermediate states as well as all closed channels, and so on. The problem, of course, is to find an approximate representation of the integral operator on the right-hand side which has three properties:

(1) It should not involve explicit summations over states of the Q manifold.

(2) It should be calculable from the known properties of the potential energy surface.

(3) It should be as simple as possible (preferably a local operator) and reasonably accurate.

The resolution of the question of whether or not this will be feasible is certainly in the future, but several approaches have been tried, for the case of elastic scattering only, which seem hopeful. Perturbation theory can be applied to the righthand side only, yielding the equation

$$(E^+ - PHP)P\psi^+ = PV(E^+ - QH_0Q)^{-1}QVP\psi^+ \tag{32}$$

That this may be a reasonable approximation is suggested by the recent work of Roberts and Ross[51] in which they evaluated the imaginary portion of the semiclassical phase shift in elastic scattering by perturbation theory.

Their results agree well with the exact numerical calculations of Marriott and Micha.[52]

Two other alternatives are to use adiabatic or impulse approximations in the equivalent potential. These have been discussed briefly by Mittleman[44] and Micha.[47] The other possibility, that of replacing the nonlocal equivalent potential by a local imaginary potential has been attempted also in the elastic scattering case, but it is not at all clear whether a reasonably accurate and general prescription for this procedure can be given. In any case, further developments in these areas are of great interest.

A rather different approach has recently been suggested by Pechukas[53] who investigated semiclassical scattering theory from the viewpoint of Feynman's path integral formulation of quantum mechanics. Although this was developed with applications to elastic and inelastic collisions in mind, the general ideas and methods are certainly applicable to reactive scattering as well. The idea presented is that one ought to be able to treat the translational motions of atomic and molecular systems rather well semiclassically since the WKB method works well. However, in the presence of quantized internal degrees of freedom, a "nonclassical" change in the translational energy (and thus the trajectory) must occur if an inelastic process occurs. For such a process, no unique "classical" path exists, but the probability amplitude of the event can be defined for an arbitrary "classical" path and these summed, in the Feynman sense, to yield the overall result.

The applicability of this method has not yet been demonstrated for the problems of interest here, but it is quite suggestive. Presumably it offers a correct prescription for "mixing" the classical and quantum aspects of the problem and of looking at particular internal quantum transitions one by one. It possesses certain clear formal advantages over the normal semiclassical methods in which a classical trajectory is fixed and then time dependent quantum theory is used to evaluate transition probabilities. These formal advantages of Pechukas' theory include the conservation of energy and angular momentum and the dependence of the trajectory on the changes in the internal quantum state of the system.

Finally, the recent work of McCullough and Wyatt[54] on the solution of the time dependent Schrödinger equation needs mention. They used the formal solution of the Schrödinger equation in the form

$$\psi(t + \Delta t) = \exp\left\{-\frac{i\,\Delta t}{\hbar} H\right\}\psi(t) \tag{33}$$

to follow the motion in time and space of an initial wave packet. A finite difference approximation for H was used at each time step and approximately

200 time steps were required. Although the method, applied to the Porter–Karplus[38] potential energy surface was not fast (about 1 hour of computer time for each initial wave packet), the results are fascinating. The initial wave packet, a product of a Morse-type vibrational wave function and a Gaussian translational wave packet, moves into the interaction region and essentially splits into two portions, one transmitted, one reflected. The amplitude is largest well outside the reaction path, illustrating specifically the effects of the "centrifugal" forces. Finally, the separated portions move toward the asymptotic region and the transmission probability is computed from the fluxes. Thus, this method gives a unique physical interpretation of the dynamics of the quantum mechanical collision event. As mentioned earlier, however, it is not at present suitable for large scale computations due to the large computational requirements of the method.

Finally, there exist other methods of approach to this problem which have not been dealt with here. These include the use of the Fadeev equations, quantum statistical approximations, and density matrix methods. These have not yet been applied to chemical reactions in any detail, and their potential value is, at this time, rather hard to evaluate.

References

1. *Advances in Chemical Physics*, Ed. John Ross, **10**, *Molecular Beams*, Wiley, New York, 1966.
2. Disc. Faraday Soc., **44**, *Molecular Dynamics of the Chemical Reactions of Gases*, 1967.
3. *Advances in Chemistry Series*, **58**, *Ion-Molecule Reactions in the Gas Phase*, Ed. R. F. Gould, American Chemical Society, 1966.
4. R. D. Levine, *Quantum Mechanics of Molecular Rate Processes*, Oxford University Press, 1969. This seems to be the best reference for work up to 1968.
5. R. G. Newton, *Scattering Theory of Waves and Particles*, McGraw-Hill, New York, 1966.
6. M. Goldberger and K. Watson, *Collision Theory*, Wiley, New York, 1964.
7. K. J. Laidler and J. C. Polanyi, in *Progress in Reaction Kinetics*, Ed. G. Porter, Pergamon Press, Oxford, 1965.
8. J. C. Light, *Disc. Faraday Soc.*, **44**, 14 (1967).
9. (a) J. C. Polanyi and S. D. Rosner, *J. Chem. Phys.* **38**, 1028 (1963); (b) P. J. Kuntz, E. M. Nemeth, J. C. Polanyi, S. D. Rosner, and C. E. Young, *J. Chem. Phys.* **44**, 1168 (1966); (c) K. G. Anlauf, J. C. Polanyi, W. H. Wong, and K. B. Woodall, *J. Chem. Phys.* **49**, 5189 (1968); (d) P. J. Kuntz, E. M. Nemeth, and J. C. Polanyi, *J. Chem. Phys.* **50**, 4607 (1969); (e) P. J. Kuntz, M. H. Mok, and J. C. Polanyi, *J. Chem. Phys.* **50**, 4623 (1969).
10. (a) N. C. Blais and D. L. Bunker, *J. Chem. Phys.* **37**, 2713 (1962); (b) N. Blais, *J. Chem. Phys.* **49**, 9 (1968).
11. (a) M. Karplus, R. N. Porter, and R. D. Sharma, *J. Chem. Phys.* **40**, 2033 (1964); **43**, 3259 (1965); (b) M. Godfrey and M. Karplus, *J. Chem. Phys.* **49**, 3602 (1968).
12. M. G. Evans and M. Polanyi, *Trans. Faraday Soc.* **35**, 178 (1939).

13. D. J. Kouri, *J. Chem. Phys.* **51**, 5204 (1969).
14. W. N. Sams and D. J. Kouri, *J. Chem. Phys.* **51**, 4809 (1969).
15. D. Micha, *Arkiv. För Fysik* **30**, 425 (1965).
16. M. Karplus and K. T. Tang, *Disc. Faraday Soc.* **44**, 56 (1967).
17. H. Hulbert and J. Hirschfelder, *J. Chem. Phys.* **11**, 276 (1943).
18. K. Tang. B. Kleinman, and M. Karplus, *J. Chem. Phys.* **50**, 1119 (1969).
19. P. D. Robinson, private communication.
20. D. R. Dion, M. B. Milleur, and J. Hirschfelder, private communication.
21. D. W. Jepsen and J. Hirschfelder, *J. Chem. Phys.* **30**, 1032 (1959).
22. E. M. Mortensen and K. S. Pitzer, *The Transition State*, Chem. Soc. (London), Spec. Publ. **16** (1962).
23. S. Sato, *J. Chem. Phys.* **23**, 592 (1955); **23**, 2465 (1955).
24. E. M. Mortensen, *J. Chem. Phys.* **48**, 4029 (1968).
25. D. J. Diestler and V. McKoy, *J. Chem. Phys.* **48**, 2941 (1968); **48**, 2951 (1968).
26. D. J. Diestler, *J. Chem. Phys.* **50**, 4746 (1969).
27. E. M. Mortensen and L. D. Gucwa, *J. Chem. Phys.* **51**, 5695 (1969).
28. W. Kohn, *Phys. Rev.* **74**, 1763 (1948).
29. R. A. Marcus, *J. Chem. Phys.* **45**, 4493 (1966).
30. R. A. Marcus, *J. Chem. Phys.* **49**, 2610 (1968).
31. S. Glasstone, K. Laidler, and H. Eyring, *The Theory of Rate Processes*, McGraw-Hill, New York, 1941.
32. C. C. Rankin, Thesis, University of Chicago, 1968.
33. C. C. Rankin and J. Light, *J. Chem. Phys.* **51**, 1701 (1969).
34. S. Chan, J. Light, and J. Lin, *J. Chem. Phys.* **49**, 86 (1968).
35. R. Gordon, *J. Chem. Phys.* **51**, 14 (1969).
36. P. Pechukas and J. Light, *J. Chem. Phys.* **44**, 3897 (1966).
37. R. E. Wyatt, *J. Chem. Phys.* **51**, 3489 (1969).
38. R. N. Porter and M. Karplus, *J. Chem. Phys.* **40**, 1105 (1964).
39. G. Miller, private communication.
40. H. S. Johnston and D. Rapp, *J. Am. Chem. Soc.* **83**, 1 (1961).
41. D. Russell, private communication.
42. D. Russell and J. Light, *J. Chem. Phys.* **51**, 1720 (1969).
43. D. J. LeRoy, B. A. Ridley, and K. A. Quickert, *Disc. Faraday Soc.* **44**, 92 (1967).
44. M. H. Mittleman and R. Pu, *Phys. Rev.* **126**, 370 (1962).
45. C. Schwartz, *Ann. Phys.* (New York) **16**, 36 (1961).
46. W. H. Miller, *J. Chem. Phys.* **50**, 407 (1969).
47. D. A. Micha, *J. Chem. Phys.* **50**, 722 (1969).
48. J. L. F. Rosenfeld and J. Ross, *J. Chem. Phys.* **44**, 188 (1966).
49. H. Y. Sun and J. Ross, *J. Chem. Phys.* **46**, 3306 (1967).
50. C. Nyland and J. Ross, *J. Chem. Phys.* **49**, 843 (1968).
51. R. E. Roberts and J. Ross, *J. Chem. Phys.* **52**, 1464 (1970).
52. R. Marriott and D. Micha, *Phys. Rev.* **180**, 120 (1969).
53. P. Pechukas, *Phys. Rev.* **181**, 166 (1969); *ibid.*, **181**, 174 (1969).
54. E. A. McCullough, Jr., and R. E. Wyatt, *J. Chem. Phys.* **51**, 1253 (1969).

A REVIEW OF ION-MOLECULE REACTIONS*

LEWIS FRIEDMAN AND BRYAN G. REUBEN†

Department of Chemistry, Brookhaven National Laboratory, Upton, New York

CONTENTS

* Work supported by the U.S. Atomic Energy Commission.
† On leave from: Department of Chemistry, University of Surrey, Guildford, Surrey, England.

33

I. INTRODUCTION

The contribution of ion-molecule reactions to the publication explosion is shown in Figure 1 which illustrates the exponential increase in number of papers published in this field during the past two decades. The data presented in this figure were obtained from two bibliographies,[1,2] one of which is a continuing operation published biannually in a joint effort of the National Bureau of Standards and Oak Ridge National Laboratory, under the masthead AMPIC (Atomic and Molecular Processes Information Center). Data are also presented for the financial support given by the United States federal government to basic scientific research projects from 1956 onwards.[3] The increase in money available for basic research in the United States has until recently been large, and the increase, though not the gross amount, has been matched in some other countries. It is worth observing, however, that an astonishly high proportion of research on ion-molecule reactions is performed in the United States, and workers in this field would be sanguine indeed if they were not to wonder if the flattening of the budgetary curve will not soon be matched by a similar reduction in the rate of increase of publication.

Such a moment seems appropriate for taking stock of the work which

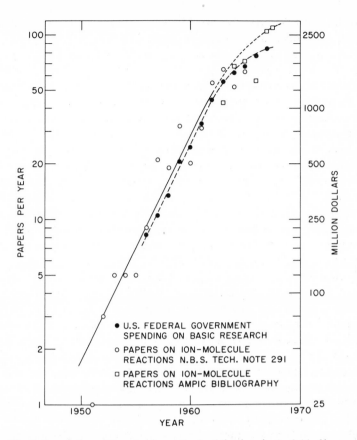

Fig. 1. Correlation of the rate of growth of publications in the field of ion-molecule reactions with growth of U.S. Federal support of basic research.

has gone on in this field in the past, for trying to assess its overall contribution to scientific knowledge, and for discussing the sort of questions on which we would hope future research might throw light.

It is necessary at this stage to define what we include in the term "ion-molecule reactions" and hence have incorporated in Figure 1. In general we have concerned ourselves with thermodynamic and kinetic aspects of ion collision processes. We have broadened this somewhat to include certain other aspects of ion thermochemistry, in view of the close relationship between the thermodynamics and kinetics of isolated reacting systems. One may think of the former as establishing necessary conditions which must be satisfied if a chemical change is to be observed in a collision process, and of the latter as providing information which can lead to data on the thermochemistry of gaseous ions.

Returning to Figure 1, the period prior to 1950 may be thought of as the "one paper a year" era when the experimental facilities required for the study of ion collision processes were limited to a very few laboratories. In those days, mass spectrometers were used mainly by experimental physicists who were concerned primarily with mass determination and isotope analysis. Many physicists were interested in electrical phenomena in gases, the conduction of electricity through gases, and the related ion collision processes, but the emphasis was on the "electrical nature of the phenomenon" rather than the chemical nature of the ions and molecules involved.[4] It is this latter aspect which has occupied many chemical physicists for the past two decades and with which we are here concerned. Nonetheless, a great deal of gaseous ion chemistry, treated from the other standpoint, is to be found in reports of these studies, and if included in this review would inflate considerably our one-paper-a-year rate.

Furthermore, the consideration of this early work points a historical lesson. Gioumousis and Stevenson's[4] contribution to the understanding of very low velocity ion-molecule reactions was based on a paper by Langevin[5] published half a century previously, and designed to serve the needs of physicists interested in electrical phenomena in gases. Eyring, Hirschfelder and Taylor,[6] in 1936, using statistical rate theory derived a result identical with that of Langevin without reference to his pioneering effort. With the small bibliography that these workers faced, there was little reason for this oversight except for the isolation in which scientists in different disciplines tend to work.

There have been a number of review articles on ion collision processes in recent years which are cited in our bibliography but not reviewed here.[7-22] In this review we shall attempt to cover most of the literature on ion collision processes published prior to Nov. 1969. These studies commenced with the use of the ion sources of commercially available mass spectrometers as reaction chambers. This single-source technique generated data and questions which provided the impetus for development of more specialized apparatus such as the tandem mass spectrometer, and the ion cyclotron resonance and flowing afterglow techniques. Experiments based on these advances in technology have been used to develop a body of knowledge which can be divided into gaseous ion thermochemistry on the one hand and the dynamics of ion-molecule collision processes on the other, and we shall consider these separately.

We have tried, in this review, to trace the development of ideas on ion-molecule reactions, and to deal with problems rather than with experimental techniques. The bulk of publication is now so large that it is impossible, even in an article of this length, to discuss all of it at any length, and we have preferred to assess critically what we consider to have been key experiments rather than to attempt to be comprehensive.

We have begun with the historical aspect of our subject. This may well be tedious for the reader who has been intimately involved in work with ion-molecule reactions, but we feel it will be of value to the non-specialist and the student. Furthermore, we have attempted to restrict ourselves to literature dealing with quantitative measurements in systems of chemical interest. We have largely ignored qualitative observations on complex systems, and measurements on such topics as ion mobilities and atomic ion scattering processes.

To those who feel their work has been unjustly overlooked we offer our apologies.

HISTORICAL BACKGROUND

A. The Single Source Technique

The simple mass spectrometer ion source has been used both for the observation of ion-molecule reactions and, at a higher level of sophistication, for the measurement of their rates.

The observations go back to J. J. Thompson who noted the formation of protonated water and hydrogen molecules in his parabola mass spectrometer.[23] These observations served more to inhibit than to stimulate progress in the field, because they were not considered as interesting processes to be studied from the standpoint of chemical kinetics, but as obstacles to the use of mass spectrometry as an analytical tool. A review of the history of the discovery of the rare heavy isotopes of light elements is to be found in Aston's book.[24] The failure to exploit ion-molecule reactions in the decades prior to 1950 is also in part attributable to the state of the arts of vacuum techniques and ion detection. In order to follow the reduced yields of ions produced in collision processes in systems at low pressures, it was necessary to develop more sensitive ion detection techniques and to improve high-vacuum techniques. Little effort was applied to the problem of distinguishing between primary ions produced by electron impact, and ions produced by secondary processes, and indeed most of the early efforts were directed toward the suppression of secondary processes rather than their investigation. The principal criterion used to single out primary ions was their different dependence on pressure, although an alternative method was used by Hogness and Harkness[25] in which the path length of ions through un-ionized gas was varied at very low pressures. If the attenuation of primary ions is trivial, the variation of secondary ions will be linearly dependent on the primary ion concentration.

Perhaps the simplest method of distinguishing between primary and secondary ions is based on the observation that the rates of many collision processes depend on the velocity of the ion and on its residence time in the neutral gas. Washburn, Berry and Hall[26] found that they were able to suppress the H_3O^+ ion in the mass spectrum of water by increasing the ion-repeller potential used to drive ions from the ion source. Their objective was to improve the sensitivity of the analytical technique used for the determination of HDO in low concentrations of water, but the determination of the relative abundance of secondary to primary ions as a function of ion-repeller voltage forms a basis for quantitative determination of reaction rates in the single source mass spectrometer. This aspect of the technique was not recognized at the time.

The development of positive chemical interest in ion-molecule reactions was slow. N_2H^+ and HCO^+ were observed in N_2/H_2 and H_2/CO mixtures respectively, the former by Eltenton,[27] the latter by Kondrajew.[28] The observation of CH_5^+ in the mass spectrum of methane was well known in the fraternity of mass spectrocopists prior to 1950,[29] but the chemical implications were not appreciated until the work of Tal'rose in 1952,[30] and subsequent studies by Field, Franklin and Lampe,[31] and Stevenson and Schissler.[32]

By 1955, a small corpus of knowledge on ion-molecule reactions had been assembled, and workers such as Franklin, Field and Lampe, Tal'rose, and Lindholm,[33] who were to contribute so prolifically to the field in later years, had already published their first results. Nonetheless the majority of the work was qualitative. It was left to Gioumousis and Stevenson[4] to lay the basis for a quantitative treatment of ion-molecule reactions. They accepted the assumption made by Eyring, Hirschfelder and Taylor[6] that ion-molecule reactions occur at every collision, that is, every time the reactants approach within a certain critical impact parameter. This assumption was based on the idea that ions derived from neutral molecules are themselves free radicals and might be expected to react rapidly and with little or no activation energy. If this is true, then the problem of calculating a reaction rate reduces to one of calculating a collision number, and Gioumousis and Stevenson did this using the kinetic model developed by Langevin.[5]

A major contribution to the understanding of the phenomenon lay in their treatment of the kinetic energy distribution of ions in the mass spectrometer ion source. They recognized that ions formed by electron impact would initially have the Maxwellian kinetic energy distribution, with velocities determined by the temperature of the neutral gas in the source. Under the influence of the repeller field these primary ions are accelerated and reach a maximum kinetic energy by the time they arrive at the source exit slit. Beyond this there is a virtual discontinuity with a sudden increase in ion kinetic energy and decrease in neutral gas pressure. Collision processes take place with varying probability as the ion moves along its path from its point of origin in the ion beam to the source exit slit. Gioumousis and Stevenson showed how phenomenological cross sections could be derived from microscopic velocity-dependent Langevin cross sections by integration between limits set by the initial and final values of ion kinetic energy which in turn were determined by source geometry and repeller potential.

This approach was both more elegant and more successful than that of Franklin, Field and Lampe[34] who recognized the role of the ion-induced dipole interaction in determining the velocity-dependent phenomenological cross section, but chose initially to use average values of the ion kinetic energy rather than a realistic energy distribution. Stevenson and Schissler's[35] experimental investigations of the rates of reaction between ions and simple molecules were in principle well suited to the application of the Gioumousis–Stevenson (G–S) theoretical model, and in fact a high measure of agreement was found for such reactions as H_2^+/H_2 and Ar^+/H_2.

Franklin, Field and Lampe,[36] on the other hand, boldly ventured into the sphere of ion-molecule reactions in polyatomic systems, and generally found it difficult to obtain quantitative agreement between experiment and theory based on a velocity-dependent Langevin cross section. The G–S model predicted that the phenomenological cross section would vary according to the inverse square root of the repeller voltage $(E^{-1/2})$. Franklin and Field found dependence on E^{-1}. They suggested that, although their cross sections were of the order of magnitude predicted by the G–S theory, it was not a complete explanation and the finer details of the rates were determined by other unknown factors.

Hamill and his coworkers[37] investigated systems which varied in complexity from those of Stevenson and Schissler to those of Franklin, Field and Lampe. They found many reactions whose cross section showed an E^{-1} rather than an $E^{-1/2}$ dependence. They challenged the validity of the G–S model, and introduced a correction which assumed that above a certain critical energy, the ion-molecule reaction had a "hard sphere" cross section, which assumption led to the observed E^{-1} dependence.

Meanwhile, Tal'rose[38] was interested in the chemistry of protonated species. He proposed a method of bracketing bond energies in ions based on success or failure in the observation of ion-molecule reactions. He assumed that if a reaction were exothermic, the product ions could be readily observed, whereas if it were endothermic, no reaction could take place. For example,

$$H_2S^+ + H_2O \longrightarrow H_3O^+ + HS$$

occurs, whereas

$$C_2H_2^+ + H_2O \longrightarrow H_3O^+ + C_2H$$

does not so that if $PA(H_2O)$ represents the proton affinity of water, 163 kcal $\leqslant PA(H_2O) \leqslant$ 172 kcal. The weaknesses in the Tal'rose approach are of two types. With exothermic reactions, certain expected products may not appear because other more favorable reaction channels are open, or because there is no way of dissipating the heat released in the reaction; furthermore in single source experiments some exothermic processes may be masked by the presence of more abundant primary ions. Endothermic processes, on the other hand, may take place with the energy deficiency supplied from the kinetic energy of the reactant ion, or the internal excitation associated with the primary ionization process. Nevertheless, at the risk of some speculation, Tal'rose obtained useful information on proton affinities which, while not precise, was not otherwise available, and where more rigorous and conservative methods had failed. His innovation was to derive quantitative limits for thermochemical quantities from qualitative observations on the gross magnitudes of the rates of ion-molecule reactions, and without reference to quantitative rate studies.

The early experiments in the field of ion-molecule reactions described above were attractive not only on theoretical and chemical grounds but also because certain experimental difficulties were absent or could be neglected. It is worth outlining the factors in favor of these experiments in order to gain some insight into the difficulties facing later workers studying more complex systems. They were as follows:

(1) The proton affinity experiments of Tal'rose, and the proton or hydrogen ion transfer experiments of other workers involved the measurement of collected ion currents due to primary and secondary ions which differed in mass, according to whether hydrogen or deuterium was used, by only one or two mass units. The only interference with secondary peaks at these mass numbers comes from parent-plus-one and parent-plus-two isotopic contributions and these can be easily calculated and subtracted from apparent secondary ion intensities.

(2) The small mass difference between primary and secondary ions means that mass discrimination between them in the mass spectrometer will be negligible, as will differences in multiplier response. Thus the relative

ion currents recorded at the collector will represent accurately the relative ion concentrations in the source.

(3) If the secondary ion in an ion-molecule reaction is formed with a considerable amount of kinetic energy, it will be discriminated against strongly in the mass analyzer, and thus collected ion currents might not represent ion concentrations. It appears, however, that if there is indeed any significant liberation of kinetic energy in these processes (and this is very doubtful) then it tends to be deposited in the neutral species rather than with the ion.[39,40]

(4) A more compelling reason for the examination of simple reactions involving monatomic or diatomic ions or molecules is that the number of reaction channels available is extremely limited. While Stevenson and Schissler did not explicitly indicate their concern about competitive and consecutive processes, they were certainly aware of a better correlation in simpler systems between the measured cross section for appearance of product, and the calculated cross section for disappearance of reactant. Many of the problems of other workers arose from neglect of alternative reaction channels.

(5) Simple systems offer the best approximation to the Langevin assumption of a point charge approaching a polarizable neutral molecule with a negligibly small "hard core" cross section.

(6) Simple systems tend to have widely spaced energy levels and few internal degrees of freedom. The partition functions of the reactants and of a loosely bound activated complex will therefore be very simple and of a form which validates the approximations made by Eyring, Hirschfelder and Taylor in their statistical calculation of the rates of ion-molecule reactions, and which they indeed carried out for the specific case of the H_2^+/H_2 reaction.

We shall discuss more of the results obtained by the single source technique and the various modifications which it has undergone at a later stage.

B. Further Development of the Single Source

The single source technique, which was initially used to build up a body of descriptive chemistry of gaseous ions which included information on limiting values of heat of formation of product ions, moved in two radically different directions during the early part of the last decade. Techniques of high pressure mass spectrometry were developed which permitted the study of relatively high kinetic order reactions. These studies will be discussed in one of the following sections of this review. The other direction of investigation, the low pressure single source study, had as its goal the quantitative absolute reaction rate measurement and correlation of absolute rate data with theoretical models. This work was built on foundations established by Gioumousis and Stevenson, with major contributions by Franklin, Field and colleagues, and by Hamill and his students at the University of Notre Dame.

The quality of ion-molecule reaction rate coefficients or cross sections obtained in single source experiments depends to a large extent on the validity of the assumption that product ions are collected with approximately the same efficiency as reactants. This assumption appears to be relatively sound for reactions in which a hydrogen atom is transferred from a hydrogen molecule to a heavier ionic species yielding an XH^+ product, where X might be Ar^+, N_2^+, CO^+, and so on. The possible perturbation on the X^+ energy distribution is small, and even if the reaction were strongly exothermic and a considerable portion of the heat of reaction were deposited as kinetic energy in the products, most of this energy would be carried off by the neutral H atom. Stevenson and Schissler[35] were properly not concerned with this source of error in experimental measurements of phenomenological cross sections for most of the inert gas or diatomic molecule/hydrogen reactions. On the other hand, the remarkable agreement between calculated and measured values of rate coefficients in the hydrogen molecule ion reactions suggested to Reuben and Friedman[39] that either very little kinetic energy was deposited in the products in the course of reaction or that Stevenson and Schissler's data were in fortuitous agreement with theory. A third possibility seriously considered at that time was that the theoretical model of Gioumousis and Stevenson[4] was of limited reliability.

Isotopic hydrogen ion-molecule reactions were used by the Brookhaven group to test the hypothesis that indeed very little internal to kinetic energy transfer took place in these exothermic reactions. Reuben and Friedman[39] studied the kinetic energy distributions in D_3^+ and D_2^+ by careful peak shape measurements and independently by retarding ions in the source with a reversed repeller technique. Their results showed no significant different between D_3^+ and D_2^+ kinetic energy distributions. This was consistent with predictions made for simple reactions of this type by J. C. Polanyi.[40] Cross sections for the isotopic hydrogen ion-molecule reactions were found to be in excellent agreement with results of Stevenson and Schissler and the G–S theory. The only case that provided Stevenson and Schissler with some difficulty was the HD^+/HD system, where they found that the sum of the yields of H_2D^+ and D_2H^+ were considerably larger than theoretically predicted. The results obtained at Brookhaven did not support these earlier findings and suggested an interference of D_2^+ in the determination of H_2D^+.

The hydrogen ion-molecule reactions stand as a remarkable quantitative confirmation of theory with rate coefficients at relatively low velocities and intermolecular isotope effects both in precise agreement with the predictions of the G–S model.

The omission of reference to the helium/H_2 system was conspicuous in the Stevenson and Schissler paper dealing with rare gas/H_2 ion-molecule reactions. A hint of anomalous behavior with respect to theory was given in their analysis of the Ne/H_2 results. The Ne/H_2 and He/H_2 systems are special cases because the reactions of hydrogen molecule ions with these rare gases are not exothermic. Stevenson and Schissler incorrectly assumed that NeH$^+$ was formed by a reaction of Ne$^+$ ions with H_2, because this was the only available exothermic reaction channel. The Brookhaven group, concerned with the problem of energy transfer, investigated these systems in greater detail to study the possibility of driving the endothermic reaction channels:

$$H_2^+ + He \longrightarrow HeH^+ + H$$
$$H_2^+ + Ne \longrightarrow NeH^+ + H$$

with kinetic energy supplied to the H_2^+ ions from the repeller voltage applied in the single source experiment.[41,42] The large difference between the appearance potentials of H_2^+ and He$^+$ or Ne$^+$ made identification of reactant ions simple. Investigations of appearance potentials of the product ions immediately showed that, contrary to expectation, reaction was proceeding in both cases through the "endothermic" channels. A peculiar energy dependence was observed for these reactions with some evidence of an energy barrier or threshold, and at higher ion kinetic energies, a rather rapid decrease in rate with increasing repeller voltage. This decreasing rate with increasing repeller voltage had been previously observed in the isotopic hydrogen ion-molecule reactions. The overall energy dependence in these systems is complex, and has so far not been quantitatively explained. The reactions show clearly the effect of vibrational energy deposited in the hydrogen molecule ions in the electron impact process by which they are formed, and these excited ions appear to have a greater probability of reaction. The assumption that one is dealing with ground state ions in a single source mass spectrometer experiment is destroyed by these studies because the bulk of the processes observed involve reactions of excited ions. The discrepancies between the experiment and theory are minimized if one considers the populations of excited ions produced by electron impact.

In the case of reactions which are endothermic for ground state ions and exothermic or thermoneutral for excited ions, it is found that reaction cross sections for excited ions are significantly larger than those for ions which have insufficient internal energy for reaction but a surplus of translational energy. This shows that the role of translational energy in driving an endothermic reaction is a minor one.

Intramolecular isotope effect studies on the rare gas/HD systems[43] show that in reactions of HD^+ with helium or neon, kinetic energy is converted to internal energy which eventually destroys the XH^+ or XD^+ ion-molecule reaction product. The inert gas/hydrogen ion-molecule reactions proved to be a rich source of data showing that with exothermic reactions in the Ar^+/H_2, and Kr^+/H_2 systems, the predicted G–S velocity dependence was readily observed. With endothermic H_2^+/He and H_2^+/Ne systems, where kinetic energy plays a significant role in bringing the reactants into sufficiently close contact to bring about reaction, one cannot expect to observe phenomenological cross sections varying with the reciprocal velocity of reactant ions (i.e., with $E^{-1/2}$). Nor can one expect cross sections equal in magnitude to the Langevin cross section if only a fraction of the ions have sufficient energy to react in isolated collisions. For exothermic processes, the G–S model provides a useful upper limit for low velocity (< 10 eV laboratory kinetic energy) ion-molecule reaction cross sections.

The failure to observe reaction through the strongly exothermic He^+/H_2 and Ne^+/H_2 channels provides an interesting example of reactions which are prohibited primarily because energy surfaces for initial and final states appear to intersect only in regions where distances between He^+ and H_2 and Ne^+ and H_2 respectively are very small. The energy of these reactions must be disposed of as kinetic energy in the products because there are no accessible internal energy states available in the neutral helium or H^+ products, and H atoms do not appear to be a suitable energy sink for the amount of energy that has to be dissipated. The conversion of internal to kinetic energy was shown again to be a process requiring very small impact parameter collisions.

The single source work on low energy elemental ion reactions which gave experimental results in very good agreement with G–S theory was not initially accepted without reservation because of difficulties encountered by many workers in their attempts to carry out single source experiments. The H_2^+/H_2 and rare gas/H_2 systems appeared to be exceptions to general observations. Phenomenological cross sections in more complex systems were frequently significantly smaller than those predicted by theory, and the dependence of rate on repeller voltage was in many cases an inverse energy (E^{-1}) rather than an inverse square root energy dependence

($E^{-1/2}$). In many cases no simple correlation of phenomenological cross section with repeller voltage could be made. Henchman in his 1965 review[15] cautiously noted his general reservations on repeller studies in the single source.

The Brookhaven group attributed these difficulties to problems associated with conversion of translational to internal energy in systems which were produced originally with non-Maxwellian internal energy distributions and to the limitations of the single source experiment in detecting all channels of reaction. Guidoni and Friedman[44] selected the methane system as an exhaustively studied case testifying to the general weakness of the G–S model, and reinvestigated it (Figure 2). They measured the disappearance cross section of CH_4^+ in a single source experiment and compared the velocity dependence of this cross section with the phenomenological formation cross section for formation of CH_5^+. The principal channel of ion-molecule reaction in the methane system had been assumed

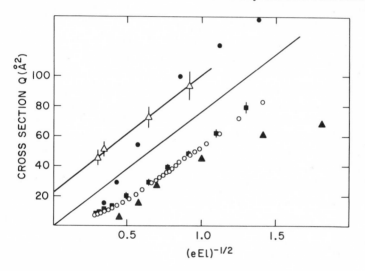

Fig. 2. Single source studies of the $CH_4^+ - CH_4$ system. The solid line through the origin is a plot of the G–S reaction cross section showing the linear dependence of cross section with reciprocal ion velocity. CH_5^+ formation cross sections obtained by Stevenson and Schissler are given by solid shaded circles ●; Kubose and Hamill, open circles ○; Field, Franklin and Lampe, solid triangles ▲; Guidoni and Friedman ■. The CH_4^+ disappearance cross sections obtained by Guidoni and Friedman are presented with the open triangles △. These points were uncorrected for collision induced dissociation of CH_4^+ in the mass spectrometer analyzer tube but show the expected linear inverse velocity dependence predicted by the polarization model.

to be $CH_4^+ + CH_4 \rightarrow CH_5^+ + CH_3$. The CH_4^+ disappearance cross section was found to vary linearly with reciprocal average ion kinetic energy, in excellent agreement with theory. The magnitude of the disappearance cross section was systematically larger than the G–S theoretical cross section. These differences in cross section were explained by postulating collision induced dissociation of CH_4^+ outside the mass spectrometer ion source. When suitable correction was made for processes which contributed to a net loss of CH_4^+ outside the ion source, there was excellent agreement between theory and experiment. The CH_3^+ data reported by Guidoni and Friedman were consistent with results previously published by Kubose and Hamill.[37g] Guidoni and Friedman showed that by systematically varying the ionizing electron energy and the repeller voltage, cross sections for CH_5^+ formation could be obtained which were in quantitative agreement with G–S theory. The problem was simply one of reducing the internal and translational energy in CH_4^+ to a level that would not permit either the direct collision-induced dissociation of CH_4^+ or the subsequent reaction channel $CH_5^+ \rightarrow CH_3^+ + H_2$.

The $CH_4^+ + CH_4$ study pushed the single source technique to its limits and demonstrated a need for the recognition of energy transfer processes in opening up competitive and consecutive reaction channels. The success of the single source technique was best established with systems which were capable of only a very few (or one) channels of reaction. The major advantage in the single source technique is the general advantage of low resolution measurements where problems in collection efficiency and sensitivity are minimized, and quantitative results can be obtained for comparison with theoretical models. The methane study demonstrated the care required for the proper application of the single source technique and served to dispel many of the doubts and reservations about the G–S theory. These conclusions were strongly reinforced subsequently by the use of more sophisticated techniques particularly the flowing afterglow experiments (Section IV-E-3).

C. The Development of the Tandem Mass Spectrometer

1. Lindholm

The tandem mass spectrometer offers many advantages over the single source technique for the study of ion-molecule reactions. It provides a means by which a beam of ions of a particular species can be selected from the mixture of ions produced in the ion source, and injected with controlled kinetic energy into a collision chamber.

However, the earliest tandem mass spectrometers were not built with this purpose in mind. A two-stage machine was constructed by Inghram and

Hayden, in the years prior to 1954,[45] designed to facilitate the detection of rare isotopic species, and in 1954 Lindholm[46] reported results obtained on a tandem mass spectrometer designed with what has come to be known as perpendicular geometry. Ions were produced and mass-analyzed in an 18-cm radius, 180° mass spectrometer. The resultant beam was decelerated by a series of retarding slits and directed into a collision chamber. Ions were extracted from the collision chamber at right angles to the incident beam and mass analyzed by a 25-cm radius mass spectrometer. The perpendicular geometry discriminates against secondary ions which have acquired appreciable forward momentum from the primary ions in the incident beam. The whole apparatus was enclosed in an evacuated bell jar 1.3 meters in diameter and 1.1 meters high which made alignment of the sections easier, but resulted in problems due to lack of efficient differential pumping.

The initial purpose of this machine was the measurement of the internal energy distributions of ions using the concept of recombination energy (RE). This is defined as the energy given by an ion to a neutral species when the ion is neutralized by it. In certain cases (for example, the hydrogen atom) the RE is equal to the ionization potential (IP). In the process

$$H^+ + M \longrightarrow H + M^+$$

(where M is a neutral molecule) in the absence of low-lying metastable states of hydrogen, the energy given to M must be 13.6 eV, the IP of hydrogen.

The difference between the recombination energy and the ionization potential, however, is significant. The ionization potential is the energy of transition from the ground state of the molecule to the ground state of its ion, although this ideal has probably not been achieved in all measurements. Recombination energy does not necessarily correspond to a $0 \leftarrow 0$ transition, and frequently measures an energy considerably smaller than the ionization potential, because of the high probability of recombination to an excited state of a neutral. The concept is especially important in the case of ions such as ArH^+ and H_3^+ which do not exist as neutral molecules.

Lindholm argued that if he directed a beam of ions into a collision chamber of neutral gas, and collected product ions at right angles to the beam, the only product ions he would see would be those formed by accidentally resonant charge exchange with no transfer of momentum, and an amount of energy exactly equal to the RE of the ion would be deposited in the neutral molecule. From this he hoped to estimate distribution of energy states in the ion beams with which he dealt.

For example, he bombarded hydrogen with atomic oxygen ions obtained variously from carbon monoxide, carbon dioxide, and nitrous oxide and measured the relative cross sections for production of H_2^+ and H^+.*[46c] These values are shown in Table I-A. In all cases, H^+ was obtained. Since the process

$$O^+ + H_2(^1\Sigma_g^+) \longrightarrow O + H^+ + H \tag{1}$$

TABLE I

(A) Relative Cross Sections for Bombardment of Hydrogen with O^+ Ions from N_2O, CO or CO_2.

		Relative cross sections									
		Production of H_2^+					Production of H^+				
Energy of bombarding ions, eV		25	50	150	400	900	25	50	150	400	900
Source of bombarding ions (created by 100 volt electron impact)	CO	—	9	9	9	9	—	0.5	0.25	0.15	0.05
	N_2O	4.2	4.5	6.0	8.0	7.0	0.65	0.45	0.18	0.12	0.05
	CO_2	1.2	2.4	3.0	4.0	4.0	0.40	0.30	0.15	0.09	0.04

(B) Proportions of Various Excited States of O^+.

State of ion	State of Product oxygen atom	RE (eV)	Proportion of each state of O^+ in beam obtained by 100 eV electron impact		
			CO	CO_2	N_2O
^4S	^3P	13.62	30%	60%	
^2D	^3P	16.94	30%	30%	Intermediate between CO and CO_2
	^1D	14.98			
^2P	^3P	18.64	40%	10%	
	^1D	16.67			
	^1S	14.45			

* In some early work, Lindholm quoted absolute cross sections, but ceased doing this when it became apparent that the efficiency of ion extraction from the collision region was unknown.

requires the deposition in the hydrogen molecule of at least 18.05 eV, it is clear that atomic oxygen ions from all sources must contain the ^2P state of O^+ which is known from spectroscopic data to have an RE of 18.64 eV and is the only one with a sufficiently high RE to give H^+. From similarity of the relative cross sections in Table I it would appear that the ^2P state is equally abundant in the O^+ beams from both CO and N_2O but less abundant in the O^+ beam from CO_2.

The ^2P state of O^+, however, can lead not only to H^+ by process (1) but also to H_2^+ by the process:

$$O^+ (^2P) + H_2(^1\Sigma_g^+) \longrightarrow O(^1D) + H_2^+(^2\Sigma_g^+) \qquad (2)$$

The oxygen ion neutralization makes available 16.67 eV which is sufficient to ionize the hydrogen. The fact that the relative cross sections for H_2^+ production differ using O^+ ions from N_2O and CO shows that another state of O^+ with similar RE is present. Lindholm identifies this with the ^2D state which he claims is more abundant in O^+ from CO than from N_2O or CO_2. From this and similar measurements he endeavors to deduce the proportions of the various excited states of O^+ present in his various beams (Table I-B).

Lindholm's technique is useful for detecting the existence of metastable states of ions in beams. It is not sensitive to small variations in recombination energy, however, and the measurements of the amount of energy deposited are not precise. Furthermore, it can be criticized on the grounds that the cross sections for two different energy transfer processes may be radically different, and therefore the relative populations of excited states deduced from these cross section measurements may be grossly distorted. Thus the information obtained by this technique is largely qualitative.

An unresolved question emerging from this work is the problem of which recombination processes will be favored when more than one is possible. Lindholm[46f] shows that for the two REs of Xe^+ in methanol (12.13 and 13.44 eV) the former is twice as important as the latter, and a similar situation obtains with Kr^+ (14.00 and 14.67 eV). It is interesting to speculate as to whether or not this is a general phenomenon and that processes involving the transfer of a minimum amount of energy are favored.

Having obtained tables of recombination energies and estimated the proportions of excited ions in his beams, Lindholm devised a novel extension of his technique by bombarding polyatomic molecules (which may be assumed to have quasi-continuous energy levels and therefore to be able to undergo accidentally-resonant charge exchange with any bombarding species) with ions of known recombination energy. For example, by bombarding methanol with ions of steadily increasing recombination energies, he was able to deposit in the methanol steadily increasing known amounts of energy, and thus to obtain mass spectra of methanol for various energy insertions.[46f] For the detailed results of these experiments the reader is referred to the bibliography in a review by Lindholm,[18] but it is interesting to note, for example, that when methanol is bombarded with H^+ ($RE = 13.6$ eV), the fragment CHO^+ is not formed, whereas with F^+ ($RE \approx 20$ eV) it is the most abundant ion in the spectrum.

Lindholm's experiments illuminate the mechanism of the dissociation processes of energetic ions and point to a refined technique for the application of mass spectrometry to structural investigations. Electron bombardment is a crude way of obtaining a mass spectrum, in that a spread of energies is deposited. Bombardment by simple monatomic ions can deposit precise amounts. (Lindholm's early work was with monatomic and diatomic species because these have fewer degrees of freedom to accept energy in the recombination process, and less possibility of an excited neutral species being produced with a consequent error in the supposed amount of energy deposited.) This emphasized the point that electron bombardment mass spectra are only one of a whole class of mass spectra, and indeed Lindholm's ion bombardment mass spectra represent a sort of *controlled* chemi-ionization.

Several aspects of Lindholm's work are, in retrospect, open to criticism. First, his assumption that by collecting secondary ions at right angles to the primary ion beam, he was discriminating totally against secondary ions with kinetic energy, turned out not to be valid. In particular, his work supported values of 7.373 and 9.605 eV for the bond dissociation energies of nitrogen and carbon monoxide respectively, whereas spectroscopic investigation later supported higher values of 9.6 and 11.1 eV. Giese and Maier[47] suggested that in spite of his precautions, Lindholm's second mass spectrometer must have accepted ions formed partly by kinetic to internal energy transfer.

Second, the obtaining of mass spectra of polyatomic molecules by ion impact raises the question of how energy is transferred. Does it happen via a Franck–Condon transition? Does energy transfer take place more readily if the molecule has an appropriate energy level which can take up the energy to be deposited, and is it necessary to consider the mechanism of fragmentation? If ethylene is bombarded with H^+, 13.6 eV is deposited but two processes could take place:

$$H^+ + C_2H_4 \longrightarrow H + C_2H_4^+ \longrightarrow C_2H_3^+ + H + H$$

or

$$H^+ + C_2H_4 \longrightarrow C_2H_5^+ \longrightarrow C_2H_3^+ + H_2$$

There is a problem as to the exact significance which can be attached to the RE required to cause the appearance of $C_2H_3^+$.

Third, Lindholm's task was made much more difficult by experimental problems in obtaining ion beams of known internal energy distribution, so that he was not always certain what he was starting with, and his experimental design to estimate the amounts of excited species in beams tended to produce results which were not very accurate. For example, he considered his H_2^+ beams as having REs of 16.4 to 17.4 eV and possibly from about 13 to 14 eV, which is crude compared with the later estimates of Leventhal and Friedman.[48] Nonetheless, we should not minimize Lindholm's achievement in pioneering the identification of internal energy states in ion beams and the examination of the mass spectra of molecules in which precise amounts of energy had been deposited.

2. Cermak and Herman

Two mass spectrometers which might be classified as semi-tandem machines were built by Cermak and Herman in the late 1950s.[49] Both were Nier mass spectrometers with ion sources modified to give the effect of perpendicular and longitudinal geometry.

In the "perpendicular" ion source,[49a] the electrons from the filament were not initially given sufficient energy to ionize neutral gas in the source and they thus travelled across the source without producing ions. On leaving the source, however, they were accelerated by a potential on the electron trap, and acquired sufficient energy to ionize some of the gas in the region between the trap and the electron exit slit in the source. These ions, being positively charged, were repelled by the trap and travelled backwards into the source parallel to the electron beam. There they were able to undergo ion-molecule reaction and the products could be analyzed in the mass spectrometer. As in Lindholm's experiments, the geometry of the system discriminated strongly against the products of processes where momentum was transferred, and the apparatus was therefore of greatest value in investigating charge transfer and related processes. However, Cermak and Herman did attempt to measure certain absolute cross sections for the reactions of excited ions in N_2, O_2, CO, SO_2, CO_2, COS and CS_2. Their values are very small, and as the primary ions have ill-defined energies it is difficult to assess their significance. They are probably open to the same objections which caused Lindholm in his later work to quote only relative cross sections.

The Cermak and Herman technique was later applied by Henglein and Muccini[50] to electron and proton transfer reactions in which little transfer of momentum would be anticipated. For example, with methane in their source, they observed the species CH_4^+ and CH_5^+. The former is clearly produced by charge exchange but the latter could be formed either by H^+ transfer from a primary ion to a neutral CH_4 molecule, which process could be assumed to involve little momentum transfer (that is, a "stripping" mechanism), or by the conventional CH_4^+/CH_4 reaction involving a secondary CH_4^+ ion. On the basis of graphs of CH_4^+/CH_5^+ ratios vs primary ion energy, which drop steeply as energy increases, Henglein and Muccini conclude that a stripping model is a more satisfactory explanation of their results than complex formation, at any rate for primary ions with more than 5-eV energy, since if the major reaction taking place were between secondary CH_4^+ ions and CH_4 molecules, then no variation of this ratio would be expected. On this interpretation, the variation in the CH_5^+/CH_4^+ ratio is attributed to the effect of energy on the cross sections of the charge exchange and proton transfer processes, though it is also possible that some momentum transfer takes place in the proton transfer reaction and at high energies this leads to a proportion of CH_5^+ ions being lost. Neither of these hypotheses is a satisfactory explanation of the measured isotope effect (CD_5^+/CD_4^+ is about one-fifth CH_5^+/CH_4^+) and this may be related to recent observations by Mahan et al.[121c] on the N_2^+/CH_4 system where a similar effect is observed.

Cermak and Herman also succeeded in adapting a conventional ion source to act as a "longitudinal" tandem mass spectrometer.[496] In effect, they built a reaction chamber on the front of their ionization chamber. Ions formed in the usual way were drawn into the reaction chamber by a small potential difference. Secondary ions formed in this region were extracted, and passed into the mass analyzer together with unreacted primary ions. The apparatus permitted easy detection of charge transfer reactions, but was in fact used mainly for chemi-ionization and Penning ionization studies.

3. Giese and Maier

The innovation introduced by Giese and Maier in the field of tandem mass spectrometry was the use of longitudinal as opposed to perpendicular geometry, that is, their second mass spectrometer accepted ions travelling out of the collision chamber more or less parallel to the primary ion beam. This had the effect of increasing their intensities of secondary ions, and of transmitting unscattered primary ions. The point of Giese and Maier's early work was to avoid the ambiguities associated with the nature and energies of the primary ions involved in ion-molecule reactions. For example, in a single source it is difficult to distinguish between

$$Ar^+ + H_2 \longrightarrow ArH^+ + H$$

and

$$Ar + H_2^+ \longrightarrow ArH^+ + H$$

and not always easy to work out the rates of reaction of ions of a particular energy from the results of experiments on ions with a spread of energies.

Their instrument consisted of a 1-in. radius, 90° source mass spectrometer, which could operate at very low ion accelerating voltages. The primary spectrometer exit slit was followed by a field-free region and then by a collision chamber. Product ions emerged from the collision chamber and passed through a grid and a series of rings designed to provide the boundary conditions for a uniform field, followed by a quadrupole lens and two electrodes to vary the direction of the beam. By means of these the beam could be focused on the entrance slit of the second, 12-in. radius 60° mass spectrometer.

Giese and Maier used this apparatus to try to measure absolute reaction cross sections. For a particular ion-molecule reaction in their system, and a particular primary ion energy E, the cross section $\sigma(E)$ is given by

$$\sigma(E) = \frac{N_s}{N_p\, nl} \tag{3}$$

where N_s is the number of secondary ions formed per unit time. N_p is the number of primary ions incident on the target gas per unit time, n is the concentration of target molecules and l is the length of the reaction path. For all the reactions they studied, $N_s \ll N_p$, whence

$$\sigma(E) = \left(\frac{I_s}{I_p\, nl}\, \frac{A_s}{A_p}\right) K_1 K_2 \tag{4}$$

where I_s and I_p represent the recorded primary and secondary ion currents. A_p and A_s are the areas under the primary and secondary ion profiles respectively, which they obtained by scanning the ion beam across the object slit using the aforementioned electrodes. This was supposed to allow for the fact that secondary ions can acquire transverse components of momentum greater than those of primary ions, and they will therefore come to a broader focus. In the case of heavy ions striking light molecules, for example the reaction Ar^+/H_2, the primary and secondary ion profiles were similar, but when the reactants were of similar mass, for example the reaction H_2/H_2^+, the secondary peak was considerably broadened.

The factor K_1 corrects for the fact that the response of secondary electron multipliers is mass dependent. K_2 is another correction factor based on the scattering of secondary ions. Giese and Maier claimed that the geometry of their apparatus was such that all primary ions entering the reaction region emerged again unless they had interacted with a neutral molecule. What happened to the secondary ions, and hence the value of K_2, depended on the ratio γ of the speed of the center of mass of the system in the laboratory to the speed of the reactant ion in center of mass coordinates. For the impact of a heavy ion on a light neutral molecule, γ is large, and vice versa.

If γ is small, many secondary ions are widely scattered and do not emerge from the collision chamber; if it is large some secondary ions formed in the field-free region in front of the collision chamber will be carried through and eventually detected. In their first paper, Giese and Maier calculated the factor K_2, but in their second they assumed it to be unity.[51] Initially they looked at two groups of reactions:

$$Ar^+ + D_2 \longrightarrow ArD^+ + D$$
$$N_2^+ + D_2 \longrightarrow N_2D^+ + D$$

and

$$D_2^+ + N_2 \longrightarrow N_2D^+ + D$$
$$H_2^+ + He \longrightarrow HeH^+ + H$$
$$D_2^+ + Ar \longrightarrow ArD^+ + D$$
$$H_2^+ + H_2 \longrightarrow H_3^+ + H$$

For the Ar^+/D_2 and N_2^+/D_2 systems, their measured cross sections were significantly above those predicted by the Gioumousis-Stevenson theory for all energies of bombarding ions. In the cases of H_2^+/H_2, D_2^+/N_2, and D_2^+/Ar, the cross sections were much higher than G–S values at low energies, but the experimental values dropped faster than the theoretical ones and the high energy values were much lower than G–S predictions. The H_2^+/He cross sections were all lower than theoretical, but at low energies they were still higher than the values reported by von Koch and Friedman.[41]

It is legitimate to examine Giese and Maier's technique more closely because neither the magnitudes of their cross sections nor their dependence on ion velocity agreed with the work of previous investigators or with the predictions of available theoretical models.* The potential function $V(r)$ for the interaction of an ion and a molecule separated by distance r is normally obtained by the superposition of a Lennard-Jones interaction on the interaction of a point charge with a polarizable molecule which gives

$$V(r) = \frac{a}{r^{12}} - \frac{b}{r^6} - \frac{c}{r^4} \tag{5}$$

* A subsequent investigation of the H_2^+/H_2 system by Vance and Bailey[52] showed an energy dependence of the reaction similar to that found by Giese and Maier, but absolute cross sections were not measured.

where a, b, and c are positive constants. The G–S theory is considered to be valid if r is sufficiently large for the first two terms to be negligible in comparison with the last, and this results in the observed dependence of cross section on $E^{-1/2}$. At high energies and small impact parameters, the r^{-6} and r^{-12} terms can no longer be ignored and the cross section would be expected to depend inversely on a power of E less than a half, that is, it begins to approximate to a "hard core" cross section, more or less independent of energy. Giese and Maier's cross sections at high energies behave in the opposite fashion, varying sensitively with E.

If we assume that the G–S theory gives the energy dependence to which Giese and Maier's results should conform, the following explanations for the discrepancies might be advanced: Giese and Maier assume, in deriving their correction factor K_2, that "there is no connection between the initial direction of the reacting primary ion and the direction of the observed secondary in center of mass coordinates", that is, the decomposition of the transition complex is isotropic. Henglein[53] has pointed out that if the reaction is partly of the "stripping" type where no true transition complex exists, then the yield of products in the forward direction would be abnormally large. Giese and Maier's data, taken on an apparatus with longitudinal geometry would then be grossly overcorrected, and the fact that their cross sections at low energies were abnormally high would be explained.

The explanation of the low cross sections at high energies for certain of the reactions is less clear. It is possible that the discrepancy is due to some form of discrimination against secondary ions which Giese and Maier did not consider. It is noticeable that the discrepancy occurs in the reaction systems where a light ion strikes a heavy molecule, e.g. D_2^+/N_2, D_2^+/Ar, but not in systems where the reverse applies (N_2^+/D_2, Ar^+/D_2). A more likely explanation, however, is that Giese and Maier neglected consideration of other reaction channels. In the "light-on-heavy" reactions, a substantial amount of energy is put into the center of mass and is therefore available to drive endothermic processes. The obvious alternative channel is the collision-induced dissociation of the impacting hydrogen or deuterium molecule ions. In the cases of N_2^+/D_2 and Ar^+/D_2, however, far less energy is put into the center of mass in Giese and Maier's experiments and therefore the endothermic reaction channel is of less significance and the apparent cross section does not drop below the G–S value.

Giese and Maier undoubtedly recognized the above points and in a much later paper Maier[54] estimates K_2 by an experimental method and investigates the following reactions:

$$He^+ + O_2 \longrightarrow O^+ + O + He$$
$$He^+ + N_2 \longrightarrow N^+ + N + He$$
$$He^+ + N_2 \longrightarrow N_2^+ + He$$

In these systems, where he studies all possible channels and makes the appropriate allowance for non-isotropic scattering in the center of mass, Maier obtains cross sections and energy dependences which agree remarkably well with the results of other workers and with the predictions of the G–S theory.

Before Maier's work on helium ion reactions, however, Giese[55] had queried the validity of the G–S model. Following some earlier work by Durup,[56] Giese considered the importance of quantum mechanical resonance forces in ion-molecule reactions. He altered the Langevin r^{-4} potential by considering quantum mechanical resonance forces and the r^{-6} term in the Lennard-Jones interaction. He predicted enhanced rates for ion-molecule reactions at high energies when the ionization potential of the neutral molecule approached the recombination energy of the reactant ion.

To sum up, Giese and Maier pioneered the longitudinal tandem mass spectrometer. They were able to point out sources of error in Lindholm's work, but their own early measurements of absolute cross sections appear to be vitiated somewhat by their failure to consider non-isotropic scattering and alternative reaction channels.

III. THERMOCHEMISTRY OF IONS

A. Introduction

Ions are closely related to free radicals in their reactivity and indeed many of them are in fact free radicals. For example, there is a formal similarity between free radical anionic and cationic polymerization, both in conventional polymerizations and in high pressure source experiments. A fundamental experimental difference between ion-molecule reactions and thermal reactions involving neutral species is that neutral species normally have a Maxwell–Boltzmann energy distribution, and excited states, when they occur, are readily recognized and taken into account. Ions formed by electron impact frequently are generated with a broad distribution of excited states. Their thermochemistry from the point of view of experiment is correspondingly more difficult. Heats of formation of ions, for example, may be estimated from ionization potential measurements, but if the ion and its parent molecule differ markedly in structure, it is likely that Franck–Condon transitions from the ground state of the molecule will give a vibrationally excited state of the ion, and it might not be possible at all to detect ions formed by the $0 \rightarrow 0$ transition. In this case, the electron impact ionization potential will be in error by the extent of the excitation energy. Nonetheless, if gaseous ions and their reactions are to be integrated into the corpus of general chemistry, it is necessary to know about their thermochemistry. Furthermore, information on ionization potentials, recombination energies and bond strengths in ions is of importance not only from the point of view of providing information which will enable us to predict necessary conditions for reaction, but also to provide a body of information which may be used to estimate the chemistry of ions in terms of general chemical affinities, and to predict the stability and properties of reactive intermediates which cannot be isolated for independent study.

In the following section, we discuss data on the thermochemistry of ions obtained by methods involving collision processes. We propose in addition to list sources of data obtained by other methods including theoretical calculations,[57] and to use results obtained in these ways. We discuss briefly the experimental methods available and in more detail the application of these to thermochemical problems such as the measurement of the proton affinities of hydrogen, water, and ammonia.

B. Ionization and Appearance Potentials

Heats of formation of stable ions which can be formed from stable molecules by primary ionization can be calculated if the ionization or appearance potentials are known; that is, in addition to the normal contributions to ΔH_f it is necessary also to take the ionization energy into account. Franklin and Field[57a] were among the first to tabulate ionization and appearance potentials and to deduce heats of formation from them which could be used in Hess' law calculations. These data were obtained mainly by electron impact and are subject to experimental error and to problems of energy distribution in ionization products. Misleading ionization potentials may be obtained if ions are formed in high vibrational states, and appearance potential measurements carry the additional risk that the neutral fragment might be formed with kinetic energy. Stevenson's rule,[58] based on a study of paraffin hydrocarbons, suggests that with a molecule R_1R_2 where $I(R_1) < I(R_2)$, the process $R_1R_2 \rightarrow R_1^+ + R_2$ will take place to give R_1^+ and R_2 in their ground states; on the other hand the process yielding R_2^+ will occur with excitation.

Appearance and ionization potential measurements by electron impact also suffer from errors due to the spread of energies in electron beams obtained by conventional methods. Various techniques have been adopted to overcome this, and these have been reviewed by McDowell.[59] They fall into three main classes. Three other methods, not involving electron impact, are also summarized here.

(1) Graphical and deconvolution techniques, for example, the methods of McDowell and Warren,[60] Lossing,[61] and Morrison.[62] These accept

that a spread of electron energies is inevitable, and attempt to allow for it by mathematical treatments of varying degrees of sophistication of experimental ionization efficiency curves. The existence of on-line computers has led to a revival of interest in these techniques, but it is difficult to know how much reliance can be placed on fine structure developed by double differentiation of apparently smooth ionization efficiency curves.

(2) Monoenergetic electron techniques, for example, Marmet and Kerwin's,[63] and Clarke's[64] electron energy selectors. Ionization efficiency curves are obtained using electrons of uniform energy generated by some design of electron monochromator. Marmet and Kerwin claimed that their instrument would resolve the $2P_{3/2}$ and $2P_{1/2}$ states of argon separated by 0.18 eV, but its design involved overcoming serious problems of space charge and perturbation of electron orbits. In spite of great ingenuity lavished on its construction, Marmet and Kerwin were able to obtain beams of only 10^{-7} amp with a 100-mV energy spread.

(3) Retarding potential difference method, for example, the methods of Fox et al.[65] and Cloutier and Schiff.[66] These methods involve the use of a retarding potential so that the energy spread of a conventional electron beam is cut off sharply at an energy chosen to correspond roughly to its maximum intensity. Variation of the retarding potential by a small amount ΔV allows through, an additional "slice" of the electron energy distribution curve which results in an increase ΔI in the ion current. This difference is due to electrons lying within the energy band ΔV, and the plotting of ΔI against different values of V gives an ionization efficiency curve equivalent to that which would have been obtained with monoenergetic electrons. The retarding potential difference methods seem to give slightly lower values than other methods, and are possibly influenced by the build-up of surface charges, etc., in the source. There is some evidence that they are subject to systematic errors of unknown origin.[67,68]

(4) Ion impact spectroscopy, for example, Lindholm.[46] As discussed previously, the condition for charge exchange in Lindholm's apparatus, if he is to detect the product ion, is that RE (ion) = IP (neutral). Lindholm's results provide a way of estimating the lowest appearance potential of a molecule, and in some cases throw light on higher ionization potentials.

(5) Photoionization. In the most modern applications of this technique, a vacuum monochromator is arranged to irradiate material in the source of a mass spectrometer.[69,70] The wavelength of light at which any given ion appears can be determined and hence its IP or AP calculated. The method offers very good energy resolution (about 0.001 eV) and should, in theory, be the complete answer to the problems associated with IP and AP measurement. The main drawback of the technique is the very low ion currents it is possible to produce.

(6) Photoelectron spectroscopy. This is another photon impact method. Material is irradiated with light of very short wavelength (usually the helium resonance line at 584 Å = 21.21 eV) and the energy spectrum of the photoemitted electrons is scanned. It should be possible to obtain both adiabatic (0 → 0 transition) and vertical (Franck–Condon) ionization potentials, the former from the highest electron energy within a group, the latter from the energy of the most abundant electrons. In general, the results obtained have agreed well with values obtained from, for example, Rydberg series. There have been a few disagreements with electron impact values, which were ascribed, among other things, to autoionization in the latter case.[71,72]

C. Kinetic Approach to Thermochemistry

The above methods, particularly those involving photon impact, make it possible to obtain values for ionization potentials of stable molecules, and hence ionic heats of formation of a high proportion of fragments arising from them which exist in the gas phase. Ion impact spectroscopy can also be used to give relatively crude values for recombination energies of such molecules as H_3^+ and ArH^+, but in general the thermochemistry of protonated molecules and exotic species must be determined by other methods. These may be illustrated by recapitulation of the continuing history of the experimental determination of three outstanding thermochemical quantities, the proton affinities of hydrogen, water, and ammonia, followed by a review of related experiments.

1. Proton Affinity of Hydrogen

The proton affinity of hydrogen in the gas phase was measured nearly thirty years ago by Russell, Fontana, and Simons,[73] (See Table II) who studied the scattering of low velocity protons in hydrogen and calculated a potential function and a proton affinity from their attenuation data. Their value of $PA(H_2) = 80.7$ kcal bears comparison with many modern values.

Another early measurement of the proton affinity of hydrogen was made by Tal'rose by the method already described.[38] He was able to observe the reaction:

$$H_2^+ + H_2 \longrightarrow H_3^+ + H$$

but not

$$H_2^+ + C_2H_2 \longrightarrow H_3^+ + C_2H$$

whence he deduced that 61 kcal $\leqslant PA(H_2) \leqslant$ 79 kcal on the assumption that the former reaction must be exothermic and the latter endothermic.

TABLE II

Experimental Values of Proton Affinities

	Worker	Method	Proton affinity (kcal/mole)		Refer-ence
			Mini-mum	Maxi-mum	
HYDROGEN	Russel, Fontana, and Simons	Low velocity ion scattering	80.7		73
	Tal'rose and Frankevich	Sucess or failure in observing ion-molecule reactions	61	79	38
	Aquilanti and Volpi	Success in obser-ving reactions	—	85	74, 75
	Chupka, Russell, and Refeay	Photoionization	78.3	—	80
	Leventhal and Friedman	Collision-induced dissociation	98.5	101.5	83
WATER	Simons, Francis, Mushlitz, and Fryburg	Low-velocity ion scattering	150		87
	Tal'rose and Frankevich	Success or failure in observing ion-molecule reactions	163	172	28
	Van Raalte and Harrison	Appearance potentials	151 ± 3		88
	Chupka and Russell	Photoionization	161	—	89
	Beauchamp and Buttrill	Ion cyclotron resonance	160	168	90
	De Pas, Leventhal and Friedman	Collision induced dissociation	179 ± 7		97
	Haney and Franklin	Appearance potentials with calculation of excitation	165		98
	Vetchinkin, Pshenichov, and Sokolov	Crystal lattice energies	171		85b
	Kondratyev and Sokolov	Crystal lattice energies	187		85a
	Lampe and Futrell	Crystal lattice energies	187		86

TABLE II—CONT.

	Worker	Method	Proton affinity (kcal/mole) Minimum	Maximum	Reference
AMMONIA	Russell, Fontana, and Simons	Low velocity ion scattering	217		73
	Chupka and Russell	Photoionization	196	—	89
	Haney and Franklin	Appearance potentials with calculation of excitation	207	—	98
	Vetchinkin, Pshenichov, and Sokolov	Crystal lattice energies	216	—	85b

Aquilanti and Volpi[74,75] were able to observe the reaction

$$H_3^+ + CH_4 \longrightarrow (CH_5^+)^* + H_2 \longrightarrow CH_3^+ + 2H_2$$

and on the basis of the overall reaction being exothermic, they concluded that $PA(H_3^+) \leqslant 85$ kcal. This method of fixing an upper limit by observation of a reaction rather than by failure to observe one is clearly superior to Tal'rose's method. Nonetheless, it must be emphasized that endo- or exothermicity of a reaction is not a sound criterion for success or failure in the observation of a reaction. For example, $H_2^+ + He \rightarrow HeH^+ + H$ ($\Delta H = +1.1$ eV) is observed in a conventional mass spectrometer ion source and is formed from excited H_2^+ [41] while the reaction $He^+ + H_2 \rightarrow HeH^+ + H$ ($\Delta H = -8.3$ eV) is not observed[76] because the HeH^+ ion has no way of dissipating the heat generated in the reaction. Similarly, while protonated methane (CH_5^+) is well known[36e,57a,77] and easily generated, protonated ethane $(C_2H_7^+)$, propane $(C_3H_9^+)$,[77] and others, have been observed only with great difficulty. It is unlikely that the reactions forming them are endothermic and it is possible that these protonated species are formed but that they decompose instantaneously to give ethyl and propyl ions and hydrogen. Protonated benzene $(C_6H_7^+)$ is a common intermediate in liquid phase organic chemistry, but efforts to detect it in mass spectrometric collision processes by bombarding benzene with H_2O^+ [50b] or H_2^+ [78] have failed, and the ion when produced in the mass spectrum of ethylbenzene loses molecular hydrogen to give the phenyl ion,[79] emphasizing the importance of this competitive reaction channel.

It is interesting to note that while mass spectrometrists rely on the observation that the majority of the heat of reaction in an exothermic ion-molecule reaction is retained as internal energy of the products rather than appearing as kinetic energy, they are reluctant to consider that this energy may well be capable of driving the reaction in a consecutive channel.

Chupka,[80] in his experiments to measure the proton affinity of hydrogen, faced the problem of selecting definite energy states for his reactants. He constructed a photoionization chamber in which he could generate ions in different internal energy states by irradiating the neutral molecules at different wavelengths.[81,82] These ions could go on to react, and reactants and products could be detected in an attached mass spectrometer.

The $D\ ^1\pi_u$ system of hydrogen shows vibrational bands with an onset of continuous absorption at 14.676 eV corresponding to the dissociation

$$H_2(^1\pi_u) \longrightarrow H(1s) + H(2s\ or\ 2p)$$

Chupka found that even when generated with thermal energies, the excited hydrogen atoms are able to undergo the reaction:

$$H(2s\ or\ 2p) + H_2 \longrightarrow H_3^+ + e^-$$

As the energy states of both reactants are clearly specified, the criterion of exothermicity is valid, and $\Delta H_f(H_3^+)$ must be less than or equal to $\Delta H_f[H(2s\ or\ 2p)] = 286.8$ kcal, whence $PA(H_2) \geqslant 78.3$ kcal.

The above experiments all aim at establishing upper or lower limits for thermochemical quantities. Friedman and Leventhal[83] devised a direct method of measurement using a tandem mass spectrometer with a high pressure source. D_3^+ ions are generated from deuterium in the ion source, mass selected with a conventional 60° 12-in. radius magnetic mass spectrometer, decelerated, and allowed to collide with inert gas molecules in a differentially pumped collision chamber. They looked for the kinetic energy threshold for the reaction

$$D_3^+ + X \longrightarrow D^+ + D_2 + X$$

analyzing their D^+ product ions in a longitudinally oriented quadrupole mass spectrometer (X = inert gas). The threshold, in terms of center of mass energy, was the same no matter which inert gas they used as target, showing the absence of systematic error in the determination of the D_3^+ kinetic energy (Figure 3). If the pressure in the ion source was lowered, a different threshold was obtained, about 2 eV lower, showing that D_3^+ was formed in an excited state in the initial reaction but that equilibration of this internal energy occurred by relatively few collisions with D_2 molecules. The threshold for production of D^+ was at about 4.8 eV, in

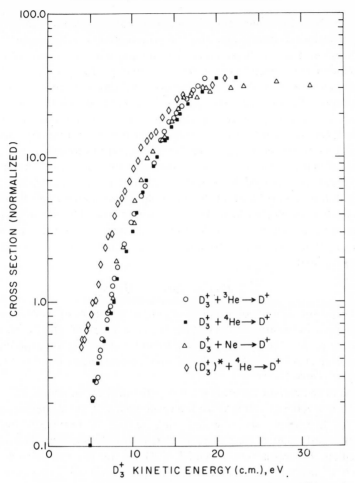

Fig. 3. Normalized cross sections for production of D^+ from collisions of D_3^+ with ^{3}He, ^{4}He, and Ne plotted versus the center-of-mass energy of the colliding system. Naturally occurring neon ($\sim 10\%$ ^{22}Ne and 90% ^{20}Ne) was used. However, the difference in the center-of-mass energy with these two isotopes is not sufficient to affect the results in these experiments. $(D_3^+)^*$ refers to D_3^+ ions having excess internal energy generated at source pressures less than 5 μ.[83]

close agreement with the most recent theoretical work.[84] While there is no certainty that the D_3^+ in these experiments was completely de-excited, the agreement with theory is impressive and leads to a proton affinity of hydrogen in the range 4.27 to 4.40 eV (98.5 to 101.5 kcal). Extrapolation of the data for excited D_3^+ suggests that D_3^+ ions produced at low pressures contain about 2 eV internal excitation and that this accounts for the previously estimated lower values for the proton affinity of hydrogen.

The observation of threshold energies for collision-induced dissociations in a tandem mass spectrometer as described above is a very useful technique. It offers a chance of observing an unobscured threshold process, without interference by consecutive or competitive processes in the vicinity of threshold, providing of course that the various thresholds are widely spaced. As this technique has been used extensively by Friedman and his coworkers, it is appropriate to stress the assumptions involved. First, it is assumed that the internal energy of the projectile ion can be specified, or that it can be modified by repeated collisions to give the ground state. Second, it is assumed that, at the threshold, the available channel of reaction leads to products which are in the ground state. That is, no center of mass energy is converted to kinetic or internal energy. This is the equivalent of saying that the collision-induced dissociation has zero activation energy near threshold.

2. Proton Affinity of Water

The proton affinity of water is an important quantity because of its significance in solvation studies. It has been the subject of rather more work than the proton affinity of hydrogen. In addition to collision experiments, there have also been values deduced from measurements of crystal-lattice energies (Table II).[85,86]

One of the earliest experiments was performed by Simons et al.[87] who obtained a value of 150 kcal from measurements of the scattering of low velocity hydrogen ions in water. Tal'rose and Frankevich[28] obtained another early value. (See Section II-A.) Von Raalte and Harrison[88] measured the appearance potential of H_3O^+ by electron impact on a variety of alcohols, esters, and ethers. They obtained consistent values for $\Delta H_f(H_3O^+)$ from the different compounds, and considered this to be a justification for their technique. On the other hand, the self consistency of their values was partly due to arbitrary choices of the neutral products from the fragmentations. Chupka[89] has pointed out that all but one of the processes investigated by these workers involve a sequence of two decompositions producing a total of three fragments and that the appearance potential for such processes is almost always considerably higher than the thermochemical threshold. Chupka and Russell[89] determined the proton affinity of water by attempting to observe the reaction

$$NH_3^+ + H_2O \longrightarrow NH_2 + H_3O^+ \tag{6}$$

in their photoionization mass spectrometer. Their photoionization efficiency curve did not show a clear threshold, but with ground state ammonia ions the reaction

$$NH_3^+ + H_2O \longrightarrow NH_4^+ + OH \tag{7}$$

went at least 100 times as fast as (6). Chupka and Russell obtained a probable threshold for reaction (7), and from it deduced $PA(H_2O) = 161$ kcal, which value they considered to be a lower limit.

Beauchamp and Buttrill[90] applied the technique of single and double ion cyclotron resonance to the determination of the proton affinities of water and hydrogen sulfide. This technique has been developed mainly by Wobschall, Graham, and Malone,[91] Llewellyn,[92] and Baldeschwieler[93] and his students,[94] among whom Beauchamp and Buttrill[95,96] are numbered. For a detailed account of the operation of an ion cyclotron resonance spectrometer, the reader is referred to the above literature; suffice to say that ions are produced in a conventional electron impact source and permitted to drift into a cell where ions of a particular mass are trapped by the combined effect of a radiofrequency field and a perpendicular uniform magnetic field. If the cyclotron frequency of an ion ($\omega = eB/m$, where e is the charge of the ion, m its mass and B the strength of the magnetic field) is equal to that of the applied radiofrequency field, it will absorb energy continuously until the ion collides with another molecule or a confining wall. The rf oscillator thus loses energy, and this energy absorption by the ions can be measured by a modulated-phase-sensitive detector. The mass spectrum can be scanned either by varying the magnetic field or the frequency of the rf field, and the primary and secondary ions present can be identified. In a reaction of the type $A^+ + B \rightarrow C^+ + D$, both A^+ and C^+ will be detected by this method, but it will not be certain whether they are primary or secondary ions, and whether A^+ gives rise to C^+. It is found, however, that if a second rf field is applied at the cyclotron frequency of A^+, while the main field is set for resonance of C^+, a significant change in the yield of C^+ may occur (since A^+ is being accelerated, and the cross sections of most ion-molecule reactions vary with ion velocity) indicating that A^+ is indeed a reactant ion. This is called the double resonance technique.

It permits the unambiguous identification of reactants and products, and also gives a sign to $(dk/dE_{ion})°$, the variation of rate constant with ion kinetic energy at vanishing ion energies, which is the same as the sign of the pulsed double resonance contribution from the reactant ion to the observed product. If this is negative, the reaction *must* be occurring at thermal energies and is therefore exothermic or thermoneutral. A reaction which gives a positive pulsed double resonance signal may be endothermic or exothermic but *must* give a positive signal near threshold, reflecting the kinetic → internal energy conversion needed to push the reaction. This is an elegant technique and is discussed in more detail in Section IV-F-4. It suffers from the drawback that if a reaction takes place exothermally through internally-excited species, the sign of the pulsed double resonance signal is negative even though the reaction of the ground states may be endothermic.

Nonetheless, Beauchamp and Buttrill's technique, which is still essentially a "bracketing" method, represents a considerable advance over that of Tal'rose. They were able to observe the reactions

$$H_3O^+ + CH_2O \longrightarrow H_2O + CH_2OH^+ \quad [\Delta H_f(H_3O^+) \geqslant 142 \text{ kcal}]$$

and

$$C_2H_5^+ + H_2O \longrightarrow C_2H_4 + H_3O^+ \quad [\Delta H_f(H_3O) \leqslant 149 \text{ kcal}]$$

both of which had a negative $(dk/dE_{ion})°$. They also observed

$$H_3O^+ + C_2H_4 \longrightarrow H_2O + C_2H_5^+ \quad [\Delta H_f(H_3O^+) \geqslant 149 \text{ kcal}]$$

also with a negative $(dk/dE_{ion})°$, but they were able to produce evidence that the H_3O^+ in this reaction possessed vibrational excitation and the lower limit derived from it was therefore in error. They recognized the possibility of internal excitation in their reactants, and operated with electron energies as low as possible. After an intelligent and careful study, they concluded that $\Delta H_f(H_3O^+) = 143 \pm 4$ kcal, whence $PA(H_2O) = 164 \pm 4$ kcal, a result in reasonable accord with crystal lattice measurements.

In spite of this, De Pas, Leventhal and Friedman,[97] using the technique they had applied to hydrogen, were able to show that in a source containing water at fairly low pressures, an abundance of excited species was produced. These proved more stable and difficult to deactivate than the excited states of H_3^+, and the workers were able to derive a relatively imprecise result of $PA(H_2O) = 179 \pm 7$ kcal; again in reasonable agreement with values based on crystal lattice energies.

Haney and Franklin[98] attempted to calculate the excess energy liberated when molecules were formed by electron impact. They found a semi-empirical relationship between the excess energy released in an ionic fragmentation and the relative translational energy of the products, as follows:

$$E^* = (0.44)N\bar{\varepsilon}_t \tag{8}$$

E^* is the difference between the experimental appearance potential and the appearance potential calculated from ground state heats of formation of products and reactants. N is the number of oscillators present in the reactant ($3n - 6$ for a non-linear molecule) and $\bar{\varepsilon}_t$ is the average value of translational energy produced in the fragmentation. Because of experimental difficulties in measuring small translational energies, the method is limited to molecules with a small number of oscillators ($\gtrsim 25$), and Haney and Franklin therefore used ethanol as their source of H_3O^+. They measured the appearance potential of H_3O^+ from it by the retarding potential difference method and also the average translational energy of the ions produced. They discussed the following possible modes of fragmentation of C_2H_5OH

$$C_2H_5OH + e^- \longrightarrow C_2H_5O^+ + H \xrightarrow{m^*} H_3O^+ + C_2H_2$$
$$\longrightarrow H_3O^+ + C_2H_3$$
$$\longrightarrow H_3O^+ + C_2H_2$$

and concluded that their appearance potential (which at 14.3 eV was more than 2.3 eV higher than previous values) corresponded to the last of these reactions and that 14.8 kcal excess energy was liberated in the fragmentation. This gave a value of 165 kcal for $PA(H_3O^+)$, a figure higher than that of von Raalte and Harrison by approximately the calculated amount of excitation energy.

The relatively minor experimental discrepancy in the proton affinity of water (that is, the range of 165–180 kcal mole) is viewed with pessimism by many workers equipped to measure proton affinities. The existence of stable long-lived excited states, and the problems generated by these states in measurements of proton affinities, by either the observation of reactions in a single source or the measurement of collision-induced dissociation thresholds, is well established. There is little doubt that upper limits established by the latter technique are reliable and that the proton affinity of water is less than 190 kcal. But the question of how close the correct value is to 180 kcal, 170 kcal, or 165 kcal, remains open for the present and will probably not be settled without considerable improvement in available experimental techniques or by an accurate theoretical calculation. Theoretical calculations give a wide spread of values. These are discussed by Hopkinson et al.[99] whose calculations tend to favor the high value of Friedman et al.

3. Proton Affinity of Ammonia

The proton affinity of ammonia is less well established than that of hydrogen or water. The only recent crystal lattice energy value is shown in Table II.[85] Russell, Fontana and Simons[73] obtained a value of 217 eV from the scattering of low velocity hydrogen ions in ammonia in some very early work and Chupka[89] has obtained a value $\geqslant 196$ kcal based on the occurrence of the reaction

$$NH_3^+ + H_2O \longrightarrow NH_4^+ + OH$$

in his photoionization source with ammonia ions in the vibrational ground state. Haney and Franklin,[98] using the method described above, obtained proton affinities of 207 and 181 kcal from experiments with ethylamine and dimethylamine respectively. They considered the latter value to be anomalous, and accepted the former as agreeing reasonably well with crystal lattice and theoretical values. There is clearly scope for further work in this area.

4. Proton Affinities of Other Molecules

Several measurements have been made of the proton affinity of hydrogen sulfide. Kiser[100] has obtained a value of 200 kcal based on the electron impact method, and Beauchamp and Buttrill[90] have found

$$PA(H_2S) = 179 \pm 2 \text{ kcal}$$

using their ion cyclotron resonance technique to bracket the value. Haney and Franklin[98] obtained a slightly higher value of 179 kcal by applying the method already described to dimethyl sulfide and ethanethiol.

Other recent estimates of proton affinities and related thermodynamic quantities include the work of Harrison and his coworkers[101] on carbonyl compounds of the types RCHO and RCOOH, of Munson[102] on RCOOH and CH_3OH (R = H, CH_3, C_2H_5), and of Haney and Franklin.[98c] All groups used appearance potential measurements to obtain their data.

The proton affinity of methane has attracted little attention, though a lower limit of 111 kcal can be[36e,77] deduced from the occurrence of an ion-molecule reaction in methane giving CH_5^+. For carbon dioxide values of 110[57g] and 121 kcal[98c,103] can be deduced from appearance potential data on carboxyl compounds. Moran and Friedman[104] have calculated the proton affinities of a number of atoms (including the rare gases) using the Platt electrostatic model.

Many values have or can be obtained for proton affinities of molecules and heats of formation of ions by appearance potential measurements and observation of ion-molecule reactions, and some of these can be found or deduced from the sources already quoted. Although data obtained by these methods are usually the best available, they should be treated with caution until further work has elucidated the role of excitation in ions formed by electron bombardment.

5. Threshold for the Reaction $C^+ + D_2 \longrightarrow CD^+ + D$

In theory it is possible to measure quantities other than proton affinities by the above methods, but so far, little work has been done. Maier[105] has attempted to measure the threshold and variation of cross section with ion energy of the reaction

$$C^+ + D_2 \longrightarrow CD^+ + D$$

The importance of this reaction is that its threshold energy is known from standard thermochemical data to be 0.40 eV, and the experiment provides a method of comparing a result obtained in a longitudinal tandem mass spectrometer system with a known value. This kind of experiment, designed to validate an experimental technique, has been conspicuously absent from a great deal of work on ion-molecule reactions, and Maier deserves considerable credit for attempting it. His experience exemplifies several of the problems which can interfere with experiments using a tandem mass spectrometer.

He found that at low ion accelerating voltages his source mass spectrometer transmitted a significant current of mass 14 when set for maximum mass 12 transmission. Primary ions having laboratory kinetic energies below about 5 eV were obtained by retarding the ions when they were travelling between the source mass spectrometer exit slit and the reaction chamber, and even so the magnetic field had to be varied around the position for maximum mass 12 transmission in order to allow the background to be estimated. Maier considers this to have been the reason for his obtaining an unreasonably low value for the threshold energy. He shifted all his data by a constant increment so that his measured threshold energy agreed with the computed value, and then showed that the shape of his cross section versus barycentric energy curve near the threshold was in good agreement with the theory of Pechukas and Light.[106]

The lack of resolution in his source mass spectrometer emphasizes the problem of obtaining intense beams of primary ions at low energies because of the tendency of the ion beam to "explode" due to mutual interionic repulsions. Furthermore, Maier took his threshold for CD^+ production at a point where the secondary ion current was about 3 % of its maximum. A drop-off of several orders of magnitude would have defined a threshold more satisfactorily, and the data suggest that there were problems in obtaining sufficient intensity of secondary ions. A possible explanation of the low threshold, apart from the one given by Maier, is that a proportion of his carbon ions which were obtained by electron bombardment of carbon monoxide were themselves excited, thus lowering the apparent threshold energy. Maier was aware of this problem and used low energy ionizing electrons to minimize the possibility.

6. N_3^+, C_2O^+, O_3^+

The ion-molecule reactions in nitrogen, carbon monoxide, and oxygen are formally similar to the reactions in hydrogen:

$$N_2^+ + N_2 \longrightarrow N_3^+ + N$$
$$CO^+ + CO \longrightarrow C_2O^+ + O$$
$$O_2^+ + O_2 \longrightarrow O_3^+ + O$$

and Cermak and Herman[107] observed these products, although in yields well below those predicted by the G–S theory. They found maxima in the X_3^+ ionization efficiency curves not observed with the primary X_2^+ ions, and concluded that only primary ions in the quartet metastable excited states underwent reaction. This is not unreasonable, since the reaction of ground state ions is certainly endothermic. Friedman and Leventhal[108] subsequently showed that N_3^+ and C_2O^+ were only minor products of the above collision processes and that the major products arising from N_2^+/N_2 and CO^+/CO collisions were N^+ and C^+ ions. These had escaped detection in previous experiments because they were masked by ions from the primary mass spectrum of CO and N_2. The major reactions producing atomic ions were thought to be

$$N_2 + N_2^+ \longrightarrow N^+ + N_3$$
$$CO + CO^+ \longrightarrow C^+ + CO_2$$

since the collision-induced dissociation processes were not energetically possible for low kinetic energy ions under the conditions of the reaction. The above reactions are also endothermic for ground state ions, but kinetic to internal energy conversion can occur. Friedman and Leventhal deduced that, as the charge preferred to remain with the atomic rather than the triatomic molecule, $I(N) < I(N_3)$ and $I(C) < I(CO_2)$. The latter conclusion is, of course, well established, but by a study of the energetics of the nitrogen reaction, a crude value of $D(N_2 - N)$ was obtained.

No similar production of O^+ was found in the O_2^+/O_2 system, and it was suggested that the low yields of O_3^+ were due to its decomposing to give $O_2^+ + O$, a reaction which would be concealed in single source studies by primary ions, and which would require only 0.39 eV excess energy in the O_3^+ to enable it to take place.

Maier[109] investigated the N_2^+/N_2 system in a tandem mass spectrometer and found that the formation of N^+ required more kinetic energy than had been claimed by Friedman and Leventhal. He suggested that the N^+

product arose mainly by collision-induced dissociation. The energy to bring this about in the single source experiment might arise from the primary N_2^+ being formed with greater internal excitation than Friedman and Leventhal had assumed, and thus the energy requirement of the reaction might be met from internal excitation of N_2^+ rather than from the heat of formation of N_3.

Ryan[110] carried out a single source experiment in a system similar to that of Friedman and Leventhal, but with a longer ion path length, to give any excited N_2^+ ions a longer time to decay. He found that N^+ was formed with an energy threshold intermediate between those of Friedman and Maier (whose path length in a tandem instrument was extremely long). This could indeed be due to deexcitation of N_2^+ primary ions over the longer path, but Ryan pointed out that it could also be due to a reduction in the collection efficiency of product ions brought about by the change in primary ion path length. Thus it is still an open question whether the stability and dissociation energy of N_3 can be inferred from study of ion-molecule reactions, although there is certainly spectroscopic evidence for its existence.

7. Attenuation Data and the Identification of Excited States in Beams

We have already discussed the pioneering work of Lindholm on the identification of the proportion of excited electronic states in ion beams. The states concerned were metastable states, that is, their lifetimes had to be commensurate with the ion transit times inside the first of his mass spectrometers.

Photoelectron spectroscopy can be applied to the same problem, but because it measures the energies of the electrons emitted when a substance is irradiated with light in the far ultraviolet, it gives information on the immediate products of photon impact rather than the metastable products of electron impact.

Another technique which can be used (among other things) as a diagnostic tool for the detection of long-lived excited states is that of measurement of the attenuation of ion beams passing through a gas-filled collision chamber. If two states of an ion have different collision cross sections σ_1 and σ_2 for collision with a given scattering gas, and a mixture of these of concentrations C_1 and C_2 is directed into a collision chamber of length l and containing n molecules of gas per unit volume, then the ratio of the attenuated to the initial ion beam intensity will be given by

$$\frac{I}{I_0} = C_1 e^{-n\sigma_1 l} + C_2 e^{-n\sigma_2 l} \tag{9}$$

A graph of $ln\ I/I_0$ against nl (proportional to pressure) will thus consist of two approximately linear sections, with gradients $-\sigma_1$ and $-\sigma_2$ and a curved transitional section between them. This behavior has been found by Turner, Rutherford, and Mathis[111] for the scattering of O^+ and O_2^+ in N_2, and NO^+ in H_2, N_2, Ar, O_2 and H_2O. They concluded, for example, that when O_2^+ is made by electron impact with 25, 50, and 100 eV electrons, the resultant beams contain 22, 30, and 33% of excited states.

In addition to the detection of excited states, attenuation data can be used to investigate repulsive potential energy interactions, that is, "the wall" on the potential energy curve at short interatomic distances. The bond energy and bond lengths are considered as equilibrium properties of a molecule, and it is not unreasonable for us to extend this idea to include the whole of the potential function, even though thermodynamics is normally concerned only with the potential energy well where equilibrium is a meaningful concept. Investigation of the steep section of a potential function thus forms a link between thermodynamics and kinetics, and measurement of cross sections for collisions between ions and molecules can provide structural and thermochemical information. This is potentially a fruitful field for research and the results should be of value not only for the information they provide about repulsive potentials (how far are "hard core" molecule models valid?) but also because it should be possible to use them for semi-empirical tests of models of clustered and hydrogen-bonded complexes.

The scope of such experiments was outlined in a much neglected group of papers by Simons and his coworkers[73,87,112] published during World War II. They directed low velocity beams of H^+, H_2^+, and H_3^+ into hydrogen, helium, water, and ammonia, and observed the attenuation and scattering of the beams. They obtained potential laws for the systems and were able to derive not only these potential laws, but also energies of combination and proton affinities. These last were quoted in the previous section, and stand comparison with more recent work. They comment[73] "As measurements of this kind can be made using many kinds of ions each of which can be scattered in a great variety of gases, much information of value can be obtained such as force laws, proton affinity or absolute basicity of substances, energies of solvation of ions etc." This gauntlet which they threw down in 1941 has been picked up only recently.

Mason and Vanderslice in 1959[113] reinvestigated the mobilities of H^+, H_2^+, and H_3^+ in hydrogen both by theory and by scattering experiments, but calculated only the potential law and not thermodynamic quantities. De Pas, Leventhal and Friedman[97] much more recently studied the attenuation of ionic clusters of water $D^+(D_2O)_n$ by helium in their longitudinal tandem mass spectrometer. Relative attenuation cross sections were measured as a function of n, and are shown in Figure 4. These data were more or less independent of ion energy which was varied between 10 and 60 eV suggesting "hard core" collisions over this energy range. In the case of the solvated species of D_3O^+, the main effect of the charge is to attract water molecules according to a potential in which the energy V varies as the inverse square of the distance r:

$$V(r) = -\frac{\mu e}{r^2} \tag{10}$$

The enthalpies of solvation of the various ionic clusters were deduced by Friedman et al. in a different series of experiments[97] (Section III-D). Using these values and the above equation they were able to calculate the separation between the neutral polar molecule and the center of charge in D_3O^+. From this, they computed the effective radius of the water molecule by subtracting the distance from the center of charge in D_3O^+ to the edge of the molecule ion, assuming it to be a rigid disc with radius approximately equal to the O—H bond length in neutral water (0.96 Å).

All the solvating water molecules were then considered as noninteracting rigid discs packed around D_3O^+, and average values of the geometric cross section were calculated for the various species, the disc thicknesses all being equal to the O—H bond length. The results of this admittedly crude technique are included in Figure 4. The level of agreement with experiment suggests that ion-dipole interaction is the dominant factor determining the physical geometry of these relatively weakly bonded species. The main discrepancy between this model and experiment occurs in the case of $D_{11}O_5^+$ and may reflect the fact that the model does not take into account repulsive dipole-dipole interactions between bound water molecules. These would indeed be expected to increase in importance after the first hydration shell had been filled in $D_9O_4^+$.

D. Equilibrium Approach to Ion Thermochemistry

Thermochemical information on neutral molecules is customarily gathered by the investigation of systems at equilibrium. Enthalpies are found by calorimetry and heats of reaction are found wherever possible by measurement of the variation of the equilibrium constant of a reaction with temperature. Ways in which equilibrium measurements can be applied to ion thermochemistry are not immediately obvious, since ions will always tend to discharge themselves at the walls of their container, and at equilibrium there will be none left. The only major attempt to achieve equilibrium conditions in an ion source was made by Kebarle and his coworkers[114] who carried out extensive work on the solvation of gaseous

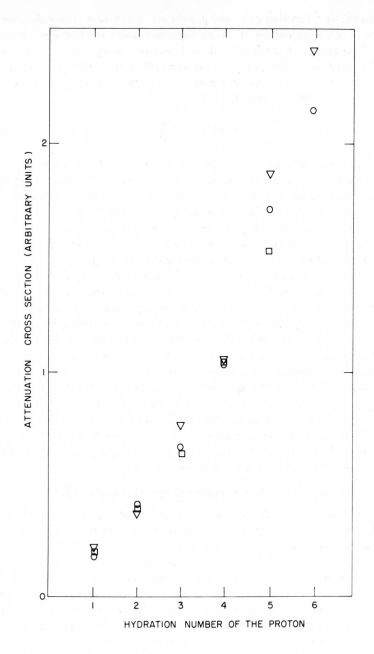

ions, that is, the formation of clusters of molecules around an ion even in the gas phase, together with other high pressure ion-molecule reaction studies.

Early work on clustering was carried out by many physicists to resolve the controversy as to whether gaseous ions were molecular clusters of large radius or charged single molecules retarded in their motion by dielectric forces in the gas. In particular, we call attention to the measurements by Munson and Hoselitz[115a] of the mobilities of alkali metal ions in inert gases at a total pressure of the order of 20 mm. They found that only Li^+ could serve as the center of a cluster, and that the greater the polarizability of the rare gas atom, the higher the temperature at which it would form a cluster with the lithium ion. An attempt made to observe a K^+/Ar cluster at 126°K was unsuccessful, whereas Li^+/Ar readily formed clusters at 195°K. We can see clearly in these experiments the opposed factors of enthalpy (ion-dipole attraction) and entropy, with the entropy factor being dominant at high temperatures. This type of experiment has been treated theoretically by Eyring, Hirschfelder and Taylor,[6] Overhauser,[115b] and Bloom and Margenau.[115c]

Unlike the workers on ion mobilities, Kebarle and his coworkers used a quasi-static system in which to create their clusters, and constructed a number of combined reaction vessels and ion sources.[114b,m] Because their aim was to achieve equilibrium, it was necessary for them to operate at very high source pressures. Ions were created inside the sources by three methods; by irradiating neutral gas with α-particles from a polonium source, by ionizing with a 25,000 eV electron beam through a nickel window, and by bombarding with a 100 keV proton beam from a Cockroft–Walton accelerator. These techniques give poorly defined energy distributions in the resultant ions, many of which are formed by secondary electrons displaced from the walls of the source, but they reduce the general

---

Fig. 4. Relative values for the attenuation of beams of hydrated protons when passed through a collision chamber containing helium at low pressures. The data are actually presented for D_3O^+, $D_5O_2^+$, $D_7O_3^+$, $D_9O_4^+$, and so on. The triangles and circles were obtained in experiments with collision chambers having radically different geometry. The squares were obtained by using a crude rigid disc model for the hydrated species in which bond lengths were assumed to be correlated with bond energies determined by ion-dipole interactions.[97a] Remarkably similar values of attenuation cross sections were observed for hydrated hydroxyl ions H_3O_2, $H_5O_3^-$, etc., that is, OH^- and H_3O^+ were found to have nearly the same attenuation value as did the isoelectronic $H_3O_2^-$ and $H_5O_2^+$.[97b]

problem of temperature control. The ions formed underwent a series of ion-molecule reactions in the source, and the mixture of ions which resulted was sampled by a cone with a leak at its apex which projected into the irradiated volume. Ions passing through the leak were accelerated, magnetically analyzed, and detected in a conventional mass spectrometer.

We shall consider the work of Haynes, Hogg and Kebarle on ammonia[114e,g] as exemplifying the experiments carried out by this group. The primary ammonia ions produced by irradiation of the source are rapidly converted by ion-molecule reactions to ammonium ions

$$NH_3^+ + NH_3 \longrightarrow NH_4^+ + NH_2$$

The ammonium ion then "solvates" with ammonia molecules:

$$NH_4^+ + NH_3 \longrightarrow NH_4^+(NH_3)$$
$$NH_4^+(NH_3) + NH_3 \longrightarrow NH_4^+(NH_3)_2$$
$$. \quad . \quad . \quad . \quad .$$
$$NH_4^+(NH_3)_{n-1} + NH_3 \longrightarrow NH_4^+(NH_3)_n$$

Kebarle et al. obtained clusters of this type up to $n = 20$. By varying temperature and pressure they were able to obtain a series of equilibrium constants $(K = [NH_4^+(NH_3)_n]/[NH_4^+(NH_3)_{n-1}][NH_3])$ from which the entropies and enthalpies for each successive solvation step could be derived.

Some of the results obtained are shown in Figure 5. The associated entropy changes are of the expected magnitude; the enthalpy values show

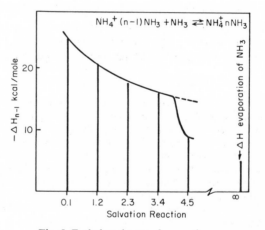

Fig. 5. Enthalpy changes for reactions:
$$NH_4^+ \cdot (n-1)NH_3 + NH_3 \longrightarrow NH_4 \cdot nNH_3 (n-1, n).[114e,g]$$

a sharp drop after four ammonias have been bound to the central ammonium ion, suggesting that the innermost solvation shell is completed with four ammonia molecules. Kebarle et al. failed to find a point, which could be expected at about $n = 16$, corresponding to the completion of the second shell, but problems due to lack of sensitivity and adiabatic cooling (see next paragraph) were becoming acute by this stage.

These values depend on two basic assumptions; first, that the various clusters in the ion source are in equilibrium with each other and with the neutral gas molecules, and second, that the measured ion intensities are truly representative of concentrations of ion clusters in the source.

Kebarle et al. considered that equilibrium in the ion source would be established if the ions underwent a reasonable number of collisions before being discharged. They estimated a typical lifetime for their ions of 0.1 to 1 msec before discharge, and showed that at 10 torr they would undergo about 10^4 collisions in this period. Experimental support for the establishment of equilibrium came from the fact that the measured equilibrium constants were indeed independent of pressure, and that no change in the ion ratios was observed when the α-particle beam was moved away from the sampling leak so that the reaction time was increased. Ion ratios in ethylene (a non-equilibrium system) changed dramatically when this was done.

Many of the possible errors in sampling would affect both the $NH_4^+(NH_3)_{n-1}$ and $NH_4^+(NH_3)_n$ peaks by similar amounts, for example, discharge of ions at the source walls, mass discrimination in the mass spectrometer focusing system, and fractionation during flow through the leak. The most serious source of potential error was felt to be the cooling effect due to adiabatic expansion of gas through the sampling leak. This might result in additional clustering, and was never completely eliminated, although the residual effect was thought to be small. To guard against this, it was necessary to have a sampling leak small enough for the flow through it to be molecular, and to keep the partial pressure of neutral gas sufficiently low for the mean free path to be much larger than the diameter of the sampling leak.

Kebarle et al. also studied the competition between water and ammonia molecules for places in the solvation shells.[114d] They found a strong preference for ammonia in the inner shell (ammonia was from 10–25 times as likely to be taken up as water) and a similarly strong preference for water in the outer shell (the probability of water being taken up was greater by a factor of almost 1000). The effect is so marked that, in an approximately equimolar mixture of water and ammonia at 1.8 torr, $NH_4^+(NH_3)_4$ is more abundant than any other ion with four solvating molecules, $NH_4^+(NH_3)_5$

is virtually absent, and $NH_4^+(NH_3)_4(H_2O)$ is the most abundant of the ions with five solvating molecules.

In another paper, Kebarle, Searles, Zolla, Scarborough and Arshedi[114i] reported the clustering in water to give ions of formula $H^+(H_2O)_n$. Some of their values for the enthalpy of solvation are shown in Table III. They

TABLE III

Hydration Enthalpies (eV) of D_3O^+: $D_3O^+(D_2O)_{n-1} + D_2O \longrightarrow D_3O^+(D_2O)_n$

$n-1, n$	From threshold for detachment of 1 D_2O	From difference between thresholds for detachment of 2 D_2O and 1 D_2O	From difference between thresholds for detachment of 3 D_2O and 2 D_2O	Enthalpy value considering the association of D_2O molecules	Kebarle's data
0, 1	1.4	1.1	1.15	1.4	1.56
1, 2	0.2	0.75	0.70	1.0	0.97
2, 3	0	0.45	0.43	0.74	0.74
3, 4	0	0.40	—	0.70	0.67
4, 5	0	—	—	—	—

found no distinct completion of an inner shell as there had been with ammonia, but some sort of "crowding" or starting of a new shell occurred after $n = 4$.

In addition, Kebarle, Haynes and Collins have studied the competitive solvation of the hydrogen ion by water and methanol,[114h] and Kebarle, Arshadi and Scarborough[114l] have observed the solvation by water of negative ions F^-, Cl^-, Br^-, I^-, O_2^- and NO_2^-. Work has also been carried out on the solvation of sodium ions,[114m] and the comparative solvation by water of Cl^-, BCl^-, and B_2Cl^-. In addition $D(Ar^+ - Ar)$, $D(Ne^+ - Ne)$,[114k] and $D(O_2^+ - O_2)$[114n] have been determined in this type of apparatus, although the first two were calculated by a method involving recombination energies rather than equilibria.

De Pas, Leventhal and Friedman[97] have studied clustering in water, using a tandem mass spectrometer fitted with a high pressure source, and we shall consider their work here because of its relevance to the work of Kebarle et al., in spite of its not being an equilibrium method. The objective of Kebarle et al. had been to measure equilibrium constants at different temperatures for the systems

$$H^+(H_2O)_{n-1} + H_2O \longrightarrow H^+(H_2O)_n$$

Friedman et al. adopted a different approach. They tried to generate beams of $D^+(D_2O)_n$ clusters which had been thermalized in the ion source by repeated ion-neutral collisions. These were then mass analyzed and introduced with controlled kinetic energy into a collision chamber containing helium. The thresholds of collision-induced processes of the types

$$D^+(D_2O)_n + He \longrightarrow D^+(D_2O)_{n-1} + D_2O + He$$
$$\longrightarrow D^+(D_2O)_{n-2} + 2D_2O + He$$
$$\longrightarrow D^+(D_2O)_{n-3} + 3D_2O + He$$

were measured. From these thresholds, they obtained evidence similar to that found by Leventhal and Friedman in the D_3^+ system[83] for the existence of excited states of D_3O^+, $D_5O_2^+$, $D_7O_3^+$, and so on, which appeared to survive at pressures of up to 1 torr for periods in excess of 10^{-4} sec. In addition, their work on attenuation cross sections for water clusters (See Section III-C-7) suggests that in Kebarle's apparatus the large clusters would be less mobile than the small ones, and hence that the latter's assumption of a valid measurement of equilibrium was open to question from two standpoints, that is, possible existence of long-lived excited water molecules, and errors in sampling. Friedman et al. derived enthalpy values from their threshold data and these are shown in the next to last column of Table III. There, evidence is also presented for the idea that when two molecules of water are detached from a solvated ion, they break off as a dimer with 0.25–0.30 eV energy of dimerization. It will be seen that in spite of the disparity between the two methods of measurement (that is, threshold data and Kebarle's equilibrium studies), the results obtained are in remarkable agreement, and unless this is fortuitous, which seems unlikely, we must conclude that Kebarle's temperature variation experiments produce a shift in the steady state concentration of ions which indeed accurately reflect the enthalpies of association.

De Pas, Guidoni and Friedman[97b] have studied the solvated hydroxyl ion systems and found that enthalpies of hydration of $H_3O_2^-$, $H_5O_3^-$, and $H_7O_4^-$ were remarkably similar to enthalpies of solvation of the corresponding hydronium ions, that is, $H_5O_2^+$, $H_7O_3^+$, and $H_9O_4^+$ respectively. Ion attenuation studies gave results which indicated near identity in elastic scattering cross sections for the respective H_3O^+ and OH^- solvated species. These results were used as a basis for estimating the total hydration of the proton in aqueous solution to be 12.0 ± 0.3 eV. This value compares well with the Halliwell–Nyberg[97c] value of 11.35 ± 0.15 obtained from treatment of a variety of thermodynamic data consisting mainly of heats of solution. The result of De Pas, Guidoni and Friedman is derived mainly from data on gaseous ions taken with well established values of the heat of neutralization of acids in aqueous solution and the heat of vaporization of water. This work indicates a direction that might be followed to establish a *solid* connection between gaseous ion chemistry and chemistry of ions in aqueous solution.

E. Discussion

Insofar as values of proton affinities, bond strengths, and recombination energies are used as tools for the understanding of general chemistry, any future work must take as its prime objective the establishment of more reliable and more accurate data. There has indeed been a very real improvement in the quality of data available in the past five years. This may be expected to continue, and photoionization and photoelectron spectroscopy, along with the older techniques, will certainly produce further information about electronic and vibrational energy levels in stable ions. Collision-induced dissociation and attenuation methods appear to have great potential for measurements on more exotic species. The major problem to be faced is the fact that many processes appear to lead to excited species, and the importance of these must be recognized so that thermodynamic values can be obtained unambiguously for ground and well defined excited states.

De-excitation phenomena themselves also merit attention in the future. From the work of Friedman and Leventhal, it appears that vibrationally excited D_3^+ can be deactivated by relatively few collisions with D_2 molecules, but that excited water clusters are much more stable to collision.[97] How readily can excited ions be de-excited either by radiation or by collision? Are the same criteria valid as those which apply to neutral molecules?

It is, incidentally, worthy of note that theoretical methods of calculating thermodynamic properties of ions have been improved to the point where errors due to the approximations which have to be made are of the same order of magnitude as errors due to possible lack of consideration of excited species.

The implications of the clustering experiments of Kebarle and Friedman and their co-workers seem even more far reaching. Apparently they form a bridge between solution and gaseous chemistry, and permit us to examine in the gas phase the sort of systems whose existence is postulated in the liquid phase. For many years, species such as $H^+(H_2O)_n$ and $NH_4^+(NH_3)_n$ have been assumed to exist in solution, in order to explain mobilities, transport properties, etc. It is now possible to examine in isolation the bond energies and structures of these clusters and to discuss them on a molecular rather than a macroscopic basis.

The sorts of questions which might arise from these considerations are illustrated by the following speculative argument; it is believed that in solution, multi-charged species exist, and that these form multi-charged coordinated complexes e.g. $[Fe(H_2O)_6]^{3+}$ or $[Cu(NH_3)_4]^{2+}$. In the gas phase, on the other hand, although multi-charged metal ions, for example, Hg^{2+}, are well known, we are unable to find an observation of a multi-charged gas phase coordinated complex produced in a high pressure ion source, either in our own work or in the literature. This is only negative evidence and might reflect merely that no one has looked hard enough. Alternatively, it might reflect a high cross section for the process $M^{2+} + OH^- \rightarrow M^+ + OH$ since OH^- is produced by electron impact in water vapor in comparable amounts with M^{2+} ions. Meanwhile, it is interesting to consider the systems in which it is worthwhile to look for doubly charged solvated species. If, for example, we produce a doubly charged copper ion and allow it to react with water, we might expect to see

$$Cu^{2+} + H_2O \longrightarrow [Cu(H_2O)]^{2+}$$

arising from the strong ion-dipole interaction, and this would additionally be favored by the possibility of spreading the charge over a larger volume. However, $I(H_2O) = 12.6$ eV, and the first and second ionization potentials of copper are 7.68 and 20.34 eV respectively, so that the process

$$Cu^{2+} + H_2O \longrightarrow Cu^+ + H_2O^+$$

would be exothermic to the extent of about 0.06 eV and would probably take place in preference to the hydration reaction. The condition for this is that the electron affinity of the multi-charged ion be greater than the ionization potential of water. It is satisfied by a number of elements (for example, silver, argon, aluminium) which would not therefore be expected to form multi-charged gas phase complexes with water. Many elements, however, do not satisfy this condition (for example, Barium, $I^+ = 5.19$ eV, $I^{2+} = 9.95$ eV; Fe, $I^+ = 7.83$, $I^{2+} = 16.16$ eV) and they should presumably form complexes $[Ba(H_2O)]^{2+}$ and $[Fe(H_2O)]^{2+}$. These however would be likely to react rapidly with another molecule of water

$$[Fe(H_2O)]^{2+} + H_2O \longrightarrow [Fe(H_2O)_2]^{2+}$$

Now, although $I(H_2O)$ is too high for this to lose H_2O^+, $I(H_3O)$ is about 5.6 eV, and so this might decompose:

$$[Fe(H_2O)_2]^{2+} \longrightarrow FeOH^+ + H_3O^+$$

If this condition is not satisfied, further condensations can take place, and since $I(H_3O) > I(H_5O_2) > I(H_7O_3) \cdots$ there will ultimately come a point at which exothermic charge exchange can take place, and the doubly charged ion will then be destroyed. It thus seems that doubly charged gas phase solvated ions are least likely to survive if they are large, and the search for them should be confined to small clusters around metals where the difference between the first and second ionization potentials is small. These considerations also lead to speculation about the nature of multi-charged ions in solution, and the extent to which it is accurate to depict cupric and ferric ions in solution as Cu^{2+} and Fe^{3+} when a high proportion of charge probably resides in the solvation shell.

IV. KINETICS OF ION-MOLECULE REACTIONS

A. Introduction

In the previous section of this review we considered the methods used to establish the thermochemistry of ions, and some of the results obtained. The following section deals with kinetics rather than thermodynamics. The questions which can be formulated about rate processes are of two kinds, which in the final analysis merge. First, what are the rates and velocity dependencies of ion-molecule reactions, and second, what is the mechanism of the collision process? The experimental problems associated with the first question are ones of detecting the products formed, identifying the reactant ions and their energy states, and establishing the reaction channels. The second question demands that we construct microscopic "models" of the energy and momentum interchange during reaction and the nature of the collision process itself. Two models on which attention is currently focused are reaction through complex formation and through stripping. In the former case, which corresponds in some respects to classical transition state theory, the reactants are assumed to form a complex whose lifetime is sufficiently long for equilibration of vibrational energy throughout the complex to take place. In the latter case, the reaction is supposed to take place without complex formation and without any transfer of momentum to the neutral species, that is, in the reaction

$$N_2^+ + H_2 \longrightarrow N_2H^+ + H$$

the H_2 molecule is thought of as being composed of separate atoms so that after collision, the neutral H atom, known as the "spectator", has not gained energy or shared in the reaction exothermicity.

These problems are summarized in the flow sheet in Figure 6. Clearly, most of the answers in which we are interested are not accessible by the simple single source technique described in our historical introduction. The various methods designed to overcome the above problems are listed in Figure 7. We shall now discuss some of the problems and by this means illustrate the application of the various experimental methods.

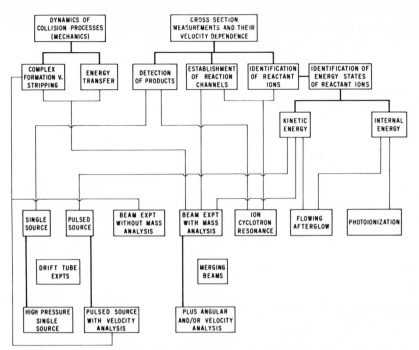

Fig. 6. Flow sheet indicating genetic relationships between problems investigated in the area of ion-molecule reaction kinetics, specific questions asked and techniques developed for the experimental attack on these problems and questions. For example, in the dynamics of collision processes the question of complex formation has been studied in beam experiments without mass analysis, in pulsed source experiments with velocity analysis, and in more sophisticated beam experiments with mass analysis and angular and/or velocity analysis. The general questions concerned with kinematic studies and cross section measurements are identified, but their isolation is obviously arbitrary. The relationships shown are designed to indicate the response of the experimentalist to questions of current concern in the field of ion-molecule reaction kinetics. Drift tube experiments and merging beams are not connected in this scheme although the latter are definitely related to kinematic studies while the former play a role in mobility measurements which for the present are related to cross section studies.

B. Complex Formation versus Stripping

1. Stripping Mechanisms

The idea of "stripping" reactions in ion-molecule systems was suggested by Henglein and his coworkers on the basis of their extensive measurements of the velocities of reaction products.[18,50,116] They constructed an apparatus which measured velocity spectra and which was entirely lacking in any facility for mass analysis.[116b,e] Primary ions such as Ar^+,

N_2^+, and CO^+ were produced by electron impact in a conventional ion source and accelerated by voltages between 20 and 200 volts. The ion beam entered a collision chamber containing H_2 or D_2, and the resulting mixture of primary and secondary ions entered a Wien filter which allowed ions of equal velocity to reach the collector. Henglein thus obtained velocity spectra for a number of reactions exemplified by

$$Ar^+ + H_2 \longrightarrow ArH^+ + H$$

His spectra showed a sharp peak due to the transmitted Ar^+ ions, plus a more diffuse peak at lower velocities which he assigned to ArH^+. He was able to derive reaction cross sections by measuring the relative areas under these peaks, and to calculate where the secondary peak should be on the hypotheses of complex formation or stripping. His results showed that, in the range of ion energies with which he was concerned, the stripping process was always observed. He confirmed this result by showing that the peaks due to ArH^+ and ArD^+ appear in the same position irrespective of whether they are produced by Ar^+/H_2 or Ar^+/HD collisions in the one case, or Ar^+/D_2 or Ar^+/HD collisions in the other.[116a] This was an elegant series of experiments, particularly in view of the simplicity of the apparatus employed, and Henglein and his coworkers must receive considerable credit in addition for application of the idea of stripping reactions to ion-molecule reaction mechanisms.

Henglein et al. also measured ArH^+/ArD^+ isotope effects. They predicted that the cross section for a stripping reaction would become zero at the point where the internal energy developed in the product exceeded its dissociation energy. As this point occurred at a higher incident ion energy for ArH^+ than for ArD^+, the isotope effect ArH^+/ArD^+ should become infinite after the ArD^+ critical energy had been reached (94 eV). By extrapolation of the data at lower energies, they were able to obtain moderate agreement with prediction.[116c,d] The validity of this extrapolation can be questioned, and Futrell and Abramson,[18] using a tandem mass spectrometer, found an increase of ArH^+/ArD^+ above 50 eV incident ion energy considerably smaller than that calculated by Henglein, Lacmann and Knoll. This was confirmed in another perpendicular tandem mass spectrometer experiment by Berta, Ellis and Koski[18] who measured isotope effects and absolute cross sections for the reaction of HD^+ with rare gases. They studied both the reaction leading to protonated rare gas ions, and the collision-induced dissociation. They obtained a value lower than that of Henglein et al. for the ArH^+/ArD^+ ratio at 40 eV, and also showed that in HD^+/Ar collisions, product ions exist at much higher ion bombarding energies than are predicted by the stripping model. They concluded that, at these high energies, the noncolliding partner of the diatomic ion was no longer a spectator, and carried away some of the energy which would otherwise be sufficient to break the ArH^+ or ArD^+ bond.

Henglein's work on ion-molecule reactions and the increasingly interesting results obtained from crossed molecular beam studies of reactions between neutral molecules stimulated interest in beam studies in general, and several groups developed tandem mass spectrometer systems in which it was possible to measure both the energy and the angular distribution of the product ions.

2. Development of Angular and Velocity Distribution Measurements

Before the complete energy and angular distribution measuring systems had been devised, Stebbings et al.[117] investigated the angular distribution of secondary ions from the N_2^+/D_2 reaction and disregarded their velocities, in contrast to Henglein et al. who measured velocity but not angular distributions. They managed to obtain reaction cross sections as a function of scattering angle at various impact energies, and the total reaction cross section as a function of impact energy. The results showed "peaking" of secondary ions in the forward direction, but the limitation imposed by the measurement of only angular product distributions made it impossible to choose between complex formation and stripping.

Much of the pioneering work on tandem mass spectrometers with facilities for measurement of angular and velocity distributions was carried out by Bailey and his coworkers. Initially Vance and Bailey[52] used a tandem instrument which measured only velocity spectra, and studied charge transfer and collision-induced dissociation in N_2^+/H_2 and H_2^+/H_2 systems. They also examined the kinetics of H_3^+ formation, where their cross sections, measured under similar conditions to those of Giese and Maier,[51] showed the same energy dependence.

Subsequently, Doverspike, Champion and Bailey[118,119,120] used the fully developed instrument, built by Leventhal, Doverspike and Champion in Bailey's laboratory, to study the collision-induced dissociation of D_2^+ by Ar and N_2, and the ion-molecule reaction systems Ar^+/D_2, N_2^+/D_2, D_2^+/D_2, D_2^+/H_2, and H_2^+/H_2. In the case of the Ar^+/D_2 and N_2^+/D_2 reactions, they found something approximating to spectator stripping (which they and some other workers describe as a "pick-up" mechanism) for primary ion energies between 2 and 100 eV.[118] There were, however, a few discrepancies. First, the change of velocity of Ar^+ and N_2^+, after picking up a deuterium atom, was less than would have been expected on a pure stripping model (complex formation would have involved a larger change). This effect was only small but agreed, with slight deviations, in the results of Henglein et al.

Second, the N_2D^+ and ArD^+ velocity profiles differed from each other in a way which indicated that a small amount of momentum was being transferred to the spectator atom, and third, Doverspike et al. were able to observe "back scattering", that is, product ions which had been formed in head-on collisions with neutral molecules and where the reactant ions had lost energy on a scale to suggest that an intermediate complex might have been formed. This indication of complex formation was noticeable at low ion kinetic energies but diminished rapidly as these were raised, and Mahan and Wolfgang[121,122] do not consider this back scattering to be evidence for complex formation, although its magnitude was rather large to be interpreted purely as a consequence of head-on collisions.

In the D_2^+/D_2 and D_2^+/H_2 systems, Doverspike et al.,[119] following up some of the work on the same systems by Vance and Bailey,[52] observed a stripping mechanism with, again, slightly too little energy being lost in the collision process.

The angular distribution spectrum for D_3^+ showed two peaks which appeared to correspond to two reaction channels, transfer of D^+ to the target D_2 molecule, and spectator stripping of a D atom from the neutral D_2, the latter being the more likely. It might be supposed that as the pick-up of a deuterium atom by D_2^+ involves no transfer of momentum to the spectator atom, the same would also be true for deuteron transfer to the neutral molecule. This was not found to be so. At collision energies below about 5 eV, the reaction was completely inelastic, and all the collision energy appeared as excitation energy of the product. These conclusions were confirmed by experiments on the D_2^+/H_2 system.

3. Ion Intensity Maps

Mahan and his co-workers[121] also constructed a tandem mass spectrometer in which a beam of mass- and energy-selected ions was directed into a collision chamber and the emergent ions analyzed by mass, energy, and angle.[121d] They used it to study the dynamics of the reactions of N_2^+ with H_2, D_2 and HD[121a,b] and, in later experiments, with CH_4 and CD_4.[121c]

They analyze their results in terms of center of mass coordinates and refer to the phenomenon of a reactant ion not losing as much energy on collision as would be expected if a complex were formed as "forward scattering". These terms derive from nuclear physics and are becoming more and more widely used in connection with the dynamics of ion-molecule collision processes, along with particle counting techniques of nuclear physics which make possible many of the experiments described here.

Mahan et al. represent their results by constructing contour maps of measured ion intensities.[121b,c,e] These are plotted in polar coordinates in the center of mass system. The original direction of the ion beam is taken as the zero angle, and other angles represent the scattering of ions in that direction relative to the center of mass. The distance of a point from the origin represents the velocity of an ion, and thus the "height" of a contour at a particular point (r, θ) represents the number of ions of velocity r scattered through an angle θ.

If we were to plot such a diagram for a system in which elastic scattering takes place (for example, Ar^+ on He), we would expect the velocity of our ions in center of mass coordinates to remain unaltered so that the final velocity vectors should lie on a circle called the elastic circle whose radius is equal to the initial velocity of the ion in center of mass coordinates. Any ion undergoing a collision in which kinetic energy is absorbed will give a point inside the circle, and a translationally exothermic collision will result in a point outside it.

If we consider a reaction such as

$$N_2^+ + D_2 \longrightarrow N_2D^+ + D$$

which is exothermic, there will be a "superelastic" circle corresponding to all the heat of reaction going into the kinetic energy of N_2D^+, and conservation of energy demands zero intensity of N_2D^+ ions beyond this circle. Furthermore, there will be another circle of smaller radius such that any N_2D^+ corresponding to a point inside the circle would have to store so much energy internally that it could not exist and would dissociate into N_2 and D^+. We should therefore expect all our N_2D^+ ion intensity to fall inside the concentric ring formed by these two circles.

Figure 7 shows the results obtained by Mahan et al.[121] for the above reaction. The shaded areas are the regions of velocity space forbidden by the above considerations, and the fact that some product intensity is found in these regions is probably a consequence of stray electric fields and finite energy and angular resolution of the apparatus.

If an intermediate complex were formed which decayed randomly after several vibrations and revolutions, Figure 7 would be a series of concentric circles, or at any rate symmetric about the $+90°$ to $-90°$ axis. The fact that it is not, and that the velocity vectors of product ions are concentrated in the forward region of the diagram, shows that a long lived intermediate complex is not being formed, and suggests that spectator stripping is occurring. Furthermore, the fact that the peak of the distribution occurs near the inner limiting circle shows that the N_2D^+ carries with it a large

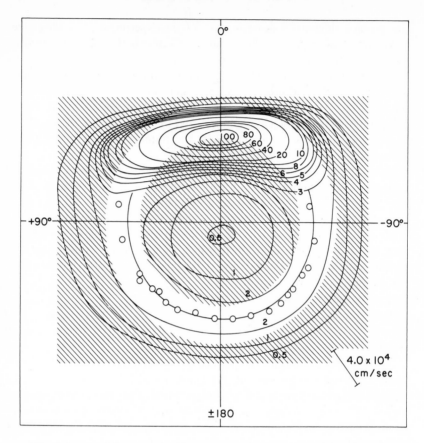

Fig. 7. A map of the normalized intensity distribution of N_2H^+ in the center-of-mass coordinate system. The relative energy of N_2^+ and H_2 reactants was 5.6 eV. The outer shaded area represents values of Q greater than $+1.0$ eV, and in the inner shaded area Q is less than -2.5 eV. The circled points represent the actual maxima in the intensity which were located in the energy and angular scans.

amount of excitation energy, and indeed if this amount were increased only slightly, it would dissociate. Similarly, the majority of backward scattered ions are excited, but less so than those which are scattered forward.

These considerations are discussed in more detail by Mahan.[121e] His results and those of his co-workers show that the N_2^+/D_2, H_2 and HD reactions occur by a stripping mechanism but with the product having slightly more forward velocity than the ideal model predicts. Internal excitation of products is high, but decreases somewhat with increasing scattering angle. For HD, there is an isotope effect which favors N_2H^+ by large factors at small scattering angles and N_2D^+ by smaller factors at large angles.

Their results from a study of the reactions

$$N_2^+ + CH_4 \longrightarrow N_2H^+ + CH_3$$
$$N_2^+ + CH_4 \longrightarrow N_2D^+ + CD_3$$

accord very well the ideal stripping model. Although no momentum is imparted to the neutral methyl radical, it is left with a substantial amount of internal energy. A large isotope effect is found favoring N_2H^+. The N_2D^+ product is both of lower intensity and confined to smaller scattering angles.[121c]

4. Crossed Beam Experiments

Herman, Rose, Kerstetter, and Wolfgang[122] developed a tandem mass spectrometer which differed in two major respects from that of Mahan et al. They were able to operate at lower ion bombarding energies (as low as 0.1 eV in the center of mass) and the neutral gas, instead of filling the collision chamber, was produced as a pulsed molecular beam at 55°C and was assumed to have a Boltzmann distribution of energies. As a consequence of this, the center of mass zero angle was displaced by a few degrees from the direction of the reactant ion beam.

Wolfgang et al. studied the reactions Ar^+/D_2 and N_2^+/D_2 and, contrary to the suggestions of earlier workers, were unable to find complex formation even at the lowest energies. They postulated the dominance of long range forces at all energies. At low reactant energies, part of the energy of reaction appeared as translational energy, but at high energies, kinetic to internal energy conversion took place. Forward scattering was again in excess of what would be predicted on the ideal stripping model.

Whatever the discrepancies between the data obtained in the above experiments, there is virtual unanimity that an excess of forward scattering over what would be expected from an ideal stripping model is found in the reactions of various rare gas and nitrogen ions with deuterium, and this result was further confirmed by Ding, Henglein, Hyatt and Lacmann[23] (v. infra) in experiments on reactions of the type $X^+ + D_2 \rightarrow XD^+ + D$ ($X^+ = Ar^+$, N_2^+, CO^+) for high energies of bombarding ions.

Mahan et al. suggested that this effect was due to kinetic and internal energy of the neutral target molecules, and developed a modified stripping theory to allow for it.[121a] Henglein et al. tested this theory by changing the temperature of their target gas between -190 and $20°C$, and showed that it had no effect on their ion distributions.[123] They therefore supported the theory put forward by Wolfgang et al., which was also a modification of stripping theory and which took into account that the polarizability of the neutral reactant molecule is greater than that of the spectator atom which remains after reaction has taken place.[122]

Consider, for example, an Ar^+ ion approaching a D_2 molecule. As these reactants approach, they are accelerated toward one another by the ion induced dipole potential

$$\alpha_{D_2} \frac{e^2}{r^4}$$

and at the moment of reaction the Ar^+ ion will be travelling faster than its original speed because of this attractive potential. It will then pick up a deuterium atom and the product ArD^+ ion will start to recede from the spectator D atom. As it does so, it will be decelerated by the ion-dipole interaction. Were the polarizability of D the same as that of D_2, the acceleration and deceleration would cancel out, and the net effect would be zero. α_D however is smaller than α_{D_2} and therefore the deceleration as the product ArD^+ ion separates from the spectator D atom is less than the acceleration of the reactant. Thus, the forward scattering will be greater than the pure stripping model would predict, and the neutral species will carry energy away from the collision but in the direction opposite to that of the approach of the Ar^+ ion. In center of mass coordinates it has been scattered strongly backwards, and Wolfgang et al. therefore named their model "recoil" stripping.

5. Evidence for Complex Formation

A noteworthy feature of the experiments described above and others related to them is that the stripping mechanism appeared to be widespread if not universal. Mahan commented that "The occurrence of the stripping process in this series of hydrogen abstraction reactions is an interesting unifying feature. However, the prevalence of this mechanism at this time must not be overinterpreted. The grazing or large-impact parameter collisions associated with small angle scattering and stripping produced a large total reaction cross section. In selecting systems for our first experiments, we naturally pick reactions which have large cross sections so that products will be easily observed. This process almost automatically selects reactions that display something close to the stripping phenomenon."[212]

Wolfgang et al. attached greater significance to the prevalence of the stripping mechanism, which they succeeded in observing at lower energies than Mahan et al. They commented "At lower energies, however, it has been widely thought that ion-molecule reactions in general and these processes in particular occur via an intermediate complex of sufficiently long lifetime to undergo normal unimolecular decomposition, that is, having a lifetime of many vibrations. The evidence for such a belief is, however, circumstantial being based largely on a proposed interpretation of observed isotope effects."[122a]

The circumstantial nature of the isotope effect evidence will be outlined in a subsequent section of this review; it is certainly difficult to rationalize it without the assumption of an intermediate complex. More cogent arguments can, however, be adduced against the viewpoint of Wolfgang et al. and indeed they have themselves since abandoned it. Persistent complexes in ion-molecule reactions have been observed by Pottie and Hamill,[37c] Lifshitz and Reuben,[124] Field,[36f] and Henglein[125] himself.

Matus, Opauszky, Hyatt, Masson, Birkinshaw and Henchman[126,127] have used a pulsing technique for the study of ion-molecule reactions at energies of about 1 eV. Short high voltage pulses applied to a mass spectrometer ion source repeller plate produced reactant ion beams which traversed the source at constant velocity. Reactant and product ions were velocity analyzed by a time of flight technique involving pulsed defocusing of the beam. The technique makes it possible to distinguish between collisions in which there is little momentum transferred to the neutral reaction products, and those in which momentum is distributed among the nuclei involved in the reaction. It was found that the reactions

$$CD_3^+ + CD_4 \longrightarrow C_2D_5^+ + D_2$$

and

$$HD^+ + HD \underset{\displaystyle \searrow}{\overset{\displaystyle \nearrow}{}} \begin{array}{c} H_2D^+ + D \\[2mm] HD_2^+ + H \end{array}$$

involved complex formation at laboratory energies of 1.7 and 0.36 eV respectively. The nonreactive processes

$$He^+ + H_2 \longrightarrow He^+ + H_2$$
$$C_2H_4^+ + C_2H_6 \longrightarrow C_2H_4^+ + C_2H_6$$

were also studied. Complex formation and back reaction are observed in the latter case but not in the former. The reaction

$$CD_4^+ + CD_4 \longrightarrow CD_5^+ + CD_3$$

takes place at low energies through complex formation in the view of Henchman et al. and Henglein and Muccini.[50] On the other hand, isotopic experiments show that in this reaction, the identity of the collision partners is not lost,[128] and it appears that the complex must be bound strongly enough to permit momentum distribution and energy equilibration, but loosely enough to prevent proton and deuteron scrambling or the electron transfer which would result in loss of identity.[103]

A different kind of evidence on complex formation emerges from a study by Leventhal and Friedman of the collision induced dissociation of D_3^+ in a tandem mass spectrometer.[83] The energy threshold for the process

$$D_3^+ + X \longrightarrow D^+ + D_2 + X$$

was the same for $X = {}^3He$, ${}^4He$, or Ne, indicating that the whole of the energy in the center of mass was being used to cause dissociation of the D_3^+. In order for this energy to be made available, some sort of XD_3^+ complex would have to exist for sufficiently long for energy transfer to occur.

6. Transition from Stripping to Complex Formation

There was thus a substantial body of evidence that under certain circumstances ion-molecule reactions could take place through complex formation, and under other circumstances they undoubtedly took place by a stripping mechanism or some modification of it. Few workers, however, had succeeded in observing a smooth transition from complex formation at low energies to a stripping mechanism at higher energies until this was achieved by Henglein, Lacmann et al.[123,129]

These workers added a quadrupole mass spectrometer to their previous velocity analysis apparatus. Initially, they looked at the reactions of Ar^+ and N_2^+ with H_2, D_2, and HBr at ion energies between 5 and 20 eV. (5 eV Ar^+ ions striking stationary hydrogen molecules correspond to an energy of about 0.25 eV in the center of mass.) No transition from stripping mechanism to complex formation could be found.[129a]

Henglein et al. then studied the classic CD_4^+/CD_4 reaction, and found evidence for complex formation. At 7 eV laboratory energy, they obtained two peaks in the CD_5^+ velocity spectrum, one in the predicted position for stripping and the other in the position for complex formation, suggesting that these were competitive processes.[129b]

The reaction $CD_3^+ + CD_4 \rightarrow C_2D_5^+ + D_2$ involves the breaking and formation of two bonds, and might therefore be expected to proceed through complex formation, and this was indeed found by Henchman et al.[127] Henglein et al. on the other hand found that their results were better explained by the stripping mechanism.[129c] For the first time, Henglein et al. expressed their results in terms of forward scattering, that is, they showed that their product ion distribution in center of mass coordinates was forward of what would be expected from complex formation.

Ding, Henglein, Hyatt and Lacmann[123,129d] then attempted to observe the whole range of complex formation, stripping etc. in a single reaction at different ion bombarding energies. Initially they considered the reactions $X^+ + D_2 \rightarrow XD^+ + D$, and $X^+ + CD_4 \rightarrow XD^+ + CD_3$ ($X^+ = Ar^+$, N_2^+, CO^+), and found that at high energies, as already mentioned, additional forward scattering of product ions was observed. They interpreted this in terms of recoil stripping. At low energies (~ 1 eV), behavior intermediate between complex formation and spectator stripping was observed. They postulated an intermediate XD_2^+ complex with a life shorter than half a period of rotation. This rapidly equilibrated excess energy among its vibrational degrees of freedom, so that when it broke up, although the product was still scattered forward because insufficient time had elapsed for rotation, some energy was still transferred to the neutral species so that spectator stripping could not be said to occur. Henglein et al. pointed out that their use of the word "complex" means a transition state which decomposes isotropically in center of mass coordinates, and should not be confused with the statistical mechanical transition complex which is sufficiently long-lived for complete equilibration of vibrational energy. This being so, it is perhaps not strictly accurate for Henglein et al. to say that their XD_2^+ complex undergoes energy equilibration but rather that it exists for long enough for the D atom to act as a "sink" for excess energy. This, of course, would not require such a stable complex as would complete equilibration.

The reactions of Ar^+ and N_2^+ with CD_4 fitted the stripping model over a wide range of energies. At low energies, no preferential forward scattering was found as in the corresponding reactions with D_2, and there was evidence for strong interaction between the incident ion and the CD_3 group. The formation of an intermediate XCD_4^+ ion which would decay isotropically was expected at energies of a few tenths of a volt but such low primary ion energies were not accessible.

In an effort to achieve such an isotropically decaying complex, Henglein et al. decided to study reactions in polyatomic molecules where the complexes would presumably be longer-lived.[129e] They chose two reactions of methanol

$$CH_3OH^+ + D_2 \longrightarrow CH_3OHD^+ + D$$

and

$$CH_3OH^+ + CD_4 \longrightarrow CH_3OHD^+ + CD_3$$

and measured the distribution of kinetic energy of the product ions at various energies of bombarding ions. At low energies, they obtained complexes which decayed isotropically; at high energies there was a transition to a stripping model. The reaction with D_2 was endothermic and had a threshold energy of 0.2 eV in the center of mass system. The complex formed at bombarding energies slightly above this would be expected to decay without the product having appreciable kinetic energy, since all the center of mass energy is needed to drive the reaction, and this was in fact observed.

Having found a system which formed complexes in an accessible region with respect to ion bombarding energies, Henglein and his coworkers studied the transition between the complex and stripping mechanisms. They proposed a model for D atom transfer which took account of both the collision time and energy transfer from the critical degree of freedom of the transition species into other internal degrees of freedom. They found experimentally that for the CH_3OH^+/D_2 reaction, the transition from complex formation to stripping occurred at CH_3OH^+ bombarding energies of about 40 eV. This corresponds to an ion velocity of about 1.6×10^6 cm sec^{-1}. Assuming that energy interchange can take place at distances of up to 3 Å, the collision process can be thought of as lasting for 2×10^{-14} sec. In the case of CH_3OH^+/CD_4, a similar argument leads to a collision time of 6×10^{-14} sec, and Henglein et al. conclude that internal energy relaxation in polyatomic species is a very fast process, and that the time for partition of vibrational energy among the separate degrees of freedom is of the order of a single vibrational period.

As we have observed previously, the time for deposition of energy in an energy "sink" is likely to be much shorter than that required for equilibration of vibrational energy; nonetheless the collision times estimated by Henglein et al. are astonishingly small, and the implication of their work is that energy transfer takes place not by motion of atomic nuclei but by motion of electron clouds. When a complex of this type is formed, it cannot be one in which bond lengths adjust and vibrational frequencies change from their values in the reactants. It must rather be one in which reaction causes a very rapid redistribution of electron clouds, leaving new potential wells into which the heavy nuclei can fall at their leisure as the aftermath of the actual collision process.

Soon after Henglein and his coworkers had reported on the transition between stripping and complex formation, similar transitions were reported by the workers equipped with instruments for velocity and angular distribution analysis of product ions.

Mahan et al.[130] observed complex formation in the endothermic reaction

$$O_2^+ + D_2 \longrightarrow DO_2^+ + D \quad (\Delta E_0{}^0 = -1.9 \text{ eV})$$

at a relative bombardment energy of 3.86 eV. The "contour map" in the center of mass system showed a series of concentric rings centered at the origin. As the relative energy was increased, the product distribution ceased to be isotropic, and eventually displayed the forward peaking characteristic of the stripping mechanism. While the symmetry of the intensity about the barycentric angle of $\pm 90°$ indicated the occurrence of a collision complex which lasted several rotational periods, the isotropy suggested high internal excitation and low angular momentum in the complex, so no definite conclusion was reached as to its lifetime.

Mahan et al. suggested two reasons why the reaction proceeded through a complex. First, the ground state of the $D_2O_2^+$ intermediate is 2.6 eV lower in energy than the reactants; second, the overall reaction is 1.9 eV endothermic. Consequently, there is a high threshold for dissociation to products. Furthermore, for the reaction to proceed by a stripping mechanism, the bombardment energy relative to *one atom* would have to be greater than the endothermicity, whereas reaction by complex formation can occur for a similar bombardment energy relative to the whole molecule, and this of course is achieved at lower laboratory energies.

Wolfgang et al. then produced their first results on complex formation.[131] They agreed with Henglein et al.[129c] that the CD_3^+/CD_4 reaction proceeded through a stripping mechanism but found evidence for complex formation at low energies in the reaction:

$$C_2H_4^+ + C_2H_4 \longrightarrow [C_4H_8^+] \longrightarrow CH_3 + C_3H_5^+ \longrightarrow C_3H_3^+ + H_2$$

The $C_3H_5^+$ and $C_3H_3^+$ product ions were both distributed isotropically about the $\pm 90°$ axis. The $C_3H_3^+$ ions, however, tended to have low and the $C_3H_5^+$ ions high velocities. Apparently $C_3H_5^+$ ions are produced initially with different ratios of internal to translational energy. Ions with high translational energy do not react further and are detected as products, while internally excited ions decompose to give $C_3H_3^+$ product ions.

7. Discussion

Clearly much work remains to be done before the roles of complex formation, "pure" stripping, "recoil stripping" etc. are fully understood. The following observations are presented not so much as conclusions as suggestions and comments on the current situation.

(1) Experiments to determine the dynamics of collision processes are extremely difficult to carry out, and they become more difficult at low ion energies which is the region in which the chemist is principally interested. To avoid the necessity of considering the thermal motion of the neutral

product molecules, it is helpful to choose reacting systems involving the impact of heavy ions on light neutrals (for example, N_2^+ on D_2). On the other hand, this results in very small angular deviations in the scattering process, and these could well be affected by stray fields etc. The idea of Wolfgang et al.[122a] of using a beam of neutral molecules instead of a gas filled collision chamber does not so far seem to offer any substantial advantages.

Some doubt is cast on the results of many of the beam experiments by the fact that the agreement between experiment and theory is neither extensive nor exact. While energy and angular distributions of primary ions are in general narrow and Gaussian, the profiles for secondary ions are frequently asymmetric and much broader than the ideal stripping model would predict, although theoretical and experimental maxima often coincide. It is uncertain whether these discrepancies are due to real deviations from the stripping model or a consequence of the experimental difficulties outlined above. Future experiments should resolve this point.

(2) In addition to experimental difficulties, there have been certain problems in connection with the transformation from laboratory to center of mass coordinates. These were noted by Wolfgang et al. in note 15 of reference 122a, and in a subsequent note Wolfgang and Cross outline a method and a coordinate system which has been accepted by a number of workers in this field.[132]

(3) A body of experimental evidence has now been gathered to confirm the intuitive feeling that in many reacting systems complex formation is likely to be important at low energies, with stripping mechanisms dominating at higher energies. The necessary condition for complex formation is presumably the existence of a minimum in the potential energy surface of the reacting system. Once this has been fulfilled, one would suppose that the possibility of complex formation would hinge on the total energy inventory of the system and the number of degrees of freedom through which it can be spread, the depth of the potential energy well for the complex and sundry considerations of endothermicity of the kind suggested by Mahan, et al.[130] Thus complex formation would occur only at very low energies in simple systems such as N_2^+/D_2 but would continue to much higher energies in more complex systems such as CH_3OH^+/CD_4.

Nonetheless, it is surprising that no complex formation has yet been found at any energy in the N_2^+/D_2 reaction. The surface must have a potential well since the ion $N_2H_2^+$ is abundant in the mass spectrum of hydrazine and must therefore have a lifetime greater than 10^{-6} sec. It is possible that $N_2H_2^+$ is not formed because there is no way in which the energy of formation can be distributed among the degrees of freedom.

Alternatively, the approach to the potential well (which probably corresponds to the structure $H-N=N-H$) from the transition state (which since the polarizability of the hydrogen molecule is a maximum perpendicular to the bond, presumably has the structure $H-H$ involves too complicated

$$\overset{|}{N_2^+}$$

a rearrangement of heavy particles for the cross section to be detectably large.

Another possibility is that the bombarding energies of the beam experiment are still not low enough for complex formation even though the center of mass energies are as low as tenths of a volt. It may be that it is not only important in complex formation for the center of mass energy of the colliding system to be low, but also for its laboratory energy to be low. If we consider a 5 eV N_2^+ ion colliding with a deuterium molecule, the center of mass energy is only 0.625 eV, which could be absorbed by a moderately stable complex, but the relative velocity of the particles is 6×10^5 cm/sec^{-1} and the particles will be within 3 Å of each other for less than 10^{-13} sec. In view of the results of Henglein et al. it is likely that the magnitude of this "contact time" is crucial.

C. Isotope Effects

1. Simple Systems

The problem of isotope effects in ion-molecule reactions is closely related to those of complex formation and energy transfer. Such effects were first studied by Stevenson and Schissler[32,35] who used the relative rates of Ar^+/H_2 and Ar^+/D_2 reactions to establish the validity of the Langevin model of ion-molecule reactions. They looked at the intramolecular isotope effect:

$$Ar^+ + HD \begin{array}{c} \nearrow ArH^+ + D \\ \searrow ArD^+ + H \end{array}$$

but had difficulty in resolving H_2D^+ from D_2^+ in the reaction system

$$HD^+ + HD \begin{array}{c} \nearrow H_2D^+ + D \\ \searrow HD_2^+ + H \end{array}$$

These reactions were later studied in a single source by Reuben and Friedman.[34] The observed intramolecular isotope effects were at least in part dependent on the kinetic energy of the reactant ion, indicating energy transfer to internal degrees of freedom in the reaction products.

Stevenson and Schachtschneider[133] pointed out the inadequacy of the Langevin theory for predicting isotope effects and suggested that competitive unimolecular decomposition of HXD^+ intermediates could account for the isotope effects observed in the Ar^+/HD reaction between 0.2 and 2.0 eV ion kinetic energy. Differences in zero point energy in the intermediates would result in the DX^+-H bond breaking more readily than the HX^+-D bond giving the observed XH^+/XD^+ ratio of less than unity. The observed H_2D^+/HD_2^+ ratio in reaction, HD^+/HD was also less than unity and could be explained in the same way.

Such an explanation is at odds with the Eyring–Hirschfelder–Taylor model of ion-molecule reactions which postulates a very loose complex in which the internal degrees of freedom of the reactants are scarcely perturbed.[6] Reuben and Friedman[34] attempted a different explanation of intramolecular isotope effects within the framework of the Eyring, Hirschfelder and Taylor theory. In this model, the reactants are attracted by a long range ion-dipole interaction between the center of charge in HD^+ and the center of polarizability in HD. This attraction is opposed by the centrifugal force of rotation of the reactants. The fact that the center of mass and center of charge in HD^+, and center of mass and center of polarizability in HD are not coincident means that there is a higher probability of $[D-H-H-D]^+$ being formed than of $[H-D-D-H]^+$ and thus, after decomposition of these intermediates to give H_2D^+ and HD_2^+ respectively, there should be an isotope effect such that $H_2D^+/D_2H^+ > 1$. The calculation was made more general by Klein and Friedman,[39] but always predicted that in X^+/HD systems, $XH^+/XD^+ > 1$. They investigated experimentally reactions of HD^+ with He and Ne, and of Ar^+ and Kr^+ with HD, and found in all cases an isotope effect favoring XD^+ at low energies and XH^+ at high energies. At high energies their results could be explained by an isotope effect involving the center of mass/center of charge displacement noted above for which $XH^+/XD^+ > 1$. At lower energies, an intermediate was thought to be formed which decomposed in accordance with predictions based on zero point energy such that $XH^+/XD^+ < 1$, and there was also a contribution in the HD^+/He and HD^+/Ne reactions at high energies from decomposition of the ionic product into atoms.

The idea that the center of mass in HD^+ differs from the center of charge has recently been applied by Light and Chan[134] to the modified stripping model of Wolfgang et al.[122a] They attempted to predict isotope effects in the HD^+/Ar system.

Moran and Friedman[135] pointed out that if the low energy isotope effects were due to formation of a transition complex, then in a highly exothermic reaction the complex might be unstable and hence no isotope effect less than unity should be found. They studied the thermochemistry of the moderately exothermic N_2/HD and the more exothermic CO/HD, O_2/HD, and CO_2/HD interactions, and showed that only in the case of N_2/HD did the XH^+/XD^+ ratio drop below unity at low ion energies.

The above work suggests strongly the existence of ion-molecule complexes at low energies. In particular, complex formation appears to occur in the cases of N_2^+/HD and Ar^+/HD, reactions which appear on the basis of angular and velocity distribution analysis to take place entirely by a stripping mechanism.

The tandem mass spectrometer offers the opportunity of distinguishing clearly between X^+/HD and HD^+/X reactions. The work of Henglein et al., Futrell and Abramson, and Berta, Ellis and Koski on reactions of Ar^+, N_2^+, and other ions with HD was discussed in the previous section. The XH^+/XD^+ ratio rose more slowly with increasing energy than was predicted by the ideal stripping model, but considerably faster than would be expected on the "displacement" theory of Friedman et al., the approximations of which are invalid at energies as high as those involved in the above experiments.

Mahan et al. measured the variation of intramolecular isotope effect with energy and scattering angle for N_2^+/HD mixtures.[121a,b] For high projectile energies there is an enormous (~ 20 at 6–12 eV) isotope effect favoring N_2H^+ at $\theta = 0$, a similar small effect at $\theta = 180°$ and an intermediate effect at $\theta = 90°$. At low velocities, the isotope effects at $\theta = 90°$ and $180°$ are in favor of N_2D^+. Mahan et al. considered the decrease in isotope effect with increasing angle to show that the energy of the projectile relative to the abstracted atom (or equivalently the internal energy of the incipient N_2H^+ or N_2D^+) becomes less critical as the impulse imparted to the freed atom increases. Thus, in grazing collisions, N_2H^+ is much favored, but in head on collisions there is only a small or even inverse effect. These and other measurements in tandem mass spectrometers[119,136] and similar ones in pulsed sources[127,137] support qualitatively the results of the single source experiments.

2. Complex Systems

Data on isotope effects in more complex systems are scattered diffusely through the literature, and little attempt has been made to correlate, let alone to interpret, them. An interesting example is to be found in the work of Bone and Futrell in a tandem mass spectrometer study of ion molecule reactions in propane.[138] The $C_2H_4^+$ ion undergoes two reactions with propane, and if $CD_3CH_2CD_3$ is used, various isotope effects are possible:

$$C_2H_4^+ + CD_3CH_2CD_3 \quad \overset{k_1}{\underset{k_2}{\rightleftharpoons}} \quad \begin{array}{l} C_2H_4D + C_3D_5H_2^+ \\ \\ C_2H_5 + C_3D_6H^+ \end{array}$$

$$C_2H_4^+ + CD_3CH_2CD_3 \quad \begin{array}{c} \overset{k_3}{\nearrow} \\ \overset{k_4}{\longrightarrow} \\ \overset{k_5}{\searrow} \end{array} \quad \begin{array}{l} C_2H_5D + C_3D_5H^+ \\ \\ C_2H_6 + C_3D_6^+ \\ \\ C_2H_4D_2 + C_3D_4H_2^+ \end{array}$$

The overall rate of the first group of reactions at near thermal energies is about twice that of the second, and increases rapidly with increasing ion energy. At low energies, $k_3/(k_4 + k_5) = 4$ and $k_2/k_1 = 10$, both values greatly in excess of the statistical values of 0.43 and 0.3 for HD^- and H^- transfer respectively. k_2/k_1 decreases rapidly with ion bombarding energy and drops as low as 0.9 at 18 eV laboratory energy. In the reactions

$$C_2H_5^+ + CD_3CH_2CD_3 \quad \overset{k_6}{\underset{k_7}{\rightleftharpoons}} \quad \begin{array}{l} C_2H_6 + C_3D_6H^+ \\ \\ C_2H_5D + C_3D_5H_2^+ \end{array}$$

$$\frac{k_6}{k_7} = 0.33 \qquad \text{(statistical value 0.33)}$$

and for

$$C_2H_5^+ + CH_3CD_2CH_3 \quad \overset{k_8}{\underset{k_9}{\rightleftharpoons}} \quad \begin{array}{l} C_2H_6 + C_3D_2H_5^+ \\ \\ C_2H_5D + C_3DH_6^+ \end{array}$$

$$\frac{k_8}{k_9} = 1.85 \qquad \text{(statistical value 3.0)}$$

The reactions of the ethyl ion can be interpreted very simply. The chances of H^- and D^- transfer appear to be similar to the proportions of H and D atoms in the propane molecule. The situation is very different for the $C_2H_4^+$ ion in which the transfer from the propane of H^- and HD^- is strongly favored. With increasing ion energy, the isotope effect drops sharply in the former case and the relative rate of the overall reaction drops rapidly in the latter. It is tempting therefore to suppose that the isotope effects arise through complex formation, especially as the perturbing effect of the long-range interaction with propane of $C_2H_4^+$ can hardly be markedly different from that of $C_2H_5^+$.

If a stable complex is formed at low energies, then fission of the C—H bond in $CD_3CH_2CD_3$ will be favored both because it is secondary and therefore about 1.5 kcal weaker than a primary C—H bond, and because of zero point energy considerations which are worth approximately 1 kcal more. Furthermore, complex formation would make easier the fission of two bonds in the hexadeuteropropane, and it seems very reasonable that hydrogen from neighboring carbon atoms should be abstracted preferentially. This hypothesis suggests that HD^- transfer can only proceed through a complex, and that H^- or D^- transfer should show a large isotope effect favoring H^- transfer at energies where complex formation takes place. At higher energies, the HD^- reaction virtually disappears, and the isotope effect in H^- transfer drops markedly as a stripping mechanism supersedes complex formation.

Friedman and Moran[135] showed that in highly exothermic ion-molecule reactions, the isotope effect appropriate to unimolecular decomposition of a complex is not found because of the large amount of internal energy which the complex must absorb in its internal modes. If this is so, then we would predict that the heats of reaction of the processes in which H^- and HD^- are transferred to $C_2H_4^+$ should be lower than that of the $C_2H_5^+$ reaction in which one would postulate that complex formation did not take place. The relative heats can be easily calculated, and are 5, 20 and 24 kcal exothermic respectively. The above analysis lends some support to the idea that it is possible to apply to the breakup of polyatomic ions with low energies the same considerations which are applied to neutral molecules in conventional isotope effect theory. If this is so, then one would expect larger polyatomic ions at thermal energies to show these effects particularly clearly. Reuben and Lifshitz have measured isotope effects at thermal energies for ten ion-molecule reactions in the benzene and hexadeuterobenzene systems.[124] They found that for addition reactions such as

$$C_6H_5^+ + C_6H_6 \longrightarrow C_{12}H_{11}^+$$

$k_D/k_H \cong 1.4$, while for reactions such as

$$C_{12}H_{11} \longrightarrow C_{12}H_9^+ + H_2$$

which involve the decomposition of these complexes, $k_D/k_H < 1$, as it is for reactions which take place with the elimination of H or H_2, for example,

$$C_2H_3^+ + C_6H_6 \longrightarrow C_8H_7^+ + H_2$$

Reactions of an intermediate character:

$$C_{12}H_{11}^+ \longrightarrow C_{10}H_9^+ + C_2H_2$$

have $k_D/k_H \cong 1$. The isotope effects in the last three of these reactions can be explained qualitatively on the basis of the effect of zero point energy considerations on the fission of C—H and C—D bonds. The inverse isotope effects in the addition reactions require for their rationalization the occurrence of some form of back or charge exchange reactions to give, for example, $C_6H_5 + C_6H_6^+$, and Reuben and Lifshitz had already postulated these to explain why their cross sections were somewhat lower than those predicted by the G–S theory.

The calculation of isotope effects in low energy ion-molecule reactions presents many problems connected with the estimation of zero point energies and frequencies in transition states, where even the data for the reactant ions are lacking. Unimolecular decomposition calculations are somewhat speculative.[139] Theoretical models, although they often predict the correct sign for an isotope effect, can suggest its magnitude only crudely. The most satisfactory of these theories is the "phase space" model of Light[106,140,141] which gives the correct energy dependence for isotope effects in the He$^+$/HD reaction, but the absolute values are between 5 and 30% too low. Stripping and displacement isotope effects should be more readily calculable since there are fewer arbitrary parameters to be dealt with, but even here few rigorous quantitative treatments have been performed and tested.

D. Long Lived Collision Complexes

1. Introduction

Long lived collision complexes, formed by addition of an ion to a molecule, without the ejection of a neutral particle, would not be expected to occur in reactions between simple molecules because of problems associated with the removal of the heat of reaction. With polyatomic molecules, however, this energy can be taken up by the many degrees of freedom in the molecule, and possibly at some stage the complex may be stabilized by a further collision. Reactions of this kind can be conveniently divided into three groups—addition reactions, clustering reactions, and polymerization reactions. Addition reactions are sometimes referred to in the literature as condensation reactions, a usage which is acceptable so long as it is realized that the term has a totally different meaning in polymer and general organic chemistry.

2. Polymerization Reactions

Gas phase ion-molecule polymerization reactions are formally analogous to the ionic polymerization processes used to prepare polymers of laboratory and commercial interest. Olefins can frequently be polymerized with the aid of a cationic initiator by a simple chain reaction, for example,

$$H^+ + CH_2{=}CHR \longrightarrow CH_3{-}CHR^+$$

$$CH_3{-}CHR^+ + CH_2{=}CHR \longrightarrow CH_3{-}CHR{-}CH_2{-}CHR^+$$

$$\cdots\cdots\cdots$$

$$CH_3{\left(CHR{-}CH_2\right)}_n CHR^+ + CH_2{=}CHR \longrightarrow CH_3{\left(CHR{-}CH_2\right)}_{n+1} CHR^+$$

In the same way that a large number of olefins can be polymerized to give products of the above type, many unsaturated compounds can undergo reactions in the mass spectrometer to give what are apparently low molecular weight polymers containing up to about ten units of monomer. For example, Kebarle and Hogg[114b] obtained a polyethylene ion containing seven ethylene moieties.

Much work has been reported in this field, and three of the papers in the 1966 ACS symposium[18] [by Wexler, Lifschitz and Quattrochi; Kebarle, Haynes and Searles; and Meisels] dealt with this topic. Volpi et al. have studied propylene, cyclopropane, butenes, and tetrafluoroethylene.[142]

Beauchamp, Buttrill and Baldeschwieler,[96] have reported on ion cyclotron resonance studies of chlorethylene. There appear to be several series of genuine condensation reactions in which HCl is eliminated. If A^+ is any of a number of ionic species in the system, the following reaction sequence is found:

$$A^+ + C_2H_3Cl \longrightarrow AC_2H_3^+ + HCl$$

$$AC_2H_3^+ + C_2H_3Cl \longrightarrow AC_2H_2 \cdot C_2H_3^+ + HCl$$

$$A(C_2H_2)_n^+ + C_2H_3Cl \longrightarrow A(C_2H_2)_{n+1}^+ + HCl$$

Ethylene polymerization was one of the earliest polymerizations to be studied,[143-145] but processes of order greater than two were not detected until measurements were extended to higher pressures.[36f,146,147] The major products are $C_nH_{2n}^+$ and $C_nH_{2n-1}^+$. The main area of disagreement in the studies of the reaction is whether the complex formed by a second order process (that is, $C_4H_8^+$, $C_4H_7^+$, or $C_4H_6^+$) is short lived and dissociates immediately, or long lived and relatively stable. In the former case, the fragment ions react further with ethylene to give higher order products;

in the latter case these are formed by reaction of ethylene molecules with the complex. The problem of stability of transition complexes is thus of importance even with high pressures and complex systems, and here too it is still unresolved.

Ion-molecule reactions in acetylene have been studied even more extensively than those in ethylene.[148-157] The complex reaction sequence required to rationalize the ionic polymerization of acetylene has been discussed in some detail by Munson,[154] by Dervish, Galli, Giardini-Guidoni and Volpi,[155] and by Wexler, Lifshitz and Quattrochi,[18] all of whom used high pressure ion sources in their experimental work. More recently, Futrell and Tiernan[156] reported work carried out on their tandem mass spectrometer.[157] Their mass spectrum of acetylene as a function of source pressure agrees qualitatively with results of Volpi et al.[155] The main primary ion is $C_2H_2^+$ and this after a single reaction step gives $C_4H_3^+$ and $C_4H_2^+$. A second step leads to $C_6H_5^+$ and $C_6H_4^+$, and a third to $C_8H_7^+$ and $C_8H_6^+$. An ion $C_6H_6^+$ is formed by higher order processes and was not observed by Melton.[153] Neglecting minor products, Futrell and Tiernan produced a mechanism to account for the polymerization which involves some 28 reactions and will not be reproduced here. The tandem study makes it possible to see which reactions may lead to a long lived 'sticky' collision complex. $C_2H_2^+$ does not react to give detectable amounts of $C_4H_4^+$, but C_2H^+ gives a trace of $C_4H_3^+$. $C_4H_4^+$ derived from benzene reacts with acetylene to give $C_6H_6^+$, but the major product is $C_6H_5^+$.

The tandem mass spectrometer technique involves ion bombarding energies of the order of volts, and it is to be expected that the cross sections for long lived complex formation will drop sharply with energy, so that Futrell and Tiernan's results do not eliminate the possibility of sticky collisions being more important at lower energies. Munson comments on the basis of his results that, after the loss of H or H_2 in the first collision process

$$C_2H_2^+ + \begin{cases} C_4H_3^+ + H \\ C_4H_2^+ + (H_2) \end{cases}$$

the reactions involve only addition of acetylene molecules. For small aggregates, he was able to show that these had to be stabilized by collision or they would decompose, for example,

$$C_4H_3^+ + C_2H_2 \longrightarrow [C_6H_5]^+ \begin{cases} \xrightarrow{C_2H_2} C_6H_5^+ + C_2H_2 \\ \longrightarrow C_6H_3^+ + H_2 \text{ or } 2H \\ \text{decompositions} \searrow \text{unimolecular} \\ C_4H_3^+ + C_2H_2 \end{cases}$$

For higher molecular weight polymers, he considered that this would not be necessary. It is of interest to consider the number of degrees of freedom required to allow appreciable amounts of a long lived complex to exist while absorbing a given amount of heat of reaction. Examination of the higher acetylene polymers might find the point at which this occurs. It is clear that in olefinic systems such a point has hardly been reached with the C_6 ions, although Volpi et al. have observed metastable peaks due to the decomposition of such complexes. Almost all the intermediates in the above polymerizations can be stabilized by collision at pressures of the order of several torr. It appears that the collision complexes in acetylene and ethylene have lives intermediate between those discussed in the previous section and those to be discussed in the next paragraph.

3. Addition Reactions

Examples of persistent ion-molecule collision complexes were observed by Pottie, Barker and Hamill[37b,c] who studied the reactions

$$C_2H_5I^+ + C_2H_5I \longrightarrow C_4H_{10}I_2^+$$
$$C_2H_5Br^+ + C_2H_5Br \longrightarrow C_4H_{10}Br_2^+$$
$$C_3H_7I^+ + C_3H_7I \longrightarrow C_6H_{14}I_2^+$$

They failed to observe ion-molecule reactions in methyl iodide and methyl bromide or propyl chloride but obtained $C_3H_8I_2^+$ from a mixture of methyl and ethyl iodides. Above a certain critical value of the ion velocity, the cross sections for these reactions fell to zero.

Addition reactions have also been observed in negative systems,[158] for example,

$$HCOO^- + HCOOH \longrightarrow HCOOHCOOH^-$$

and reactions of the type $CH_3CO^+ + CH_3COOC_4H_9 \rightarrow C_8H_{15}O_3^+$ are a well known source of confusion in the mass spectra of esters.[159]

Henglein observed several persistent complexes in his studies of ion-molecule reactions in acrylonitrile and benzene,[125,160] the most important ones being

$$C_3H_3N^+ + C_3H_3N \longrightarrow C_6H_6N_2^+$$

and

$$C_6H_5^+ + C_6H_6 \longrightarrow C_{12}H_{11}^+$$

The cross sections for these reactions decreased according to $E^{-1.2}$ and $E^{-1.4}$ respectively, where E is the repeller field.

Field, Hamlet and Libby[161] studied ion-molecule reactions in benzene, producing their ions both by electron impact and by impact of rare gas ions.

They found the $C_6H_6^+$ ion to be unreactive, but claimed that if produced by chemical ionization in xenon it would react to give $C_{12}H_{12}^+$ though still only slowly. The $C_3H_3^+$ ion seemed also to be unreactive though Beauchamp[162] had observed it to react with benzene to give $C_9H_7^+$ and H_2. The lack of reactivity of $C_6H_6^+$ agrees with the observation by Munson in his experiments on acetylene.[154]

Wexler and Clow[163] studied benzene and toluene in a high pressure source. In contrast to the above experiments, they found that addition was the most prominent reaction and also observed charge transfer. With toluene, dimeric species were also formed:

$$C_7H_7^+ + C_7H_8 \longrightarrow C_{14}H_{15}^+$$
$$C_7H_8^+ + C_7H_8 \longrightarrow C_{14}H_{16}^+$$

Giardini–Guidoni and Zocchi[165] studied benzene, toluene, and xylenes in a beta ray ion source at pressures up to 0.1 torr and found dimer ions to be the main products at the higher pressures. Like Wexler and Clow, they obtained large yields of $C_{12}H_{12}^+$ and $C_{12}H_{11}^+$ in benzene, both formed by addition reactions.

Field, Hamlet and Libby,[161b] however, have offered a different explanation to reconcile the low ratio of $C_{12}H_{12}^+/C_6H_6^+$ which they found with the much higher ratio found by Wexler and Clow. They investigated the effect of temperature on the high pressure mass spectrum of benzene, and claimed that an equilibrium:

$$C_6H_6 + C_6H_6^+ \rightleftharpoons C_{12}H_{12}^+$$

was established, and for which they measured $\Delta H = -15$ kcal/mole and $\Delta S = -23$ kcal/°/mole. It appears that their source was not designed specifically for the purpose of equilibrium measurements and one would expect the problems of sampling, and achievement of equilibrium to be more acute than those encountered by Kebarle et al. (Section III-D), although Field et al. nonetheless obtained good agreement with these workers in the experiments performed to validate their method.

Field et al. recognize that their value of ΔH is surprising. A reaction which gave the two benzene rings linked at one point would almost certainly be endothermic, and instead it is suggested that the dimer ion has a sandwich structure which is known to exist in the liquid phase. The two above hypotheses are not mutually exclusive, and the possibility exists that the formation of $C_{12}H_{12}^+$ is a "traditional" chemical reaction possessing both a finite heat of reaction and energy of activation.

Lifshitz and Reuben[124] studied benzene and hexadeuterobenzene at low pressures, both in a pulsed-source mass spectrometer and one with a conventional source. They obtained higher yields of C_9 to C_{11}^+ products than other workers, and their main reaction was addition to give $C_{12}H_{11}^+$. They were unable to detect $C_{12}H_{12}^+$ in the pulsed system in which the ions were effectively thermal, but these appeared in their other source as soon as the repeller field was raised. The reaction giving $C_{12}H_{12}^+$ appears to be endothermic and requires kinetic to internal energy transfer before it can take place. This conclusion was reached separately by Beauchamp[162] using an ion cyclotron resonance mass spectrometer.

Lifshitz and Reuben also found that their $C_{12}H_{11}^+$ was not stable, and broke up to give $C_{12}H_9^+$ and $C_{10}H_9^+$, accompanied by the appropriate "metastable" peaks. Observations of metastable peaks for transitions from ion-molecule complexes have been reported previously by Dervish et al. in acetylene, and confirm the long lived nature of these complexes. A summary of the data on $C_{12}H_{11}^+$ gives the series of reaction channels shown in Table IV. The rate constants found by various workers are shown in subsequent columns. The discrepancies are large and there are clearly problems associated with the calculation of rate constants in high pressure systems. These will be discussed at the end of this section.

The formation of $C_{12}H_{12}^+$ apparently by kinetic to internal energy conversion poses a question as to why it does not take place with benzene ions formed by electron impact, which undoubtedly have a spread of internal energies which in some cases should be sufficient to drive the reaction. Does the reaction have a translational energy barrier such that the species can only get close enough to react if one of them has sufficient kinetic energy to surmount this barrier, that is, is this a system where internal to kinetic energy conversion is necessary rather than the reverse? The role of xenon in increasing somewhat the rate of the $C_6H_6^+/C_6H_6$ reaction in the high pressure system is obscure and any attempted explanation would be entirely speculative.

The nature of the $C_{12}H_{11}^+$ ion also requires further elucidation. It is unclear from the published work whether the $C_{12}H_{11}^+$ observed is a stable ion or a genuine transition complex which, if left long enough, would react along the various available channels. The fact that Reuben and Lifshitz obtained $C_{12}H_{11}^+$ in a system at thermal energies, where collisional deexcitation is unlikely, suggests that most of the $C_{12}H_{11}^+$ ions have a similar energy inventory and if one molecule decomposes so ultimately will the others. The instability of the complex is confirmed by the observation of the "metastable" peaks already mentioned. On the other hand, observations of peak intensity versus pulse time for $C_{12}H_{11}^+$ by the same workers was of a form which suggested that almost all the $C_{12}H_{11}^+$ with sufficient energy to decompose had done so in a time span much shorter

TABLE IV

Ion-Molecule Reactions in Benzene

	Reuben[+124] and Lifshitz	Field, Hamlet[161] and Libby	Giardini-Guidoni[165] and Zocchi	Wexler*[163] and Clow	G-S
					Theory
k in cc molecule^{-1} sec^{-1} $\times$ 10^{10} Repeller field V/cm	0	1–5	4	12.6	
$C_6H_5^+ + C_6H_6 \rightarrow [C_{12}H_{11}^*]$ $\rightarrow C_{12}H_{11}^+$	1.44	25			
$\rightarrow C_{12}H_9^+ + H_2$	1.13				
$\rightarrow C_{10}H_9^+ + C_2H_2$	1.46	25			
$\rightarrow C_6H_6^+ + C_6H_5$	not determined				
Total	4.03	50	0.10	2.81[a] 4.52[b] 5.87[c]	12.03

+ All experiments except those of Reuben and Lifshitz were carried out in high pressure sources.

* Rate constants calculated from cross sections obtained at 12.6 V/cm and 3.2 mm path length.

[a] At 80 eV electron energy.

[b] At 400 eV electron energy.

[c] At 1000 eV electron energy.

than that involved in the pulsing experiments. In the high pressure experiments, of course, the $C_{12}H_{11}^+$ all had ample opportunity to become at least in part collisionally de-excited.

$C_{12}H_{11}^+$ and $C_{12}H_{12}^+$ are probably the best investigated of the relatively few known long-lived complexes. The work done so far, however, generates more questions than it solves, and it might be worthwhile for some of the beam techniques used for simpler reactions to be applied to the benzene system. Here at least there is small probability of a stripping reaction at low energies.

4. Clustering Reactions

This term covers the reactions in which polar molecules group themselves around an ion and are held in place by ion-dipole or ion-induced dipole interactions. Examples are $NH_4^+(NH_3)_n$ and $H^+(H_2O)_n$. These were discussed in the thermochemistry section (Section III-D) and we shall not refer to them further in this section.

5. Determination of Rate Constants in High Pressure Systems

Most of the experiments described in this section were carried out with mass spectrometers equipped with sources which could be operated at pressures of several tenths of a torr[36f,146,147] or even several hundred torr.[114b] Primary ionization is brought about by electron impact,[36f,166-171] glow discharge,[172,173] uncollimated alpha or beta particles and secondary electrons from radioactive deposits near to or in the source,[74,174] or a MeV proton beam from a Van de Graaff accelerator (Wexler et al.[164]).

Many workers have calculated rate constants for consecutive and concurrent reactions in these high pressure systems, and in view of the bulk of these data and discrepancies in results for the benzene system it is worth reviewing the methods used to calculate these rate constants.

Two early models to describe consecutive ion-molecule reactions in high pressure systems were proposed by Lampe, Franklin and Field[13] and Wexler and Jesse.[171] The former applied Bodenstein steady state theory to transition complexes which they assumed to be formed though they were unable to detect them. They also assumed the rate constant k of an ion-molecule reaction to be independent of the reactant ion velocity, in accordance with the G–S theory. The rate of attenuation per unit volume of a certain kind of reactant ion X^+ was taken as

$$-\frac{d[X^+]}{dt} = k[X^+]N \tag{11}$$

where $[X^+]$ and N are the concentrations of reactant ions and neutral molecules respectively.

Wexler and Jesse, on the other hand, considered the attenuation of a beam of ions of intensity I in traversing a distance y and proposed for a reaction of microscopic cross section σ

$$-\frac{dI}{dy} = IN\sigma \tag{12}$$

The theory does not allow for the formation of an intermediate complex. Wexler and Jesse assumed that σ was independent of the ion velocity, an assumption which seems very much less satisfactory than that of Franklin, Field and Lampe. Nonetheless, the two methods give similar results when applied to methane[171,175] and ethylene.[36f,147] Volpi et al[174] used Wexler and Jesse's beam method for calculating rate constants and extended it to allow for the possibility of a collision complex being formed. They also developed a different model to permit the calculation of rate constants in one of their sources which was irradiated by a radioactive emitter, and in which they assumed ions to be formed uniformly throughout.[174] The reactant ion path length is neither single nor well defined but by assuming that the rate constant was independent of ion velocity, and that neither charge exchange nor ion-molecule reactions slowed down the ion, they were able to carry out an integration which enabled them to derive rate constants. This was the method used by Giardini–Guidoni and Zocchi in the work on benzene described earlier.[165]

Hyatt, Dodman and Henchman[18] commented on the above methods and derived a theory for hydrogen transfer reactions based on three possible mechanisms, namely complex formation, proton or hydride ion stripping, and hydrogen atom stripping. Application of the theory yields information on the energy dependence of the rate constant and if this is known independently, it is possible to choose among the three mechanisms.

Szabo[177] has pointed out that most determinations of order and rate of reaction to give a particular product ion are based only on variation of pressure, and has suggested that the variation of ion path length and residence time might also be considered. He gives formulae for calculation of orders and rates by these three different methods and applies them to his own results,[178] and to those of Harrison et al.,[18] and Futrell and Miller.[157]

Field and Munson[179] have commented that the calculation of rate constants in high pressure systems requires the knowledge of ion residence times. The choice of proper values is not straightforward because the reactant ions are not formed in a well defined position, and because at high pressures the large number of collisions an ion undergoes means that the time it takes to escape from the source may not be calculable from simple electrostatics. Ion mobility considerations should be applied, but information on drift velocities of ions in the relevant systems is largely lacking. It seems likely that the residence time at constant field strength is pressure dependent.

Volpi et al. also expresses doubts about the accuracy of high pressure rate constant measurements, and suggest that for processes of higher order little more than semi-quantitative data can be expected. Szabo concurs with this and considers that rate constants of reactions of order higher than two cannot be determined because the rigorously determined formulae for ion intensities are too complicated.

The unease expressed by workers in the field about the validity of rate constants determined in high pressure experiments seems to be justified. In addition to problems associated with uncertain residence times and path lengths of reactant ions mentioned above and relating to the photoionization measurements of Warneck (see section IV-E-4) there is a problem with the internal energy of the reactant ions. Differences in energy are as important as differences in structure as far as reactivity is concerned.

Wexler et al. and Volpi et al. use methods for generating primary ions such that most of them are produced by secondary electrons of ill-defined energies. Field et al. pay some attention to this problem by the use of electron beams and by use of chemical ionization techniques to deposit better defined amounts of energy. Indeed they were able to show differences in reactivity between excited and unexcited acetylene ions. Nonetheless, even if the energy state of a primary ion is defined, the energies of secondary and tertiary ions arising from it are unknown. Thus, while results of high pressure experiments appear to provide accurate data on the order of processes leading to the various product ions, it would be unwise to attach high significance to other quantitative measurements in these systems.

E. State Selection of Reactants

1. Introduction

The cross section of a species for undergoing a chemical reaction depends not only on the molecular structure of the species but also on its internal and kinetic energy. These last two factors can be crucial. For example, in neutral–neutral reactions, the reactivity of triplet excited states and of electronically excited states in general is now well attested. Furthermore, the necessity of a molecule having a certain minimum energy inventory in order to be able to undergo reaction, is the basis of the simplest theories of chemical kinetics.[180]

In the study of ion-molecule collision processes, where there is little opportunity for thermalization of energy, it is important to define internal and kinetic energies of ions if their reaction cross-sections are to be meaningful. The efforts of Lindholm and his coworkers to estimate the proportion of metastable excited states in ion beams have already been discussed (Section II-C-1), and experiments on similar lines have been carried out for example by Lao, Rozett and Koski who used a tandem perpendicular mass spectrometer technique to estimate the relative amounts of ^2P and ^4P C^+ in a beam.[181]

There are three possible approaches to the problem of state selection. One can generate a mixture of states and try to identify the proportions of each state in the mixture; one can try to generate entirely ground state ions, or one can try to generate selected excited states. The first approach is exemplified by the work of Lindholm, the second by the pulsed source and flowing afterglow techniques, and the third by the photoionization technique.

The tandem mass spectrometer generates ions of known kinetic energy which is a form of state selection, but the application of this method was discussed in the previous section, and is not repeated here.

2. Pulsed Source Techniques

The pulsed source technique was devised in order to study the reaction of ions with thermal energies. A short pulse of electrons is admitted to a field-free ion source to produce primary ions by electron impact. A known and variable time later, a second voltage pulse is applied to an ion repeller or draw-out plate, and the ions expelled from the source are mass analyzed. In the interval between the two pulses (called the delay time) the ions are able to react, and from the variation of primary and secondary ion currents with delay time the thermal rate constants can be calculated.

The technique was first described by Tal'rose and Frankevich in 1960[182] and has since been used for the study of ion-molecule reactions in relatively low pressure systems by Harrison et al.[183] Lampe et al.[184] Hand and von Weyssenhof,[185] and Ryan and Futrell.[186] An ingenious and interesting modification of the pulsed source technique devised by Henchman et al. was described in a previous section (IV-B-5).

Harrison and his group have been easily the most prolific workers in this field. In 1966 he, Myher and Thynne[18] reviewed the previous three years' work and listed some fifty rate constants for a wide range of reactions varying from H_2^+/H_2 to CH_3CHO^+/CH_3CHO, CD_3OH^+/CD_3OH and CH_3CN^+/CH_3CN. Since then, the group has studied acetylene, acetylene/methane,[183h] ethylene,[183f] methylamine,[183f] hydrogen sulfide,[183b] propyne, allene,[183k] oxygenated species,[183l] and many other systems. Some of the earlier rate constants were in error due to incorrect measurements of ion source pressures, and Gupta, Jones, Harrison and Myher have amended and in some cases remeasured these.[183d]

Many of the amended results agree with the G–S theory. Discrepancies in some cases were attributed to alternative channels of reaction which could not be observed under the conditions of the experiment, and in others to ion-permanent dipole interactions which were treated on the basis of a modified collision theory.

Among the more recent results is the finding that the propyne ion $[HC\equiv C-CH_3]^+$ reacts more rapidly with propyne molecules than does the allene ion $[H_2C=C=CH_2]^+$ with allene molecules, possibly on account of its having a finite dipole moment. Furthermore, investigation of competitive reaction channels with labelled molecules showed that in reactions where hydrogen atoms are transferred, reaction takes place without "scrambling" of the other hydrogen atoms. If a larger species is transferred, however, (for example, $C_3H_4^+ + C_3H_4 \rightarrow C_6H_7^+ + H$) "scrambling" of hydrogen atoms does occur. Harrison et al. suggested that in the former case a loose complex is formed while the latter corresponds to an "intimate" complex. Increase of ion kinetic energy discriminates markedly against formation of the intimate complex, and to a lesser extent against the hydrogen transfer reaction. Alternative reaction channels appear to be involved but the pulsed source shares with the single source the difficulty of identifying these.

Pulsed sources have other drawbacks. They produce ions with controlled kinetic energy but offer no control over internal energy; they are liable to problems in connection with the build-up of small surface charges which prevent the ions from being truly thermal, and the investigations of processes with small cross sections is hindered by the possibility of reactions occurring during the draw-out pulse.

The problem of internal energy deposited in ions in a pulsed source has been considered carefully by Lampe and co-workers.[184] They have developed a technique which integrates a pulsed beam of ionizing electrons of controlled energy with a variable delay time extraction field. A study of the ionization efficiency curves of ion-molecule reaction products provides a means of identification of excited reactant species. In the formation of N_3^+, for example, no gross change of ionization efficiency curve with delay time was observed[184b] indicating that only one excited state in reactant N_2^+ participates in formation of N_3^+. On the other hand, in studies of the formation of Ar_2^+ and He_2^+[184c] three and four families of excited states respectively were detected. In the latter case, de Corpo and Lampe have shown that the lifetimes of the reactant species range from 0.47 to 1.33×10^{-6} seconds. Rate constants for the respective reactions yielding He_2^+ were measured and found to vary between 1.8 and 3.6×10^{-9} cm^3/molecule sec.

Lampe et al. have been most successful in their exploitation of the pulsed source technique. They have been able to show clearly the effect of electronic excitation on the "size" of reactant rare gas species. Their technique shows much promise for more detailed study of energy transfer in collisions of excited molecules and ions.

3. Flowing Afterglow

Stationary pulsed afterglow techniques have been used for the investigation of ion-molecule reactions by Sayers et al.,[187] Fite et al.,[188] and Langstroth and Hasted,[189] and these have been reviewed in reference (22). They suffer from the same problems of state selection as the single source mass spectrometer and we shall consider here only the newer flowing afterglow system which has been developed by Fehsenfeld, Schmeltekopf, Goldan, Schiff and Ferguson.[190] Ions are produced in a discharge at one end of a tube and the recombining afterglow plasma is separated spatially from the active discharge by rapid pumping. Optical spectrometers and microwave interferometers can be applied to the flowing afterglow if desired. The ions from the afterglow are carried down the tube in a stream of helium or argon past a position where a neutral reactant is added. The ion composition is monitored by a quadrupole mass spectrometer sampling the gases through a leak at the end of the tube. Measurement of reactant ion disappearance as a function of added neutral reactant leads directly to reaction rate constants. The fact that the ions spend some time in a stream of neutral inert gas before reacting makes it highly probable that they are at thermal energies. Ferguson has recently reviewed his technique[190h] in some detail. He provides an excellent bibliography and our treatment is not intended to be exhaustive.

Flowing afterglow measurements have been applied mainly to simple reactions, especially those of interest in the upper atmosphere, for example, the reactions of N_2^+, Ar^+, CO_2^+, NO^-, and O^- with diatomic and triatomic molecules.[190] Of the very large number of rate constants for positive ion reactions measured by Ferguson et al., about 60% are for charge transfer and dissociative charge transfer reactions, about 8% are for reactions of metal ions,[190o] and the remainder are almost entirely for proton or hydrogen atom transfer reactions.

A striking feature of these rate constants is their excellent agreement with the values predicted by the G–S theory, and the obtaining of such values by a technique far removed from the typical experiment in a mass spectrometer ion source provides the theory with solid support.

One case in which the agreement with G–S theory occurred unexpectedly was the reaction

$$O^+(^4S) + CO_2(^1\Sigma) \longrightarrow O_2^+(^2\Pi) + CO(^1\Sigma)$$

which apparently takes place at every collision in spite of the fact that spin is not conserved.[190c] Paulson et al.[191] have shown by isotope studies that the reactant oxygen ion always appears in the O_2^+ product. This throws doubt on a suggestion that the reaction proceeds via a quartet CO_3^+ intermediate where spin reversal can take place over a longer time span. Lipeles, Novick and Tolk[192] claim that these rules are not valid. Reuben et al.[67,193] have pointed out that the decomposition of N_2O and N_2O^+ takes place with low frequency factors and Lindholm summarizes evidence that ionization through charge exchange is governed by spectroscopic selection rules. Nonetheless, any defense of these rules for application to ion-molecule reactions must explain the above result in the O^+/CO_2 system.

The group of reactions between metal ions and ozone:

$$M^+ + O_3 \longrightarrow MO^+ + O_2 \quad (M = Mg, Ca, Fe, Na, K)$$

are interesting examples of reactions of relatively stable ions (Na^+ and K^+ have inert gas structures) with a highly reactive neutral species.[190o]

Ferguson et al. measured rate constants for charge transfer and ion-molecule reactions of negative ions.[190h] An area which they explored most carefully was that of associative detachment reactions[190c,f,g]

$$A^- + B \longrightarrow AB + e$$

For example,

$$O^- + CO \longrightarrow CO_2 + e$$

Unlike positive ion reactions where the product ions can be detected, measurements on this class of reaction depend entirely on observations of reactant ion attenuation. The rate constants found are high, being of the same order of magnitude as G–S values.

Another interesting possibility offered by the flowing afterglow technique is that of reacting ions with atoms, free radicals, and other reactive species (such as, H, O, N, O_3). For example, if the neutral gas inlet is supplied with nitrogen atoms produced in a discharge and estimated by a titration technique of the kind reviewed by Campbell and Thrush,[194] it becomes possible to look at the rates of reactions such as

$$O_2^+ + N \longrightarrow NO^+ + O$$

which are not accessible by other methods.[190b]

In addition, Ferguson et al. have been able to measure the rate of the reaction

$$O^+(^4S) + N_2(^1\Sigma) \longrightarrow NO^+(^1\Sigma) + N(^4S) \qquad \Delta H = -1.1 \text{ eV}$$

as a function of the vibrational excitation of the nitrogen molecule.[190i] Increase of vibrational temperature from 300 to 5000°K causes an approximately fifty-fold increase in reaction rate.

This reaction is interesting both from its theoretical aspects and because it is responsible for removing O^+ ions from the ionosphere after nightfall. Its rate constant at normal temperatures has been measured by several workers both in flowing afterglow,[187d] single source,[195,196] and tandem mass spectrometer systems,[197,198] and appears to be about 0.1 % of the G–S value. It increases with increasing ion kinetic energy and goes through a peak at about 10 eV lab energy. It is surprising that the rate of an exothermic reaction should be so low and that it should increase with increasing kinetic and internal energy which, on phase space theory at least, might be expected to open up alternative reaction channels. Wolf has, in fact, performed some calculations on this system.[199]

This reaction has the makings of a classic reaction kinetics problem. Three experimental facts are in need of explanation (1) the rate is extremely low for an exothermic ion-molecule reaction, (2) the reaction apparently has zero activation energy for thermal energy ions yet its rate increases with increasing ion kinetic energy, (3) the rate is also increased by increasing the vibrational excitation of the N_2.

There are three possible reasons for the low rate. First, the potential energy surfaces for reactants and products may not cross except at small impact parameters, a situation analogous to, but less marked than, that in the He^+/H_2 and Ne^+/H_2 reactions.[41,42] Second, the potential energy curves might not cross at all at accessible impact parameters, and the reaction might go by a double change of multiplicity

$$O^+(^4S) + N_2(^1\Sigma_g^+) \longrightarrow N_2O^+(^2\Pi_i) \longrightarrow NO^+(^1\Sigma_g^+) + N(^4S)$$

This explanation is developed in detail by Kaufman and Koski.[200] They construct the potential energy curves for N_2O^+ and suggest that reaction occurs by the O^+ and N_2 combining along the repulsive $^4\Sigma^-$ curve of N_2O^+. The reactants then cross by a change of multiplicity to the $^2\Pi_i$ ground state curve of N_2O^+ and thence by another spin forbidden transition to the $^4\Sigma^-$ repulsive state of N_2O^+ which separates into NO^+ and N. This reaction would require a significant activation energy because of the reactants having to climb the repulsive potential energy curve, but Kaufman and Koski point out that the slope of this curve will be much diminished by the ion-dipole interaction. The increases of rate with increased kinetic energy of O^+ or vibrational energy of N_2 are accounted for by the accessibility of other specified potential energy curves which allow reaction to take place through other reaction channels which are not spin forbidden. This is a plausible and well developed theory, but if the two spin forbidden processes are both slow, it is difficult to understand why the $N_2O^+(^2\pi_i)$ ion has not been detected. Reuben et al.[67] claim to have measured a half life of 0.54 μsec for the metastable N_2O^+ ion involved in the reaction

$$N_2O^+(^2\pi_i) \longrightarrow NO^+(^1\Sigma_g^+) + N(^4S)$$

in which case it should certainly be possible, perhaps by an experiment designed with this in mind, to detect it.

The third possibility is that the polarization forces due to the O^+ ion might be insufficient to break the strong N–N bond. The fission would be facilitated both by vibrational energy in the N_2 which would reduce the strength of the bond, and by kinetic energy of the O^+ which would permit greater perturbation of the nitrogen charge cloud. Hirschfelder[201] has pointed out the effect of the strength of the bond to be broken on the activation energy of reactions of the type $A + BC \rightarrow AB + C$ and although ion-molecule reactions are normally assumed to have zero activation energy, this is not necessarily true where a 10-eV bond has to be broken; indeed the lack of reactivity of N_2 has not gone unnoticed in other fields of chemistry.

A semiquantitative treatment of the O^+/N_2 and related reactions on approximately the above lines has been performed by Schaefer and Henis.[202] Under certain circumstances, they consider the major rate determining factor in an ion-molecule reaction to be the extent of rearrangement of electron density around the nuclear centers which is required in order for reactants to be converted into products. Their predictions of high and low rates for a series of reactions involving C, N, and O atoms, and molecules incorporating them, agrees with observation and, in particular, rationalizes the low rate of the O^+/N_2 reaction.

The flowing afterglow technique is not as simple to operate as might appear. There are uncertainties associated with hydrodynamic flow of the carrier gas and of neutral species into it. There is also the possibility of extraneous effects involving excited species other than the ion in question and/or impurities already in the reactant gases or produced as a consequence of the discharge. The latter are of importance only in the investigation of very slow reactions.[203] Nonetheless, Ferguson et al. claim only 20–30% accuracy for their rate constants depending on the reaction in question.[21]

Farragher, Peden and Fite,[204] who studied charge transfer of N_2^+, O_2^+ and NO^+ to sodium atoms in a steady state flowing afterglow, criticized the method of data analysis used by Ferguson et al., and treated their own data quite differently. It is difficult to make any assessment of the validity of this criticism. Ferguson et al. over the past five years have published an impressive body of data which makes a self-consistent pattern, agreeing in a large number of cases with the G–S theory and, where comparison is possible, with the results of other workers. It is extremely difficult to suppose that their results are subject to systematic error, and it is clear that the flowing afterglow technique provides a different and powerful means for studying ion-molecule reactions at thermal energies.

4. Photoionization

Photoionization mass spectrometry represents, in some respects, the ultimate in state selection. Its use in ion-molecule reaction studies has been pioneered by Warneck[205] and Chupka.[80–82,89] Inghram[69] and Dibeler[70] have significantly contributed to the technique but have not used photonization mass spectrometers for this purpose. Chupka and Berkowitz[81,82] built an ion source onto a mass spectrometer which they could operate up to a few mm pressure and in which they could produce ions by photon impact. The photo source was either a hydrogen or helium discharge tube coupled to a monochromator. In the region of 800 Å, the resolution could be varied between 0.05 and 0.002 eV. Under these conditions, H_2^+ ions, for example, can be produced not only in selected vibrational states but also in desired rotational states.

Chupka, Russell and Refeay studied the H_2^+/H_2 reaction cross-section as a function of vibrational and kinetic energy of the reactant ions.[80] They found that at low kinetic energies the rate decreased as vibrational excitation increased, but at high kinetic energies the reverse was true. Chupka et al. attributed this to the existence of two different reaction mechanisms, involving complex formation and stripping at low and high energies respectively.

In addition, they found that their yield of H_3^+ did not go to zero for photon energies below the H_2^+ threshold, and took this as evidence for chemi-ionization involving not only excited H_2 molecules but also metastable H atoms, possibly in the 2^2S state.

The decrease of reaction cross section with vibrational quantum number in the H_2^+/H_2 reaction was not large (the ratio of the cross sections for ions with 7 and 0 vibrational quanta was 0.83), nor was it in agreement with the conclusions of Weingartshofer and Clarke,[206] who used an electron velocity selector to control the vibrational energy of their H_2^+ ions. Chupka et al. interpreted it in terms of Light's phase space theory.[106] This said that collision complexes were formed which then decomposed along all energetically available channels, including the back reaction, to an extent governed by the relative amount of phase space available to each channel. Thus at low energies in the H_2^+/H_2 system, little back reaction would occur because it would be thermoneutral and would correspond to a small volume of phase space. At high vibrational quantum numbers, this would not be so, and a back reaction giving less vibrationally excited reactant ions would have some probability of occurring even in competition with the exothermic forward reaction. Thus the cross section for H_3^+ production would drop with increasing vibrational energy.

At high kinetic energies, Chupka et al. suggested that the stripping reaction predominated. Molecules with high vibrational energies would be more likely to participate in this since they would have greater bond lengths and be less strongly bound. Thus at high repeller potentials, the effect of vibrational energy is reversed, and it increases reaction cross sections. The only alternative reaction channel considered by Chupka et al. was back reaction, and they disregarded the possibility of collision induced dissociation, which they presumably felt to be energetically unfeasible. They report results, however, for H_2^+ ions with 7 vibrational quanta (1.75 eV) accelerated by a repeller field of 22.5 V/cm and an ion path length of about 0.5 cm. For the average ion which reacts after traversing half of the distance to the exit slit, this would suffice to put another 2.8 eV in the center of mass, giving an energy inventory well above the 2.65 eV dissociation energy of H_2^+. In view of the fact that Vance and Bailey[52] have observed collision induced dissociation as an alternative channel in the H_2^+/H_2 system, Chupka et al. should perhaps have given it some consideration. The uncertainty about H^+ production is increased by the fact that much of it might be produced with kinetic energy and thus be collected with low efficiency. This has been observed by Friedman, Irsa and Reuben[207] for D^+ in the collision induced dissociation of HeD^+.

The hypothesis of Chupka et al. concerning the occurrence of back reaction is plausible but is not supported by quantitative measurements and cannot therefore be regarded as compelling. Previous workers on the H_2^+/H_2 reaction have not found it necessary to postulate back reaction, and the process has appeared, with what limits of accuracy one cannot be certain, to take place at every collision. Chupka, Russell and Refeay have carried out a first class piece of work on a carefully and ingeniously designed apparatus, but it would have been helpful if their published results had exposed more of their data, for example, on the effect of repeller potential on the reaction cross sections.

Chupka and Russell[89] then investigated ion-molecule reactions in ammonia. They found that the cross section for

$$NH_3^+ + NH_3 \longrightarrow NH_4^+ + NH_2$$

decreased with increasing vibrational energy while that for

$$NH_3^+ + H_2O \longrightarrow NH_4^+ + OH$$

was essentially constant, which they interpreted as indicating absence of back reaction. They did not report any investigation of alternative reaction channels, and in one case even raised their pressure to a point at which 94% of their NH_3^+ ions were reacting, without noting the formation of the $NH_4^+(NH_3)_n$ clusters of the type reported by Kebarle. It is not impossible that differential rates of clustering might account for their apparent back reaction.

Chupka and Russell's third paper dealt with H_2^+/rare gas reactions.[208] ArH^+ was formed in the H_2^+/Ar reaction for all H_2^+ vibrational energies with approximately the same cross section, and in the Ar^+/H_2 reaction with Ar^+ ions in the $^2P_{3/2}$ and $^2P_{1/2}$ states. The cross section in the latter case was higher by 30%. Chemi-ionization processes leading to ArH^+ were also found.

In the reactions

$$H_2^+ + He \longrightarrow HeH^+ + H$$
$$H_2^+ + Ne \longrightarrow NeH^+ + H$$

The HeH^+ and NeH^+ ions were produced by vibrationally excited H_2^+ ions, and the thresholds for reactions at zero kinetic energy were found to be very near the $V = 3$ and $V = 2$ states of H_2^+ respectively. H_2^+ in lower states could react if the kinetic energy were high enough, and Chupka and Russell produced two very interesting graphs (Figure 8 shows their results for H_2^+/He) showing the variation of rate constants with both vibrational quantum number and repeller potential (plotted as k_a/k_c where k_a is the rate of H_2^+/inert gas reaction and k_c the rate of the H_2^+/H_2 reaction under identical conditions).

These graphs show that the role of kinetic energy in driving a reaction is less significant than that of vibrational energy. The cross sections for HeH^+ production at a repeller voltage of 20 eV differ by a factor of ten for H_2^+ with 0 and 4 vibrational quanta. Also, the smaller the amount of translational energy which has to be converted to internal energy to drive the reaction, the larger the cross section.

These graphs were the same shape as the ones obtained by von Koch, Moran and Friedman (see Figure 8),[41,42] but even at their lowest repeller potentials the rate did not drop to zero for any of the vibrationally excited states of H_2^+, though it did so for the ground state. Chupka and Russell attributed the measurement of finite reaction rates at nominally zero repeller potentials in the presence of insufficient H_2^+ excitation to make the

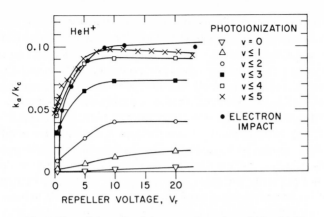

Fig. 8. Plots of the ratio of the rates of reaction of H_2^+ with He to H_2^+ with H_2. The electron impact data were obtained by von Koch and Friedman in single source experiments.[41] The photoionization data were obtained by Chupka and Russell.[208] The electron impact data were interpreted with the assumption that there was a very low kinetic energy threshold, and that only H_2^+ with sufficient internal energy to meet the requirements of the endothermic reaction, giving HeH^+, could react. Chupka and Russell did not observe a similar kinetic energy threshold in their experiment but clearly resolved the relative probabilities of reaction of H_2^+ in various quantum vibrational energy states. Reactions of H_2^+ in the $v = 0$, 1 and 2 states appear to require kinetic energy. In general the probability of a reactive collision which requires conversion of kinetic to internal energy is significantly smaller than one that does not.

reaction exothermic, to reaction just outside the ionization chamber and to possible field penetration within it. They estimated this at 0.56 volts; the exit kinetic energy of an ion would thus be 0.27 eV. The light beam however irradiated quite a large volume of the ion source and any estimates of ion kinetic energy can only be approximate.

The rates for reaction of H_2^+ with little vibrational energy at this very low effective repeller voltage were low because of the small probability of translational to internal energy conversion at such low values of translational energy. At higher vibrational energies, where such conversion was not necessary to make the overall reaction exothermic, the rate was quite high. This would not have been the case if a kinetic energy threshold of 0.35 eV had existed, as claimed by Friedman et al.

In view of the prediction by Light et al. that there would always be some translational to internal energy conversion, so that even ground state H_2^+ would have a finite probability of reacting, Chupka and Russell concluded that no kinetic energy threshold existed and that the precise agreement between theory and experiment found by Friedman et al., was fortuitous.

There are general grounds for believing that translational to internal energy conversion cannot be rigorously excluded. The work of Friedman et al. showed the earlier hypotheses on those lines to be generally but not rigorously valid, and this conclusion was confirmed by Chupka.

A photoionization mass spectrometer was used by Warneck[205] in a different way for the study of ion-molecule reactions. He employed a pulsed light source to irradiate an ion source which could operate at pressures as high as 200 microns, and contained an unpulsed repeller plate. He was not primarily interested in state selection of ions. By examining the shape of his primary and secondary ion pulses, he was able to measure the rate of ion diffusion as a function of pressure, and also to derive rate constants. He studied reactions of ionospheric interest, for example, N_2^+ ions in N_2/O_2 mixtures, the H_2^+/H_2 reaction, and the reactions of Ar^+ with a range of simple molecules. His results agree fairly well with those of other workers. A significant observation is that, at modest repeller potentials and pressures above 50 microns, the time taken for ions to emerge is consistent with their having thermal velocities. Warneck suggests that collisions are so frequent that few ions ever achieve a velocity outside the thermal range.

There is however a considerable uncertainty as to ion path length in high pressure sources and this leads to an uncertainty in interpretation of delay times when these times are found to be smaller than anticipated for the energies applied. There is no doubt that some thermalization takes place, but the same process that reduces the ion energy tends also to increase its path length, so that while the time taken for the ion to reach the slit may be characteristic of thermal velocities, it may in fact have got there by a "random walk" at a greater speed.

Photoionization experiments are difficult to carry out. Photoionization in the present state of the art produces ion currents several orders of magnitude lower than electron impact, and problems of sensitivity are acute. The ion yield can be improved by increasing the pressure of gas in the ion source, but once the point is reached at which multiple collisions are likely, some equilibration of energy can take place and the internal energy of the primary ions can no longer be considered fixed by the ionizing photon energy.

Chupka deserves great credit for managing to avoid the latter problem and still to obtain useable primary ion currents. The next stage is to find a method, perhaps a pulsing technique, of obtaining kinetic in addition to vibrational and rotational energy selection in a photoionization experiment.

F. Channels of Reaction

1. Introduction

In any investigation of ion-molecule reactions, the simplest way of measuring the rate of reaction is to observe the formation of products under conditions where the nature of the reactants is clearly defined. This is not always easy, even in the simplest reactions. In the H_2/Ne system, for example, the most obvious reaction channel leads to the product ion NeH^+. Of the two possible pairs of reactants H_2^+/Ne and Ne^+/H_2, early workers[35] thought the latter, which is exothermic, more likely than the former, which is not. This conclusion later proved erroneous.[41,42] Furthermore, there are less obvious reaction channels, such as collision-induced dissociation leading to H^+ product ions, which are now known to take place. Chemi-ionization is feasible in certain reacting systems, and "exotic" reactions such as resonant and nonresonant charge transfer may take place under certain circumstances. We shall review the methods of establishing reaction channels and, briefly, the ways in which "exotic" reaction channels may complicate the models of ion-molecule reactions.

2. Single Source and Tandem Mass Spectrometer Techniques

The earliest method used for elucidating reaction channels relied on estimates of thermodynamic and kinetic feasibility, that is, only reactions which were exothermic were considered, and primary ions of low abundance were regarded as unlikely to lead to major products since this would involve a ridiculously high rate constant.

The next method involved the measurement of the appearance potentials and ionization efficiency curves of primary and secondary ions. In a single source experiment, primary and secondary ions were expected to have identical appearance potentials and similar ionization efficiency curves. This method requires no special equipment and was widely used. It is generally applicable to exothermic reactions, but does not always produce accurate results in reactions which have an energy threshold or involve excited species. Henglein was nonetheless able to identify in this way several reactions of excited ions in CS_2 and some aromatic compounds. If the appearance potentials of possible reactants are close, ambiguous results may be obtained. For example, it is impossible by this means to distinguish between the following two reactions which may occur in the benzene system:[160]

$$C_4H_4^+ + C_6H_6 \longrightarrow C_9H_7^+ + CH_3$$
$$C_3H_3^+ + C_6H_6 \longrightarrow C_9H_7^+ + H_2$$

The tandem mass spectrometer, of course, permits unambiguous identification of reaction channels. It has been applied mainly to simple reactions where the channels are already well established.

The main proponents of the use of the tandem mass spectrometer for identification of channels are Futrell and his coworkers.[138,157,209] Only a few weeks after the single source work of Giardini–Guidoni and Friedman[44] had appeared, Abramson and Futrell published their results of a tandem mass spectrometer study of the methane system.[209b] By use of CD_4 and CH_4 they were able to distinguish between the reaction channels

$$CD_4{}^+ + CH_4 \underset{\longrightarrow}{\overset{\longrightarrow}{}} \begin{matrix} CD_4H^+ + CH_3 \\ \\ CH_4D^+ + CD_3 \end{matrix}$$

They found that at low ion energies, they recorded about half as much mass 21 (CD_4H^+) as mass 18 (CH_4D^+ or CD_3^+), and did not consider CD_3^+ a likely product on the basis of experiments with CH_4^+ and CD_4. These experiments were reported later[209c] and showed that the major product of this reaction at low energies was mass 21 (CD_4H^+) which was about three times as abundant as mass 18 (CH_4D^+ or CD_3^+). At higher energies, mass 15 (CH_3^+) predominated.

Abramson and Futrell thus confirmed Giardini–Guidoni and Friedman's assertion that reaction channels other than CH_5^+ production were important in the methane system, and that these could, to some extent, be responsible for the failure of the cross section for CH_5^+ production to reach the value predicted by the G–S theory. They differed, however, as to the nature of the channels. On the basis of the above evidence, they claimed that the major alternative channel at center of mass energies above 1 eV, was the collision-induced dissociation of methane:

$$CH_4^+ + CD_4 \longrightarrow CH_3^+ + H + CD_4$$

rather than the decomposition of a CD_4H^+ or CH_4D^+ complex as postulated by Giardini–Guidoni and Friedman on the basis of their yields of CD_2H^+ from a mixture of CD_4 and CH_4 in a single source experiment. Abramson and Futrell did show, however, that CH_5^+ formed in the reaction

$$CH_4 + CHO^+ \longrightarrow CH_5^+ + CO$$

could indeed undergo unimolecular decomposition to give CH_3^+, so that the disagreement between the two groups relates only to the importance of CH_5^+ decomposition in the CH_4^+/CH_4 system, compared with collision-induced dissociation.

The work of Giardini–Guidoni and Friedman can be criticized on the basis of the general limitations of the single source technique. They were compelled to make corrections to their data to allow for, for example, the contribution of $^{13}CH_4^+$ to their $^{12}CHD_2^+$ peaks, and their measurements of CH_4^+ disappearance are liable to the errors inherent in measuring differences between large quantities. Furthermore, their ions had a distribution of energies.

Abramson and Futrell avoided some of these problems. On the other hand, their results are most suspect at the lowest energies, where they overlap those of Friedman et al., because of the problems associated with controlling very low energy ion beams. Furthermore, it is likely that collision-induced dissociation proceeds by a stripping mechanism, which means that the CH_3^+ distribution in the tandem mass spectrometer experiment will be sharply peaked in the forward direction, while the CD_2H^+ and CDH_2^+ from the complex decomposition will be scattered isotropically in the center of mass. The proportion of CH_3^+ which is collected will thus be exaggerated.

It is tempting, therefore, to try to reconcile these sets of data by proposing that the major alternative channel in the CH_4^+/CH_4 reaction is the production of methyl ions; that at low energies these arise mainly from CH_5^+ decomposition, but that at high energies collision-induced dissociation predominates.

Futrell and his colleagues have worked mainly on ion-molecule reactions of interest in organic and radiation chemistry, and in two recent review articles[210,211] they discuss such problems as the absolute basicity of compounds (measured by their proton affinities), protonated molecules as Brønsted-Lowry acids, and carbonium ion reactions. They have studied, for example, ion-molecule reactions in propylene[209e] and isomeric butenes,[209e] in propane,[138] and of ions from propane with benzene, 1,3-butadiene, hydrogen sulfide, and nitric oxide.[156] It is interesting to note that in the final case, the ions $C_3H_8^+$, $C_3H_6^+$, $C_2H_6^+$, and $C_2H_2^+$, which have an odd electron and are therefore free radicals, undergo charge exchange with NO whereas $C_3H_7^+$, $C_3H_5^+$, $C_2H_5^+$ and $C_2H_3^+$, which have no unpaired electrons, do not react.

3. High Pressure Mass Spectroscopy

We have discussed high pressure mass spectrometry briefly in a previous section (IV-D-5), insofar as it leads to quantitative measurements. We refer to it briefly here because of the necessity for establishing reaction channels in these highly complicated systems. What is usually done is that yields of various ions are measured at various pressures. It is usually clear which ions are formed in first, second, third and higher order reactions.

By combining these observations with appearance potential measurements, it is often possible to elucidate the various channels of reaction, at any rate for the most abundant ions.

Szabo has recently pointed out that it is possible to achieve the same results by varying ion residence time or path length instead of neutral gas pressure.[177]

4. Ion Cyclotron Resonance Spectroscopy

The most recent technique for detecting channels of reaction is the ion cyclotron resonance spectrometer. The mode of operation of the ICR spectrometer has been described in detail in a review by Baldeschwieler.[93] The ability to observe reaction channels quickly and unambiguously is obviously of value, particulaily for the accumulation of a body of knowledge on the reactions which ions will undergo. Beauchamp and Buttrill observed a large number of reactions in their work on the proton affinities of water and hydrogen sulfide,[90] and their group has looked at $^3He^+/^4He$ charge exchange and the CD_4/N_2 system.[94] Gray has established about 23 reactions in acetonitrile,[212,213] King and Elleman examined reactions in hexafluorethane[214] and charge exchange in Xe/CH_4 mixtures.[215] Jennings et al.[176] investigated reactions of ethylene and vinyl fluoride, Henis[216] has looked at reactions in methanol by ICR spectroscopy, and Dunbar has applied the technique to negative ion-molecule reactions in diborane, and also to a few positive ion reactions in the same system.[217]

Baldeschwieler et al. studied the reaction channels in methyl and ethyl chlorides and used the data to give information on reaction mechanisms.[93] For example, they readily observed the protonated alkyl chloride ion formed by reactions such as

$$CH_3Cl^+ + CH_3Cl \longrightarrow CH_3ClH^+ + CH_2Cl$$

and showed that it would condense very easily with another neutral alkyl chloride molecule to give a dialkylchloronium ion and hydrogen chloride

$$CH_3ClH^+ + CH_3Cl \longrightarrow CH_3ClCH_3^+ + HCl$$

The product is isoelectronic with dimethyl sulfide, and presumably has the ether type structure shown. The double resonance technique can be operated in such a way that $CH_3{}^{35}ClH^+$ or $CH_3{}^{37}ClH^+$ can be chosen as reactant ions, while the neutral reactant molecules contain a mixture of isotopes. Thus the isotopic composition of products gives information equivalent to that obtained in a tracer experiment. In the above reaction yielding a dialkylchloronium ion, the natural $^{35}Cl/^{37}Cl$ ratio is 3 : 1, and if $CH_3{}^{35}ClH^+$ is chosen as reactant, then there are three main possibilities for the $^{35}Cl/^{37}Cl$ ratio in the product. If the chlorine atom of the neutral molecule is always retained in the product, for example, as a result of a hypothetical nucleophilic displacement

$$CH_3\underline{Cl} + H \overset{H}{\underset{H}{\diagdown}}C-ClH^+ \longrightarrow CH_3-\underline{Cl}-C\overset{H^+}{\underset{H}{\diagdown}}H + HCl$$

then $^{35}Cl/^{37}Cl$ should equal $3:1$. If on the other hand the chlorine from the ion is retained in the product, the latter should contain 100% ^{35}Cl, while if the chlorine atom comes with equal probability from either reactant, then $^{35}Cl/^{37}Cl$ should equal $7/8$. In fact, the ratio showed the third possibility to be correct, thus ruling out the nucleophilic mechanism shown.

In mixtures of methyl and ethyl chlorides, the methyl ethyl chloronium ion can be made by two different routes:

$$C_2H_5ClH^+ + CH_3Cl \longrightarrow CH_3ClC_2H_5^+ + HCl$$

or

$$CH_3ClH^+ + C_2H_5Cl \longrightarrow CH_3ClC_2H_5^+ + HCl$$

The double resonance experiments showed that in both cases the chlorine in the chloronium ion came from the methyl-containing reactant, irrespective of whether it was the ionic or the neutral species.

These data are not consistent with a nucleophilic displacement reaction, but might be rationalized by a four center mechanism involving intermediates of the types

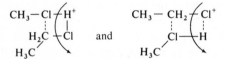

which eliminate HCl to leave the chloronium ion. To explain the above result it is also necessary for the methyl-chlorine bond to be stronger than the ethyl-chlorine bond. If this is indeed so, then the reaction

$$C_2H_5ClH^+ + CH_3Cl \longrightarrow C_2H_5Cl + CH_3ClH^+$$

should not occur, being endothermic, while

$$CH_3ClH^+ + C_2H_5Cl \longrightarrow CH_3Cl + C_2H_5ClH^+$$

should readily occur. This is in fact observed experimentally. Furthermore, at higher translational energies the proton transfer reaction is relatively favored over the four center complex reaction, which lends further support to the above mechanism

ICR spectroscopy is a powerful technique for elucidating reaction channels and the above illustration shows how observations of these channels can be related to reaction mechanisms. It would be even more impressive if it could be generally shown that absolute reaction cross sections can be obtained. In principle, these can be calculated from the shapes of the cyclotron resonance absorption peaks. The theory of the collision broadening of these line shapes has been developed by Bayes, Kivelson and Wong,[218] by Bowers et al.,[220a] and by Beauchamp,[95] and the line shapes predicted for N_2[91a] and the rare gases are in good agreement with experimental work on these molecules where the collision cross sections are well known. Bowers, Elleman and King[220] have measured thermal energy rate constants in an ICR spectrometer for the reactions of N_2^+ with H_2, HD, and D_2 to give N_2H^+ and N_2D^+ and find excellent agreement with G–S theory. They have also found a fair degree of agreement for reactions in the H_2^+/H_2, D_2^+/D_2, and HD^+/HD systems.[220b] This demonstrates the effectiveness of their method in simple systems.

Beauchamp and Armstrong have reported the development of an ion ejection technique to permit the removal of unwanted ions from the ion source region before they reach the trapping region,[221] and Futrell[222] has modified this to give what is in some respects a tandem ICR mass spectrometer, with which he claims to be able to measure rate constants at various ion energies. Early results on the Ar/H_2 and Ne/H_2 systems have shown excellent agreement with earlier work in conventional mass spectrometers.

Futrell and Clow,[222] collaborating with Tiernan and Gill[223] have cast serious doubts on some of the earlier work using the simple ICR technique. Elleman and King[214] had found a reaction between Xe^+ and CH_4 to give CH_3^+, and between CH_3^+ and Xe to give Xe^+, but no $Xe^+/CH_4 \rightarrow CH_4^+$ or $CH_4^+/Xe \rightarrow Xe^+$ reactions. Tiernan and Gill reexamined the Xe/CH_4 system in their tandem mass spectrometer and found reactions between CH_4^+ and Xe, and Xe^+ and CH_4 to give a variety of products, but no reaction between Xe and CH_3^+ at any energy. Futrell and Clow, using an ion ejection technique, showed that Xe^+ was a precursor of CH_4^+ but not CH_3^+, and that Xe^+ was produced from CH_4^+ but not CH_3^+. The rates of these two processes were approximately equal, which was why they had not been previously observed. Futrell and Clow were able to suggest why Elleman and King's results might be in error, and they point out two major problems connected with cyclotron resonance. First, it is extremely sensitive to space charge because of the trapping of low energy ions in the strong magnetic field. Second, there are ill defined ion losses at low and moderate irradiating fields, that is, the ICR cell is "leaky" because actual electrostatic fields deviate substantially from the idealized fields assumed in calculations. One consequence of this "leakiness" is that primary ions are being lost from the cell under all conditions, and this in turn invalidates the argument connecting a negative value of $(dk/dE_{ion})°$ with an exothermic or thermoneutral reaction. (See Section III-C-2.)

The ICR technique is still in its infancy. The above considerations suggest that until it is more carefully validated, quantitative results arising from it should be treated with circumspection.

A difference between ion-molecule reactions studied by ICR spectroscopy and by other techniques, where an ion may undergo several collisions, is the extremely long ion mean free path in the former case. This may turn out to be significant. The $H_5O_2^+$ ion has been observed in high pressure ion sources by numerous workers. It has not, however, been reported in ICR experiments, though H_3O^+ is well known. We may conclude that if it exists at all it is not readily detected. It is legitimate to ask why $H_5O_2^+$ should occur in one system but not in another. Wall effects may be discounted since $H_5O_2^+$ has been detected in many hydrocarbon flames. Another possibility is that $H_5O_2^+$ is formed in an excited state:

$$H_3O^+ + H_2O \longrightarrow H_5O_2^+*$$

which decomposes unless stabilized by a further collision which dissipates some of the heat of reaction. The time interval between successive collisions in the ICR experiments may be estimated from Beauchamp and Butrill's comment that "... a reaction with a rate constant of 0.3×10^{-10} cm^3 molecule^{-1} sec^{-1} will give a product ion intensity amounting to approximately 10% of parent ion intensity at 8×10^{-5} torr and with an ion transit time of 1 millisecond."[90] Since many ion-molecule reactions have rate constants of about 30 times the above, at thermal energies, the time between collisions will be of the order of 10^{-4} to 10^{-5} sec. De Pas, Leventhal and Friedman calculated that the corresponding interval in their source was of the order of 10^{-6} to 10^{-7} sec. Thus there is a certain amount of negative evidence that $H_5O_2^+$ is formed in an excited state and must be stabilized by collisions within about 10^{-5} sec, if it is not to dissociate.

The long residence time of an ion trapped in the field of an ICR spectrometer suggests the intriguing possibility of observing light emission from the decay of excited ions. This would enable accurate measurements to be made on the degree of excitation in ions formed by electron impact and ion-molecule reaction, and would clear up many of the problems noted in this review in connection with ion excitation.

5. "Exotic" Channels of Reaction

In the introductory section on channels of reaction a brief discussion was promised dealing with the ways in which "exotic" channels of reaction might complicate models of ion-molecule reactions. Many "exotic" channels have been mentioned in this review, charge transfer processes, collision-induced dissociation reactions, and others. It is clear that the definition of an exotic channel is arbitrary and was made only to provide a means of distinguishing the more common rearrangement collisions detected in early single source experiments from a variety of other processes.

In general, the two channels of reaction that must be considered separately are charge transfer processes and rearrangement collisions. This is frequently difficult if a mechanism is operative in which dissociation follows a charge transfer process which generates an excited state in the product ion. Charge transfer processes, whether accidentally or genuinely resonant processes, are frequently observed taking place with considerably larger impact parameters than rearrangement collisions,[224] and theoretical treatment of such processes should be compared only with experimental data that have been obtained with the possibility of competitive rearrangement collisions, clearly and carefully considered.

The importance of considering all possible channels of reaction is emphasized with the examination of perhaps the simplest ion-molecule reaction that is available for theoretical study, the reaction of protons with hydrogen molecules.

Csizmadia, Polyanyi, Roach and Wong[225] have recently reported preliminary results of a computer study of the dynamics of the reaction

$$D^+ + H_2 \longrightarrow DH + H^+$$

This work was cited as the first calculation of the dynamics of a chemical reaction using an *ab initio* potential energy surface. The calculation was a thorough and expensive job. They found, as expected, a deep potential well for the H_3^+ complex and with a collision energy of 3 eV, the complex was found to have a lifetime of approximately 10^{-13} seconds. More than 90% of trajectories investigated took from 0.4–3.0×10^{-13} seconds.

Reaction was observed at impact parameters >2.4 Å at both 3 and 4.5 eV. This is significantly larger than the maximum impact parameter of the Gioumousis–Stevenson polarization limit. This limit is 1.66 Å at 3 eV. Polanyi et al. noted that this should not be construed as evidence for a calculated cross section larger than one derived from the polarization model because the trajectories led to a reaction probability of 0.2 to 0.3 for $b \leqslant 2.4$ Å while the polarization model assumes a unit probability for reactive collisions with impact parameters $\leqslant$ than b_{max}, the polarization limit.

Center of mass differential cross sections were computed for this system which showed forward and backward peaked scattering. Vibrational and rotational energy distributions were estimated with a mean total value of approximately 46 kcal/mole and maximum values up to 76 kcal for 3-eV collision energies.

Approximately $\frac{1}{5}$ of the reactive collisions at 4.5 eV led to fragmentation, despite the fact that collision energies were only 0.1 eV in excess of the threshold for dissociation. At 6 eV, over 90% of the reactive collisions gave fragmentation.

The Csizmadia–Polanyi–Roach–Wong calculation clearly illustrates the complexity of this very simple system, but what remains to be considered is the question of the role of alternative channels of reaction that can take place on another potential energy surface that is thermodynamically accessible to this system with 3-eV or higher collision energy protons or deuterons. The trajectory calculations showed a high probability of conversion of translational to internal energy in reactive collisions. Molecular product was highly fragmented above threshold for molecular dissociation, and below this energy considerable energy was deposited as vibrational and rotational energy in the product molecule. The question of the possibility of the endothermic charge transfer reaction

$$D^+ + H_2 \longrightarrow H_2^+ + D \qquad (13)$$
$$\text{or} \longrightarrow HD^+ + H \qquad (14)$$

as an alternative channel of reaction is not easily attacked theoretically because a second potential energy surface for the new system $H_2^+ + D$ must be computed and the problem emerges of treating the transition from lower to upper surfaces.

This problem has been recently studied experimentally at Brookhaven,[226, 227] and results obtained show that indeed the charge transfer channels are open above the 2-eV threshold center of mass energy, and product ions from charge transfer in reaction (14), and the following processes are observed in greater abundance than the atomic ion products.

$$H^+ + D_2 \longrightarrow D_2^+ + H$$
$$\longrightarrow HD^+ + D$$
$$D^+ + HD \longrightarrow D_2^+ + H$$
$$\longrightarrow HD^+ + D$$

Experimental cross sections were found to be qualitatively in agreement with values extrapolated from the trajectories study.

The problem encountered in the theoretical study of this system of being forced to consider potential energy surface transitions in systems which are energetically capable of moving from one surface to another, gives little comfort to the theoretician faced with the task of making a rigorous analysis of reactions of energetic atoms or ions. The *ab initio* surface and trajectories on it begin to give much deeper insight into the distribution of structures that may be activated intermediates in a reactive collision. Such studies will certainly stimulate important experimental investigations, but we are left with the feeling that even with systems as simple as deuterons and hydrogen molecules there is much chemistry. With more common and complex reactions the last remark has even greater validity.

Acknowledgments

The authors wish to acknowledge permission to reproduce graphical data presented in this article from Prof. Bruce Mahan of the University of California at Berkeley, Prof. Paul Kebarle, University of Alberta, and Dr. William Chupka of Argonne National Laboratory.

References

1. F. N. Harllee, H. M. Rosenstock and J. T. Herron, *A Bibliography on Ion-Molecule Reactions*, NBS Technical Note 291, Washington, 1966.
2. *Bibliography of Atomic and Molecular Processes*, Compiled by Atomic and Molecular Processes Information Center, Oak Ridge National Laboratory. Available from Clearinghouse for Federal Scientific and Technical Information, National Bureau of Standards, U.S. Department of Commerce, Springfield, Va. 22151.
3. *Federal Funds for Research, Development and Other Scientific Activities* 1956–67, Vol. 15, p. 159, Table C53, N.S.F. 66–25, U.S. Government Printing Office, Washington, D.C. 20402.
4. See for example, *Electrical Phenomena in Gases*, K. K. Darrow, Baltimore, 1932; G. Gioumousis and D. P. Stevenson, *J. Chem. Phys.*, **29**, 294 (1958).
5. P. Langevin, *Annales de chimie et de physique*, **5**, 245 (1905). A translation can be found in *Collision Phenomena in Ionized Gases*, E. R. McDaniel, Ed., Wiley, New York 1964, p. 701.
6. H. Eyring, J. O. Hirschfelder and H. S. Taylor, *J. Chem. Phys.* **4**, 479 (1936).
7. H. Rosenstock, *U.S. Atomic Energy Commission Report*, 1959, JLI-650-3-7, TID 4500.
8. V. Tal'rose, *Pure and Appl. Chem.*, **5**, 455 (1962).
9. J. L. Franklin, F. H. Field and F. W. Lampe, *Advances in Mass Spectrometry*, Vol. 1, 1959, Pergamon Press, Oxford, p. 308.
10(a). M. Pahl, *Ergeb. Exakt. Naturw.* **34**, 182 (1962); (b) D. P. Stevenson, *Mass Spectrometry*, Ed. C. A. McDowell, McGraw-Hill, 1963, p. 589.
11. C. E. Melton, *Mass Spectrometry of Organic Ions*, Ed. F. McLafferty, Academic Press, New York, 1963, p. 65.
12. J. Durup, *Les Reactions Entre Ions Positifs et Molecules en Phase Gazeuse*, Gauthier-Villares, Paris, 1960.
13. J. L. Franklin, F. H. Field and F. W. Lampe, *Progress in Reaction Kinetics* **1**, 169 (1961).
14. J. B. Hasted, *Physics of Atomic Collisions*, Butterworth, London, 1964.
15. M. Henchman, *Ann Rep. Chem. Soc. London*, **62**, 39 (1965).
16. V. L. Tal'rose and G. V. Karachevtsev, *Advances in Mass Spectrometry*, **3**, 211 (1966).

17. C. F. Giese, *Adv. Chem. Phys.*, **10**, 247 (1966).
18. Ion Molecule Reactions in the Gas Phase, *Adv.* in *Chem. Ser.*, **58** (1966).
19. E. E. Ferguson, *Rev. Geophys.*, **5**, 305 (1967).
20. L. Friedman, *Ann. Rev. Phys. Chem.*, **19**, 273 (1968).
21. E. E. Ferguson, *Adv. in Electronics and Electron Physics*, **24**, 1 (1968).
22. V. Cermak, A. Dalgarno, E. E. Ferguson, L. Friedman and E. W. McDaniel, *Ion Molecule Reactions*, Wiley, New York, 1970.
23. J. J. Thomson, *Rays of Positive Electricity*, Longmans Green & Co., London, 1933, p. 32.
24. F. W. Aston, *Mass Spectra and Isotopes*, 2nd ed., Edward Arnold, London, 1942.
25. T. R. Hogness and R. W. Harkness, *Phys. Rev.*, **32**, 784 (1928).
26. H. W. Washburn, C. E. Berry and L. E. Hall, *Mass Spectrometry in Physics Research No.* 20, National Bureau of Standards (U.S.) Circ. 522 (1953).
27. G. C. Eltenton, *Nature*, **141**, 975 (1938).
28. V. L. Tal'rose and E. L. Frankevich, *Dokl. Akad. Nauk. SSSR*, **111**, 376 (1956).
29. G. C. Eltenton, *Monthly Reports of Shell Development Co.*, Emeryville, Calif., April 1940.
30. V. L. Tal'rose and A. K. Lyubimova, *Dokl. Akad. Nauk. SSSR*, **86**, 909 (1952).
31. F. H. Field, J. L. Franklin and F. W. Lampe, *J. Am. Chem. Soc.*, **79**, 2419 (1957).
32. D. O. Schissler and D. P. Stevenson, *J. Chem. Phys.*, **24**, 926 (1956).
33. E. Lindholm, *Proc. Phys. Soc.*, **A66**, 1068 (1953).
34. F. H. Field, J. L. Franklin and F. W. Lampe, *J. Am. Chem. Soc.*, **79**, 2419 (1957).
35. (a) D. P. Stevenson and D. O. Schissler, *J. Chem. Phys.*, **23**, 1353 (1955).
    (b) *ibid.*, **29**, 282 (1958).
36. (a) F. H. Field, J. L. Franklin and F. W. Lampe, *J. Am. Chem. Soc.*, **79**, 2665 (1957); (b) F. W. Lampe and F. H. Field, *ibid.*, **79**, 4244 (1957); (c) F. W. Lampe, F. H. Field and J. L. Franklin, *ibid.*, **79**, 6132 (1957); (d) F. H. Field and F. W. Lampe, *ibid.*, **80**, 5587 (1958); (e) F. W. Lampe and F. H. Field, *ibid.*, **81**, 3242 (1959); (f) F. H. Field, *ibid.*, **83**, 1523 (1961); (g) J. L. Franklin and F. H. Field, *ibid.*, **83**, 3555 (1961); (h) F. H. Field and J. L. Franklin, *ibid.*, **83**, 4509 (1961); (i) F. H. Field, H. N. Head and J. L. Franklin, *ibid.*, **84**, 1118 (1962).
37. (a) R. Barker, W. H. Hamill and R. R. Williams, *J. Phys. Chem.*, **63**, 825 (1959); (b) R. F. Pottie, R. Barker and W. H. Hamill, *Radiation Res.*, **10**, 664 (1959); (c) R. F. Pottie and W. H. Hamill, *J. Phys. Chem.*, **63**, 877 (1959); (d) R. F. Pottie, A. J. Lorquet and W. H. Hamill, *J. Am. Chem. Soc.*, **84**, 529 (1962); (e) N. Boelrijk and W. H. Hamill, *ibid.*, **84**, 730 (1962); (f) L. P. Theard and W. H. Hamill, *ibid.*, **84**, 1134 (1962); (g) D. A. Kubose and W. H. Hamill, *ibid.*, **85**, 125 (1963); (h) T. F. Moran and W. H. Hamill, *J. Chem. Phys.*, **39**, 1413 (1963).
38. (a) V. L. Tal'rose and E. L. Frankevich, *J. Am. Chem. Soc.*, **80**, 2344 (1958); (b) E. L. Frankevich and V. L. Tal'rose, *Dokl. Akad. Nauk. SSSR*, **119**, 1174 (1958); (c) E. L. Frankevich and V. L. Tal'rose, *Zh. Fiz. Khim.*, **33**, 1093 (1959).
39. B. G. Reuben and L. Friedman, *J. Chem. Phys.*, **37**, 1636 (1962).
40. J. C. Polanyi, *J. Chem. Phys.*, **31**, 1338 (1959).
41. H. von Koch and L. Friedman, *J. Chem. Phys.*, **38**, 1115 (1963).
42. T. F. Moran and L. Friedman, *ibid.*, **39**, 2491 (1963).
43. F. S. Klein and L. Friedman, *ibid.*, **41**, 1789 (1964).
44. A. Giardini-Guidoni and L. Friedman, *J. Chem. Phys.*, **45**, 937 (1966).
45. M. G. Inghram and R. J. Hayden, *A Handbook for Mass Spectroscopy*, National Academy of Sciences NRC Nuclear Science Series Report 14, Washington, D.C., 1954.

46. (a) E. Lindholm, *Rev. Sci. Instr.*, **31**, 210 (1960); (b) E. Lindholm, *Arkiv. Fysik*, **8**, 257 (1954); (c) E. Lindholm, *ibid.*, **8**, 433 (1954); (d) E. Lindholm, *Z. Naturforsch.*, **9a**, 535 (1954); (e) E. Gustafsson and E. Lindholm, *Arkiv. Fysik*, **18**, 219 (1960); (f) P. Wilmenius and E. Lindholm, *Arkiv. Fysik*, **21**, 97 (1962).

47. C. F. Giese and W. B. Maier II, *J. Chem. Phys.*, **39**, 197 (1963).

48. J. J. Leventhal and L. Friedman, *J. Chem. Phys.*, **48**, 1559 (1968).

49. (a) V. Cermak and Z. Herman, *Coll. Czech. Chem. Comm.*, **27**, 406 (1962); (b) *ibid.*, **30**, 169 (1965).

50. (a) A. Henglein and G. A. Muccini, *Z. Naturforsch.*, **17a**, 452 (1962); (b) *ibid.*, **18a**, 753 (1963).

51. C. F. Giese and W. B. Maier II, *J. Chem. Phys.*, **39**, 739 (1963).

52. D. W. Vance and T. L. Bailey, *J. Chem. Phys.*, **44**, 486 (1966).

53. A. Henglein, *Advances in Mass Spectrometry*, **3**, 330 (1966).

54. W. B. Maier II, *Planet. Space Sci.*, **16**, 477 (1968).

55. C. F. Giese, *Advances in Mass Spectrometry*, **3**, 321 (1966).

56. J. Durup, *ibid.*, **3**, 329 (1966).

57. (a) F. H. Field and J. L. Franklin, *Electron Impact Phenomena*, Academic Press, New York, 1957; Franklin, J. L., Dillard, J. G., Rosenstock H. M., Herron, J. T. Draxl, K. and F. H. Field, NSRDS NBS 26, June 1969; (b) K. Watanabe, T. Nakayama and J. Mottle, *Final Report on Ionization Potentials of Molecules by a Photoionization Method*, December, 1959, Dept. Army No. 5B-99-01-004, ORD TB2-0001-00R-1624; (c) R. W. Kiser, *Tables of Ionization Potentials*, U.S. Atomic Energy Commission Report TID-6142 (1960); (d) R. W. Kiser, *Additions and Corrections to Tables of Ionization Potentials* TID-6142, Dept. of Chemistry, Kansas State University, Manhattan, Kansas. Various supplements have been published; (e) A. V. Gurvich and B. A. Khachkurnzov, *Thermodynamic Properties of Individual Substances*, Izdakl'stvo Akademii Nauk SSSR, Moscow (1962), Engl. translation, U.S. Government Report FTD-HF-66-251; (f) A. Streitwieser, Jr., *Progress in Physical Organic Chemistry*. **1**, 1 (1963) (Interscience, New York); (g) R. R. Bernecker and F. A. Long, *J. Phys. Chem.*, **65**, 1565 (1965); (h) D. W. Turner, *Adv. in Phys. Org. Chem.*, **4**, 31 (1966); (i) V. I. Vedeneyev, L. V. Gurvich, V. N. Dondrat'yev, V. A. Medvedev and Ye L. Frankevich, *Bond Energies, Ionization Potentials and Electron Affinities*, Edward Arnold & Co., London, 1966; (j) Henri Lemaire, Molecular Ionization Potentials, *Comms. Energ. At. (Fr.) Serv. Doc. Ser. Bibliog.*, **81** (1967).

58. D. P. Stevenson, *Disc. Faraday Soc.*, **10**, 35 (1951).

59. C. A. McDowell, Chapter XII, *Mass Spectrometry*, McGraw-Hill, New York (1965).

60. J. W. Warren and C. A. McDonald, *Disc. Faraday Soc.*, **10**, 53 (1951).

61. F. P. Lossing, A. W. Tickner and W. A. Bryce, *J. Chem. Soc.*, **9**, 1254 (1951).

62. (a) J. D. Morrison, *J. Chem. Phys.*, **19**, 1305 (1951); (b) J. D. Morrison and A. J. C. Nicholson, *ibid.*, **20**, 1021 (1952); (c) J. D. Morrison, *ibid.*, **21**, 1767 (1954); (d) J. D. Morrison, *ibid.*, **21**, 2090 (1954); (e) J. D. Morrison, *ibid.*, **22**, 1219 (1954); (f) G. R. Herens and J. D. Morrison, *Rev. Sci. Instr.*, **23**, 118 (1952); (g) J. D. Morrison, *J. Chem. Phys.*, **22**, 1219 (1954); (h) J. D. Morrison, *J. Appl. Phys.*, **28**, 1409 (1957); (i) A. J. C. Nicholson, *J. Chem. Phys.*, **29**, 1312 (1958).

63. (a) P. Marmet and L. Kerwin, *Can. J. Phys.*, **38**, 787 (1960). (b) L. Kerwin and P. Marmet, *J. Appl. Phys.*, **31**, 2071 (1960).

64. E. M. Clarke, *Can. J. Phys.*, **32**, 764 (1954).

65. R. E. Fox, W. M. Hickam, D. J. Grove and T. Kjeldaas Jr., *Rev. Sci. Instr.*, **12**, 1101 (1955).
66. (a) G. G. Cloutier and H. I. Schiff, *Advances in Mass Spectrometry*, **1**, 473 (1959); (b) G. G. Cloutier and H. I. Schiff, *J. Chem. Phys.*, **31**, 793 (1959).
67. R. J. Coleman, J. S. Delderfield and B. G. Reuben, *Intern. J. of Mass Spec. & Ion Physics*, **2**, 25 (1969).
68. R. K. Curran and R. E. Fox, *J. Chem. Phys.*, **34**, 1590 (1961).
69. H. Hurzeler, M. G. Inghram and J. D. Morrison, *J. Chem. Phys.*, **28**, 76 (1958).
70. V. H. Dibeler and R. M. Reese, *J. Chem. Phys.*, **40**, 2034 (1964); V. H. Dibeler and J. A. Walker, *Intern. Mass Spec. Conf.*, Berlin, Photon Impact Session, September 1967.
71. D. W. Turner, "Molecular Spectroscopy," Institute of Petroleum, London, 1968, p. 209.
72. W. Price, *ibid.*, p. 221.
73. A. S. Russell, C. M. Fontana and J. H. Simons, *J. Chem. Phys.*, **5**, 381 (1941).
74. V. Aquilanti and G. G. Volpi, *J. Chem. Phys.*, **44**, 2307 (1966).
75. V. Aquilanti and G. G. Volpi, *ibid.*, **44**, 3574 (1966).
76. L. Friedman and T. F. Moran, *ibid.*, **42**, 2624 (1965).
77. M. S. B. Munson and F. H. Field, *J. Am. Chem. Soc.*, **87**, 3294 (1965).
78. B. G. Reuben, unpublished data.
79. S. Meyerson and P. N. Rylander, *J. Am. Chem. Soc.*, **79**, 1058 (1957).
80. W. A. Chupka, M. E. Russell and K. Refeay, *J. Chem. Phys.*, **45**, 1518 (1968).
81. J. Berkowitz and W. A. Chupka, *J. Chem. Phys.*, **45**, 1287 (1966).
82. W. A. Chupka and J. Berkowitz, *J. Chem. Phys.*, **48**, 5726 (1968).
83. J. J. Leventhal and L. Friedman, *J. Chem. Phys.*, **49**, 1974 (1968).
84. M. E. Schwartz and L. J. Schaad, *J. Chem. Phys.*, **48**, 4709 (1968).
85. (a) V. N. Kondratyev and N. D. Sokolov, *Zh. Fiz. Khim.*, **29**, 1265 (1955); (b) S. I. Vetchinkin, E. A. Pshenichov and N. D. Sokolov, *ibid.*, **33**, 1269 (1959).
86. F. W. Lampe and J. H. Futrell, *Trans. Faraday Soc.*, **59**, 1957 (1963).
87. J. H. Simons, H. T. Francis, E. E. Muschlitz, Jr. and G. C. Fryburg, *J. Chem. Phys.*, **11**, 316 (1943).
88. D. Van Raalte and A. G. Harrison, *Can. J. Chem.*, **41**, 3118 (1963).
89. W. A. Chupka and M. E. Russell, *J. Chem. Phys.*, **48**, 1527 (1968).
90. J. L. Beauchamp and S. E. Buttrill, Jr., *J. Chem. Phys.*, **48**, 1783 (1968).
91. (a) D. Wobschall, J. R. Graham and D. P. Malone, *Phys. Rev.*, **131**, 1565 (1963); (b) D. Wobschall, *Rev. Sci. Instr.*, **36**, 466 (1965); (c) D. Wobschall and J. R. Graham, *Bull. Am. Phys. Soc.*, **9**, 189 (1964); (d) D. Wobschall, J. R. Graham and D. P. Malone, *J. Chem. Phys.* **42**, 3955 (1965).
92. P. M. Llewellyn, *ASTM Conf. on Mass Spectrometry*, St. Louis, Mo., 1965.
93. J. L. Baldeschwieler, *Science*, **159**, 263 (1968).
94. L. R. Anders, J. L. Beauchamp, R. C. Dunbar and J. D. Baldeschwieler, *J. Chem. Phys.*, **45**, 1062 (1966).
95. J. L. Beauchamp, *J. Chem. Phys.*, **46**, 1231 (1967).
96. J. L. Beauchamp, L. R. Anders and J. D. Baldeschwieler, *J. Am. Chem. Soc.*, **89**, 4569 (1967).
97. (a) M. De Pas, J. J. Leventhal and L. Friedman, *J. Chem. Phys.*, **49**, 5543 (1968); (b) M. De Pas, J. J. Leventhal and L. Friedman, *J. Chem. Phys.*, **51**, 3748 (1969); (c) H. F. Halliwell and S. C. Nyberg, *Trans. Faraday Soc.*, **59**, 1126 (1963); (d) J. J. Leventhal and L. Friedman, *J. Chem. Phys.*, **50**, 2928 (1969).

98. (a) M. A. Haney and J. L. Franklin, *J. Chem. Phys.*, **48**, 4093 (1968); (b) M. A. Haney and J. L. Franklin, *ibid.*, **50**, 2028 (1969); (c) M. A. Haney and J. L. Franklin, *Trans. Faraday Soc.*, **65**, 1794 (1969).

99. A. C. Hopkinson, K. N. Holbrook, K. Yates and I. G. Csizmadia, *J. Chem. Phys.*, **49**, 3596 (1968).

100. R. W. Kiser, *Introduction to Mass Spectrometry and its Applications*, Prentice Hall, Princeton, N. J., 1965, pp. 202–3.

101. (a) A. G. Harrison, A. Ivko and D. Van Raalte, *Can. J. Chem.*, **44**, 1625 (1966); (b) H. Pritchard and A. G. Harrison, *J. Chem. Phys.*, **48**, 2827 (1968).

102. M. S. B. Munson, *J. Am. Chem. Soc.*, **87**, 2332 (1965).

103. (a) Wagner, C. D., Wadsworth, P. H. and Stevenson, D. P., *ibid.*, **28**, 517 (1958); (b) T. Mariner and W. Bleakney, *Phys. Rev.*, **72**, 792 (1947).

104. T. F. Moran and L. Friedman, *J. Chem. Phys.*, **40**, 860 (1964).

105. W. B. Maier II, *J. Chem. Phys.*, **46**, 4991 (1967).

106. P. Pechukas and J. C. Light, *J. Chem. Phys.*, **42**, 3281 (1965).

107. (a) V. Cermak and Z. Herman, *J. de Chimie Physique*, **56**, 51 (1969); (b) V. Cermak and Z. Herman, *Collection Czech. Chem. Commun.*, **31**, 1343 (1965).

108. J. J. Leventhal and L. Friedman, *J. Chem. Phys.*, **46**, 997 (1967).

109. W. B. Maier II, *J. Chem. Phys.*, **47**, 859 (1967).

110. K. R. Ryan, *J. Chem. Phys.*, to be published.

111. (a) B. R. Turner, J. A. Rutherford and R. F. Mathis, *J. Chem. Phys.*, **48**, 1602 (1968); (b) R. F. Mathis, B. R. Turner and J. A. Rutherford, *J. Chem. Phys.*, **49**, 2051 (1968).

112. (a) J. H. Simons, C. M. Fontana, E. E. Muschlitz Jr. and S. R. Jackson, *J. Chem. Phys.*, **11**, 307 (1943); (b) J. H. Simons, H. T. Francis, C. M. Fontana and S. R. Jackson, *Rev. Sci. Instr.*, **13**, 419 (1942); (c) J. H. Simons, C. M. Fontana, H. T. Francis and L. G. Unger, *J. Chem. Phys.*, **11**, 312 (1943); (d) J. H. Simons, E. E. Muschlitz Jr. and L. G. Unger, *ibid.*, **11**, 322 (1943).

113. E. A. Mason and J. F. Vanderslice, *Phys. Rev.*, **114**, 497 (1959).

114. (a) P. Kebarle and E. W. Godbole, *J. Chem. Phys.*, **39**, 1131 (1963); (b) P. Kebarle and A. M. Hogg, *ibid.*, **42**, 668 (1965); (c) P. Kebarle and A. M. Hogg, *ibid.*, **42**, 798 (1965); (d) A. M. Hogg and P. Kebarle, *Adv. in Mass. Spec.*, **3**, 401 (1965); (e) A. M. Hogg and P. Kebarle, *J. Chem. Phys.*, **43**, 449 (1965); (f) A. M. Hogg, R. M. Haynes and P. Kebarle, *J. Am. Chem. Soc.*, **80**, 28 (1965); (g) A. M. Hogg, R. M. Haynes and P. Kebarle, *ibid.*, **88**, 28 (1966); (h) R. M. Haynes and P. Kebarle, *J. Chem. Phys.*, **45**, 3899 (1966); (i) P. Kebarle, R. M. Haynes and J. G. Collins, *J. Am. Chem. Soc.*, **89**, 5753 (1967); (j) P. Kebarle, S. K. Searles, A. Zolla, J. Scarborough and M. Arshadi, *ibid.*, **89**, 6393 (1967); (k) P. Kebarle and R. M. Haynes, *J. Chem. Phys.*, **47**, 1676 (1967); (l) P. Kebarle, R. M. Haynes and S. K. Searles, *ibid.*, **47**, 1684 (1967); (m) P. Kebarle, M. Arshadi and J. Scarborough, *ibid.*, **49**, 817 (1968); (n) P. Kebarle, "Mass Spectrometry in Inorganic Chemistry," *Adv. in Chem. Series*, **72**, 24 (1968); (o) D. A. Durden, P. Kebarle and A. Good, *J. Chem. Phys.*, **50**, 805 (1969).

115. (a) J. R. Munson and K. Hoselitz, *Proc. Roy. Soc. (London)*, **A17**, 43 (1939); (b) A. Overhauser, *Phys. Rev.*, **76**, 250 (1949); (c) S. Bloom and H. Margenau, *Phys. Rev.*, **85**, 670 (1952).

116. (a) A. Henglein, K. Lacmann and B. Knoll, *J. Chem. Phys.*, **43**, 1048 (1965); (b) A. Henglein, K. Lacmann and G. Jacobs, *Ber. Bunsenges. Physik. Chem.*, **69**, 279 (1965); (c) K. Lacmann and A. Henglein, *ibid.*, **69**, 286 (1965); (d) K. Lacmann,

and A. Henglein, *ibid.*, **69**, 292 (1965); (e) A. Henglein and K. Lacmann, *Adv. Mass. Spec.*, **3**, 331 (1966).

117. (a) B. R. Turner, M. A. Fineman and R. F. Stebbings, *J. Chem. Phys.*, **42**, 4088 (1965); (b) B. R. Turner, R. F. Stebbings and J. A. Rutherford, *Report GA-6438*, General Atomic, San Diego (1965).

118. (a) L. D. Doverspike, R. L. Champion and T. L. Bailey, *J. Chem. Phys.*, **45**, 4377 (1966); (b) L. D. Doverspike, R. L. Champion and T. L. Bailey, *ibid.*, **45**, 4385 (1966).

119. L. D. Doverspike and R. L. Champion, *ibid.*, **46**, 4718 (1967).

120. R. L. Champion and L. D. Doverspike, *ibid.*, **49**, 4321 (1968).

121. (a) W. R. Gentry, E. A. Gislason, Yuan-tseh Lee, B. H. Mahan and Chi-wing Tsao, *Disc. Faraday Soc.*, **44**, 137 (1967); (b) W. R. Gentry, E. A. Gislason, B. H. Mahan and Chi-wing Tsao, *J. Chem. Phys.*, **49**, 3058 (1968); (c) E. A. Gislason, B. H. Mahan, Chi-wing Tsao and A. S. Werner, *J. Chem. Phys.*, **50**, 142 (1969); (d) W. R. Gentry, *Report UCRL-17691*, Univ. of California Lawrence Radiation Laboratory; (e) B. H. Mahan, *Accounts of Chemical Research*, **1**, 217 (1968).

122. (a) Z. Herman, T. L. Rose, J. D. Kerstetter and R. Wolfgang, *Disc. Faraday Soc.*, **44**, 123 (1967); (b) Z. Herman, J. D. Kerstetter, T. L. Rose and R. Wolfgang, *J. Chem. Phys.*, **46**, 2844 (1967); (c) Z. Herman, J. D. Kerstetter, T. L. Rose and R. Wolfgang, *Rev. Sci. Instr.*, **40**, 538 (1969).

123. A. Ding, A. Henglein, D. Hyatt and K. Lacmann, *Z. Naturforsch.* **23a**, 2084 (1968).

124. C. Lifshitz and B. G. Reuben, *J. Chem. Phys.*, **50**, 951 (1969).

125. A. Henglein, *Z. Naturforsch.*, **77a**, 44 (1962).

126. L. Matus, D. J. Hyatt and M. J. Henchman, *J. Chem. Phys.*, **46**, 2439 (1967).

127. L. Matus, I. Opauszky, D. Hyatt, A. J. Masson, K. Birkinshaw and M. J. Henchman, *Disc. Faraday Soc.*, **44**, 146 (1967).

128. (a) F. P. Abramson and J. Futrell, *J. Chem. Phys.*, **45**, 1925 (1966). (b) F. P. Abramson and J. Futrell, *ibid.*, **46**, 3265 (1967).

129. (a) A. Ding, K. Lacmann and A. Henglein, *Ber. Bunsenges. Phys. Chem.*, **71**, 596 (1967); (b) A. Ding, A. Henglein and K. Lacmann, *Z. Naturforsch.*, **23a**, 779 (1968); (c) A. Ding, A. Henglein and K. Lacmann, *ibid.*, **23a**, 780 (1968); (d) D. Hyatt and K. Lacmann, *ibid.*, **23a**, 2080 (1968); (e) A. Ding, A. Henglein, D. Hyatt and K. Lacmann, *ibid.*, **23a**, 2090 (1968); (f) A. Ding and A. Henglein, *Ber. Bunsenges Phys. Chem.*, **73**, 562 (1969).

130. E. A. Gislason, B. H. Mahan, C. W. Tsao and A. S. Werner, *J. Chem. Phys.*, **50**, 5418 (1969).

131. Z. Herman, A. Lee and R. Wolfgang, *J. Chem. Phys.*, **51**, 452 (1969).

132. R. Wolfgang and R. J. Cross, Jr., *ibid.*, **73**, 743 (1969).

133. D. P. Stevenson and J. Schachtschneider, *Am. Chem. Soc. Div. Phys. Chem.*, Univ. of Utah, Salt Lake City, Utah, Summer 1963.

134. J. C. Light and S. Chan, *J. Chem. Phys.*, **51**, 1008 (1969).

135. T. F. Moran and L. Friedman, *J. Chem. Phys.*, **42**, 2391 (1965).

136. C. F. Giese, *Bull. Am. Phys. Soc.*, **9**, 189 (1964).

137. A. G. Harrison, A. Ivko and T. W. Shannon, *Can. J. Chem.*, **44**, 1351 (1966).

138. L. I. Bone and J. H. Futrell, *J. Chem. Phys.*, **46**, 4084 (1966).

139. B. S. Rabinovitch and D. W. Setser, *Adv. Photochem.*, **3**, 1 (1964).

140. J. C. Light, *J. Chem. Phys.*, **40**, 3221 (1964).

141. P. Pechukas, J. C. Light and J. Lin, *ibid.*, **43**, 3201 (1965).

138   L. FRIEDMAN AND B. G. REUBEN

142. (a) V. Aquilanti, A. Galli, A. Giardini-Guidoni and G. G Volpi, *Trans. Faraday. Soc.*, **64**, 124 (1968); (b) V. Aquilanti, A. Galli, A. Giardini-Guidoni and G. G. Volpi, *ibid.*, **63**, 926 (1967).
143. J. L. Franklin, F. H. Field and F. W. Lampe, *J. Am. Chem. Soc.*, **78**, 5697 (1956).
144. D. P. Stevenson and D. O. Schissler, *J. Chem. Phys.*, **24**, 926 (1956).
145. V. L. Tal'rose and A. K. Lyubimova, *Dokl. Akad. Nauk. SSSR*, **86**, 909 (1952).
146. C. E. Melton and P. E. Rudolph, *J. Chem. Phys.*, **32**, 1128 (1960).
147. S. Wexler and R. Marshall, *J. Am. Chem. Soc.*, **86**, 781 (1964).
148. P. E. Rudolph and C. E. Melton, *J. Phys. Chem.*, **63**, 916 (1959).
149. R. Fuchs, *Z. Naturforsch.*, **16a**, 1026 (1961).
150. A. Bloch, *Advances in Mass Spectrometry*, II, Ed. R. M. Elliott, Pergamon Press, New York, 1963, p. 48.
151. E. Lindholm. I. Szabo and P. Wilmenius. *Arkiv. Fysik*, **25**, 417 (1963).
152. J. H. Futrell and L. W. Sieck, *J. Phys. Chem.*, **69**, 892 (1965).
153. C. E. Melton and W. H. Hamill, *ibid.*, **41**, 1469 (1964).
154. M. S. B. Munson, *ibid.*, **69**, 572 (1965).
155. G. A. W. Derwish, A. Galli, A. Giardini-Guidoni and G. G. Volpi, *J. Am. Chem. Soc.*, **87**, 1159 (1965).
156. J. H. Futrell and T. O. Tiernan, *J. Phys. Chem.*, **72**, 158 (1968).
157. J. H. Futrell and C. D. Miller, *Rev. Sci. Instr.*, **37**, 1521 (1966).
158. C. E. Melton, G. A. Ropp and T. W. Martin, *J. Chem. Phys.*, **64**, 1577 (1960).
159. J. H. Beynon, *Mass Spectrometry and Its Application to Organic Chemistry*, Elsevier, Amsterdam, 1960, p. 276.
160. A. Henglein, *Z. Naturforsch.*, **17a**, 37 (1962).
161. (a) F. H. Field, P. Hamlet and W. F. Libby, *J. Am. Chem. Soc.*, **89**, 6035 (1967); (b) F. H. Field, P. Hamlet and W. F. Libby, *ibid.*, **91**, 2839 (1969).
162. J. Beauchamp, *Ph.D. Thesis*, Stanford University (1968).
163. S. Wexler and R. P. Clow, *J. Am. Chem. Soc.*, **90**, 3940 (1968).
164. S. Wexler, *J. Chem. Phys.*, **41**, 1714, 2781 (1964).
165. A. Giardini–Guidoni and F. Zocchi, *Trans. Faraday Soc.*, **64**, 2342 (1968).
166. R. K. Curran, *J. Chem. Phys.*, **38**, 2974 (1963).
167. G. A. W. Derwish, A. Galli, A. Giardini–Guidoni and G. G. Volpi, *ibid.*, **40**, 5 (1965).
168. C. E. Melton, *ibid.*, **33**, 647 (1960).
169. M. Saporoschenko, *ibid.*, **42**, 2760 (1958).
170. M. Saporoschenko, *Phys. Rev.*, **111**, 1550 (1958).
171. S. Wexler and N. Jesse, *J. Am. Chem. Soc.*, **84**, 3425 (1962).
172. P. F. Knewstubb and A. W. Tickner, *J. Chem. Phys.*, **36**, 674 (1962).
173. P. F. Knewstubb and A. W. Tickner, *ibid.*, **36**, 684 (1962).
174. V. Aquilanti, A. Galli, A. Giardini–Guidoni and G. G. Volpi, *ibid.*, **43**, 1969 (1965).
175. F. H. Field, J. L. Franklin and M. S. B. Munson, *J. Am. Chem. Soc.*, **85**, 3575 (1963).
176. O'Malley, R. and Jennings, K. R., *Inter. J. of Mass Spec. and Ion Physics*, **2**, 441 (1969), **2**, 257 (1969).
177. I. Szabo, *Arkiv. für Fysik*, to be published.
178. I. Szabo, *ibid.*, **35**, 359 (1967).
179. F. H. Field and M. S. B. Munson, *J. Am. Chem. Soc.*, **87**, 3289 (1965).
180. S. Arrhenius, *Z. Physik Chem.*, **4**, 226 (1889).

181. R. C. C. Lao, R. W. Rozett and W. S. Koski, *J. Chem. Phys.*, **49**, 4202 (1968).
182. (a) V. L. Tal'rose and E. L. Frankevich, *Zh. Fiz. Khim.*, **34**, 2709 (1960); (b) V. L. Tal'rose and E. L. Frankevich, *Tr. Soveshch. po Radiacionnoi Khim*, Moscow, **13** (1967); (c) V. L. Tal'rose and G. V. Karachevtsev, *Adv. Mass. Spec.*, **3**, 211 (1966).
183. (a) A. G. Harrison, T. W. Shannon and F. Meyer, *Adv. Mass. Spec.*, **3**, 377 (1966); (b) A. G. Harrison and J. C. J. Thynne, *Trans. Faraday Soc.*, **62**, 3345 (1966); (c) A. G. Harrison and J. C. J. Thynne, *Can. J. Chem.*, **45**, 1321 (1967); (d) S. K. Gupta, E. G. Jones, A. G. Harrison and J. J. Myher, *ibid.*, **45**, 3107 (1967); (e) A. G. Harrison and J. J. Myher, *J. Chem. Phys.*, **46**, 3276 (1967); (f) E. G. Jones and A. G. Harrison, *Can. J. Chem.*, **45**, 3119 (1967); (g) J. J. Myher and A. G. Harrison, *ibid.*, **46**, 101 (1968); (h) J. J. Myher and A. G. Harrison, *ibid.*, **46**, 1755 (1968); (i) A. G. Harrison and J. C. J. Thynne, *Trans. Faraday Soc.*, **64**, 945 (1968); (j) A. G. Harrison and J. C. J. Thynne, *ibid.*, **64**, 1287 (1968); (k) J. J. Myher and A. G. Harrison, *J. Phys. Chem.*, **72**, 1905 (1968); (l) H. Pritchard and A. G. Harrison, *J. Chem. Phys.*, **48**, 5623 (1968); (m) G. P. Nagy, J. C. J. Thynne and A. G. Harrison, *Can. J. Chem.*, **46**, 3609 (1968); (n) J. A. Herman, J. J. Myher and A. G. Harrison, *ibid.*, **47**, 647 (1969); (o) J. A. Herman and A. G. Harrison, *ibid.*, **47**, 957 (1969).
184. (a) M. C. Cress, P. M. Becker and F. W. Lampe, *J. Chem. Phys.*, **44**, 2212 (1966); (b) P. M. Becker and F. W. Lampe, *ibid.*, **42**, 3857 (1965); (c) J. J. De Corpo and F. W. Lampe, *ibid.*, **51**, 943 (1969).
185. (a) C. W. Hand and H. von Weyssenhoff, *Can. J. Chem.*, **42**, 195 (1964); (b) C. W. Hand and H. von. Weyssenhoff, *ibid.*, **42**, 2385 (1964).
186. (a) K. R. Ryan and J. H. Futrell, *J. Chem. Phys.*, **42**, 824 (1965); (b) K. R. Ryan and J. H. Futrell, *ibid.*, **43**, 3009 (1965); (c) K. R. Ryan, J. H. Futrell and C. D. Miller, *Rev. Sci. Instr.*, **37**, 107 (1966).
187. (a) P. H. G. Dickinson and J. Sayers, *Proc. Phys. Soc.*, **A76**, 137 (1960); (b) P. H. Batey, G. R. Court and J. Sayers, *Planet Space Sci.*, **13**, 911 (1965); (c) J. Sayers and D. Smith, *Disc. Faraday Soc.*, **37**, 167 (1964); (d) M. J. Copsey, D. Smith and J. Sayers, *Planet Space Sci.*, **14**, 1047 (1966).
188. (a) W. L. Fite, J. A. Rutherford, W. R. Snow and V. A. J. Van Lint, *Disc. Faraday Soc.*, **33**, 264 (1962); (b) W. L. Fite and J. A. Rutherford, *ibid.*, **37**, 192 (1964).
189. (a) G. F. O. Langstroth and J. B. Hasted, *ibid.*, **33**, 298 (1962); (b) M. M. Nakshbandi and J. B. Hasted, *Planet Space Sci.*, **15**, 1781 (1967).
190. (a) F. C. Fehsenfeld, A. L. Schmeltekopf, P. D. Goldan, H. I. Schiff and E. E. Ferguson, *J. Chem. Phys.*, **44**, 4087 (1966); (b) P. D. Goldan, A. L. Schmeltekopf, F. C. Fehsenfeld, H. I. Schiff and E. E. Ferguson, *ibid.*, **44**, 4095 (1966); (c) F. C. Fehsenfeld, E. E. Ferguson and A. L. Schmeltekopf, *ibid.*, **44**, 3022 (1966); (d) F. C. Fehsenfeld, A. L. Schmeltekopf and E. E. Ferguson, *ibid.*, **45**, 23 (1966); (e) F. C. Fehsenfeld, A. L. Schmeltekopf, G. I. Gilman, L. G. Duls and E. E. Ferguson, *Bull. Am. Phys. Soc.*, **11**, 505 (1966); (f) F. C. Fehsenfeld, A. L. Schmeltekopf, and E. E. Ferguson, *J. Chem. Phys.*, **45**, 1844 (1966); (g) A. L. Schmeltekopf, F. C. Fehsenfeld and E. E. Ferguson, *J. Astrophys.*, **148**, L155 (1967); (h) F. C. Fehsenfeld, A. L. Schmeltekopf, H. I. Schiff and E. E. Ferguson, *Planet Space Sci.*, **15**, 373 (1967); (i) A. L. Schmeltekopf, F. C. Fehsenfeld, G. I. Gilman and E. E. Ferguson, *Planet Space Sci.*, **15**, 401 (1967); (j) F. C. Fehsenfeld, A. L. Schmeltekopf and E. E. Ferguson, *J. Chem. Phys.*, **46**, 2802 (1967); (k) E. E. Ferguson, F. C. Fehsenfeld and A. L. Schmeltekopf, *ibid.*, **47**, 3085 (1967); (l) F. C. Fehsenfeld and E. E. Ferguson, *Planet Space Sci.*, **16**, 701 (1968); (m) A. L. Schmeltekopf,

E. E. Ferguson and F. C. Fehsenfeld, *J. Chem. Phys.*, **48**, 2966 (1968); (n) D. B. Dunkin, F. C. Fehsenfeld, A. L. Schmeltekopf and E. E. Ferguson, *ibid.*, **49**, 1365 (1968); (o) Data of Ferguson and Fehsenfeld, see Ref. 22.

191. J. F. Paulson, R. L. Mosher and F. Dale, *J. Chem. Phys.*, **44**, 3025 (1966).
192. M. Lipeles, R. Novick and N. Tolk, *Phys. Rev. Lett.*, **15**, 815 (1965).
193. B. G. Reuben and J. W. Linnett, *Trans. Faraday Soc.*, **55**, 1543 (1959).
194. I. M. Campbell and B. A. Thrush, *Ann. Rep. Chem. Soc. (London)*, **62**, 17 (1965).
195. J. F. Paulson, E. Murad and F. Dale, *Symp. on Physics and Chemistry of the Upper Atmosphere*, Univ. of Pittsburgh, 1967.
196. P. Warneck, *J. Geophys. Res.*, **72**, 1651 (1967).
197. R. F. Stebbings, B. R. Turner and J. A. Rutherford, *J. Geophys. Res.*, **71**, 771 (1966).
198. C. F. Giese, Ref. 18, Chapter 2.
199. F. A. Wolf, *J. Chem. Phys.*, **44**, 1619 (1966).
200. J. J. Kaufman and W. S. Koski, *J. Chem. Phys.*, **50**, 1942 (1969).
201. J. Hirschfelder, *J. Chem. Phys.*, **9**, 645 (1941).
202. J. Schaefer and J. M. S. Henis, *J. Chem. Phys.*, **49**, 5377 (1968).
203. E. E. Ferguson, private communication.
204. A. L. Farragher, J. A. Peden and W. L. Fite, *J. Chem. Phys.*, **50**, 287 (1969).
205. (a) P. Warneck, *J. Chem. Phys.*, **46**, 502 (1967); (b) P. Warneck, *J. Geophys. Res.*, **72**, 1651 (1967); (c) P. Warneck, *Planet Space Sci.*, **15**, 1349 (1967).
206. A. Weingartshofer and E. M. Clarke, *Phys. Rev. Lett.*, **2**, 591 (1964).
207. L. Friedman, A. P. Irsa and B. G. Reuben, unpublished work.
208. W. A. Chupka and M. E. Russell, *J. Chem. Phys.*, **49**, 5426 (1968).
209. (a) L. W. Sieck, F. P. Abramson and J. H. Futrell, *J. Chem. Phys.*, **45**, 2859 (1966); (b) F. P. Abramson and J. H. Futrell, *ibid.*, **45**, 1925 (1966); (c) F. P. Abramson and J. H. Futrell, *ibid.*, **46**, 3264 (1967); (d) F. P. Abramson and J. H. Futrell, *J. Phys. Chem.*, **72**, 1826 (1968); (e) F. P. Abramson and J. H. Futrell, *ibid.*, **72**, 1994 (1968); (f) L. W. Sieck and J. H. Futrell, *J. Chem. Phys.*, **48**, 1409 (1968).
210. J. H. Futrell and T. O. Tiernan, *Science*, **162**, 415 (1968).
211. J. H. Futrell and T. O. Tiernan, *Fundamental Processes of Radiation Chemistry*, Ed. P. Ausloos, Wiley, New York, 1968, Chapter 5.
212. G. A. Gray, *J. Am. Chem. Soc.*, **90**, 2177 (1968).
213. G. A. Gray, *ibid.*, 6002 (1968).
214. J. King, Jr. and D. D. Elleman, *J. Chem. Phys.*, **48**, 412 (1968).
215. J. King, Jr. and D. D. Elleman, *ibid.*, **48**, 4803 (1968).
216. J. M. S. Henis, *J. Am. Chem. Soc.*, **90**, 844 (1968).
217. R. C. Dunbar, *ibid.*, **90**, 5676 (1968).
218. K. D. Bayes, D. Kivelson and S. C. Wong, *J. Chem. Phys.*, **37**, 1217 (1962).
219. M. T. Bowers, D. D. Elleman and J. L. Beauchamp, *ibid.*, **73**, 5399 (1968).
220. (a) M. T. Bowers, D. D. Elleman and J. King, Jr., *J. Chem. Phys.*, **50**, 1840 (1969); (b) M. T. Bowers, D. D. Elleman and J. King, Jr. *ibid.*, **50**, 4787 (1969).
221. J. T. Beauchamp and J. T. Armstrong, *Rev. Sci. Instr.*, **40**, 123 (1969).
222. R. P. Clow and J. H. Futrell, *J. Chem. Phys.*, **50**, 5041 (1969).
223. T. O. Tiernan and P. S. Gill, *ibid.*, **50**, 5042 (1969).
224. J. J. Leventhal, T. F. Moran and L. Friedman, *ibid.*, **46**, 4666 (1967).
225. I. G. Csizmadia, J. C. Polyanyi, A. C. Roach and W. H. Wong, *Can. J. Chem.*, **47**, 4097 (1969).
226. M. G. Holliday, J. T. Muckerman and L. Friedman, *J. Chem. Phys.* (to be published).
227. J. Krenos and R. Wolfgang, *J. Chem. Phys.*, **52**, 5961 (1970) (a similar study that came to the attention of the authors when this manuscript was in proof).

ION CYCLOTRON RESONANCE

GEORGE A. GRAY

Oregon Graduate Center for Study and Research, Portland, Oregon

CONTENTS

I. INTRODUCTION

One of the most exciting developments in mass spectrometry in the last few years has been the emergence of Ion Cyclotron Resonance (ICR) as a potent technique in the study of ion-molecule reactions. The growing interest and activity in this area is reflected in the rapid increase in published work. This review will attempt to cover the published literature up to January 1971 in a comprehensive manner as well as take advantage of the considerable amount of material submitted for publication from laboratories active in ICR research. Examination of the field will be divided into several subsections covering the basic resonance phenomenon, experimental aspects, ion-molecule chemistry in simple, typically diatomic systems, more complicated, typically organic systems, and applications to concepts important in chemical mechanisms and reactivity.

To facilitate understanding of the development of ICR, some effort will be expended on its basic principles and techniques.

Ion Cyclotron Resonance is characterized by the coherent absorption of energy by an ion from a linearly polarized rf electric field normal to an applied magnetic field. The resonance condition is fulfilled when the radio frequency matches the natural cyclotron frequency of the ion given by

$$\omega_c = \frac{qH}{mc} \tag{1}$$

where q is the charge of an ion of mass m in a magnetic field of strength H, and c is the speed of light. The absorption of energy accelerates the ions to larger orbital radii in a continuous manner until the process is terminated by either an ion-molecule collision or wall impact. The experiment is normally performed with a fixed frequency rf marginal oscillator and magnetic field-sweep. For example, at 153.57 kHz ions of mass 1 to 140 can be observed, utilizing magnetic fields of 100 to 14,000 gauss. The sensitivity of 1 to 10 ions per cubic centimeter permits high signal levels and low electron beam emission currents. The natural motion of an ion in a strong magnetic field is helical. No restraint is placed on the motion of the ion parallel to the magnetic field, but the force on a charged particle in this field, given by $\mathbf{F} = \frac{q}{c}\mathbf{v} \times \mathbf{H}$, limits the motion in the plane perpendicular to the magnetic field to a circular orbit (Figure 1). Ionic motion can be

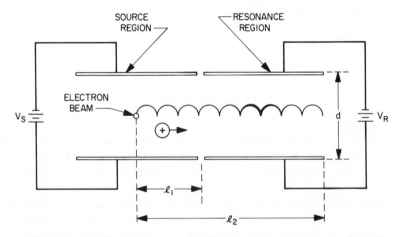

Fig. 1. Motion of an ion in a magnetic field. Superimposed on the normal circular motion is a drift motion caused by static electric fields produced by the plates shown. The magnetic field and electron beam are both parallel and perpendicular to the plane shown.

accurately controlled by suitable static electric fields applied along and normal to the magnetic field direction. The circular nature of the ion orbit is the feature which permits ICR to be useful in the study of ion-molecule reactions. This results from a basic equivalence of an ion travelling in a straight line through a gas at high pressure and a cyclotroning ion moving slowly through a dilute gas. The additional path length described by the cyclotroning ion permits a larger number of reactive events and gives a high concentration of secondary or product ions at pressures of 10^{-6} to 10^{-4} torr. A finely balanced rf oscillator detector is used to detect power absorption from the rf field and indicate the presence of ions. The possibility and frequent occurrence of complex reaction schemes illustrates the most useful aspect of ICR, that of double resonance.[1] This technique allows direct observation of ion molecule reactions and removes the normal uncertainty encountered in high pressure mass spectrometric investigation. Consider a reaction $A^+ + B \rightarrow C^+ + D$. The signal from C^+ can be continuously monitored while an irradiating radiofrequency field is swept through a range of cyclotron frequencies. When the irradiating frequency matches the natural cyclotron frequency of A^+, energy is absorbed by A^+ from the irradiating field and the translational energy of A^+ is increased. Since most reaction rate constants are dependent on the ion velocity, the number density of C^+ will change. If the irradiating field is pulsed on and off and the cyclotron resonance of C^+ is detected by a phase sensitive detector referenced to the pulsing frequency, this change in ion number density can be directly presented. If ion C^+ is observed at its cyclotron frequency while the irradiating frequency is varied, signals will be observed when the irradiating radiofrequency corresponds to the cyclotron resonance frequencies of reactant ions which yield the product C^+. This experiment can be repeated for any combination of ions, thus providing a map of the entire chemistry of a mixture of ions and neutral molecules.

Ion Cyclotron Resonance, complemented by the refinements to be detailed in the following pages, has very quickly become an important tool in the study of atomic and molecular processes.

II. EXPERIMENTAL TECHNIQUES

A. Single and Double Resonance

The first application of ICR techniques was the determination of collision cross sections for N_2^+ and Ar^+ in their parent gases by Wobschall, Graham and Malone in 1963.[2] Their analyzer cell was constructed so that ions formed by electron impact diffused along the magnetic lines of force into

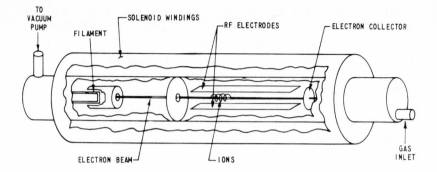

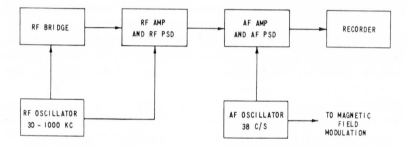

Fig. 2. ICR spectrometer of Wobschall, Graham, and Malone.[3]

a chamber consisting of a quartz tube with gold electrodes shadowed on the inside surface. The magnetic field was generated by an electromagnet. An improved version of the analyzer was later constructed using a solenoid to provide a magnetic field as shown in Figure 2.[3] The low energy ions formed in the source region drift into the analyzer region and interact with the neutral gas. The rf field is produced by a pair of electrodes which form one arm of an rf bridge. Any unbalance of the bridge is sensed and displayed. A minimum number of detectable ions was estimated to be approximately 30. General considerations of resolution, line broadening, sensitivity, effects of space change and inductive pickup of signal were examined. Ion sources, detection systems, the rf bridge and preamplifier, electrode arrangements, amplifiers and phase sensitive detectors were all described in detail.

The most widely used spectrometer system is that manufactured by

Varian,[4] the "Syrotron." Rather than a solenoid produced magnetic field, an electromagnet is employed. Figures 3 and 4 show the spectrometer system and Figure 5 the analyzer construction. Ions produced by electron impact in the source region are constrained to a circular motion in the x–y plane by the presence of the magnetic field, while simultaneously the weak potentials applied to the trapping plates restrict any substantial movements along the magnetic field lines. Drift potentials applied to plates in the x–z planes propel the ions in a direction perpendicular to both the applied magnetic field and the static electric drift fields. The ions are detected in the resonance region and are collected on the walls of the electrometer section after passing through the resonance region. Double resonance irradiating fields can be applied in either the source or resonance region.

The Varian[4] system can serve as a basis for discussion of the necessary functions of an ICR spectrometer. Essential to operation is a high vacuum capability. This requires a mechanical roughing pump and a high vacuum source such as an ion-getter pump or diffusion pump. Pressures can be monitored electronically by capacitance manometers or by ion gauges corrected for the ionization cross section for the particular gas studied. The capacitance manometer system has the advantage of obviating the latter requirement as it is sensitive to only absolute pressure. The vacuum system must be bakeable to temperatures $\geqslant 200°C$ since buildup of material on system surfaces leads to unwanted surface potentials and complicates spectral interpretation when ions appear from residual gases. The Varian system employs a Hall Effect sensor probe to regulate the magnetic field intensity provided by the 9-inch electromagnet. By setting the fixed frequency marginal oscillator to certain selected values, ion masses can be read off the magnetic field controller. The electron energy and emission current are regulated to provide a constant ion production for quantitative studies. Additional sensitivity is gained by modulating the signal output by modulation of electron energy, magnetic field, double resonance irradiating field, or drift potential. A phase sensitive detector, referenced to this modulation frequency, detects signals possessing this frequency, processes them, and outputs a d-c signal. A flange is positioned on the stainless steel can surrounding the analyzer. This allows optical experiments to be performed when the metal flange is replaced with one containing a glass window.

Double resonance frequencies can be automatically generated and synchronized with the motor used in the magnetic field experiments. This permits convenient sweeps through various ranges of atomic mass units. The signal is displayed on an x–y recorder where the x–axis is keyed to either magnetic field or double resonance sweep. Larger electromagnets

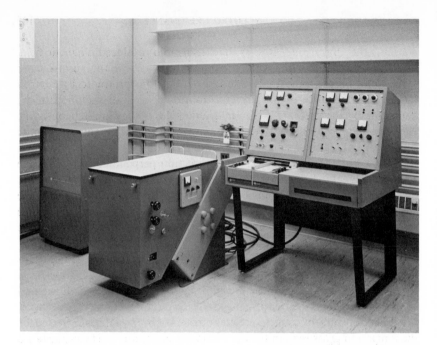

Fig. 3. Varian ICR system (courtesy Varian Associates).

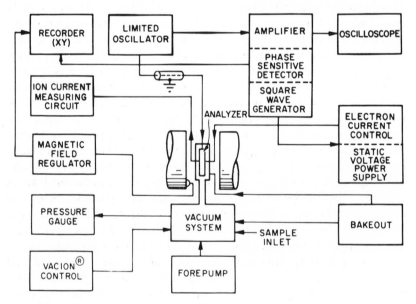

Fig. 4. Block diagram of Varian ICR system (courtesy Varian Associates).

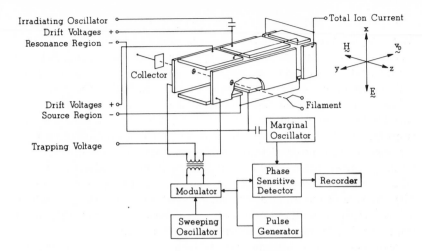

Fig. 5. Analyzer construction of ICR spectrometer (Beauchamp and Armstrong[7]).

allow larger and longer analyzer systems, more homogeneous fields, and multisectioned analyzer cells.

Various cell configurations are available depending on required properties. The most sensitive arrangement is that having a square configuration (end-on) 2.54 cm on a side (Figure 5). This cell has the largest volume and minimizes wall loss although the fields produced are not as uniform as the second type where the end-on view is rectangular. This 1 cm by 2.54 cm cell gives more uniform fields but experience has pointed out the care with which double resonance experiments must be carried out to minimize ion loss.

B. Pulse Experiments

Various adaptations have recently been formulated to extend and enhance the capabilities of the ion production and detection process.[5-10] Henis and Frasure[5] have modified the Varian electron energy circuitry to allow square wave modulation of the electron accelerating potential from just below the ionization potential to an arbitrary level above. This results in a periodic change in the ion density passing through the detector region. The resulting marginal oscillator signal is consequently modulated and is appropriately phase detected at the modulation frequency. Details of a variable d-c power supply and pulse amplifier using an external wave form generator are given. The smooth change of accelerating potential results in much better defined ion path lengths, residence times and space charge conditions during both the on and off modes of operation.

A very convenient and easily constructed modulation scheme is that of McIver's[6] for pulsed grid modulation. The grid is installed in a milled-out section of a boron nitride filament supporting block and is placed between the filament support and the trapping plate. The grid consists of a plate with a 1-mm hole to allow passage of an electron beam. The grid is d-c referenced to the filament and can be pulsed above this level. For emission currents less than 3 μamp, a pulse of 1 volt below the filament d-c level stops the flow of electrons into the cell. The reference pulse output from the spectrometer phase detector can serve without amplification. This technique gives cleaner pulses and less noise than modulation of the accelerating potential and can provide pulses as short as 100 μsec in length. The blocking of the beam during the pulse also protects from a buildup of space charge from the continuous electron beam and removes the slight trapping effect for positive ions near the electron beam.

Beauchamp and Armstrong[7] have taken advantage of the natural oscillatory motion of ions in the trapping field to develop an ion ejection technique. This natural oscillation has a frequency given by

$$\omega_T = \left(\frac{4qV_T}{md^2}\right)^{1/2} \tag{2}$$

The amplitude of these oscillations can be altered by application of an rf ejection electric field to the trapping plates (separated by distance d) as illustrated in Figure 5.[7] This field can be amplitude modulated and this modulation frequency referenced to a phase sensitive detector. Thus, direct observation of the ejected ions is possible by field sweep, or the ejection field can be swept. Alternatively, the phase sensitive detector can be referenced to a field modulation and the spectrum less the ejected ion displayed. The ejection resolution is not high and therefore is most suitable to light ions or those well separated from others in mass. The technique is very useful in eliminating one reactant in a set of concurrent reactions leading to the same product. Hence, reaction rate constants can be determined for both reactants or simply the ratio of rates. A clean ejection is pointed out in Figure 6 for the system $C_2H_4^{+\cdot}/C_2D_4^{+\cdot}$.[7]

Use of a large electromagnet enables longer cells to be constructed while retaining sensitivity and resolution. Clow and Futrell[8] used a 12-inch magnet and designed a cell which has an additional zone to which separate rf, drift, and trapping potentials can be applied. This design permits good isolation between source and detection regions and retains good ion trapping features. The electron energy is pulsed from just below the ionization potential to a few volts above at a repetition rate of 1.2 Kkz. This results in electron energy pulses of 100 μsec, after which the double reso-

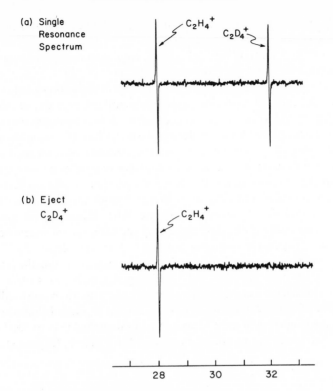

Fig. 6. (a) Single resonance spectrum of a 1:1 mixture of C_2H_4 and C_2D_4 at 14 eV electron energy and 10^{-6} torr. (b) Repeat of single resonance spectrum with simultaneous ejection of $C_2D_4^+$ at 24.4 kHz with 1.2 V trapping potential (Beauchamp and Armstrong[7]).

nance oscillator is activated for a pulse duration of 250 μsec to a known energy as the ions are rapidly drifted out of the source region. Subsequent reaction with neutrals occurs as the ions drift more slowly through the reaction zone. A static magnetic field is used and the marginal oscillator swept over the desired resonance frequencies. Currents are then measured using the ion collector.

Anders[9] has designed a similar pulse sequence for examining relative rates of reaction for CH_3^+ in methane to give $C_2H_5^+$ (exothermic by 0.9 eV) and $C_2H_3^+$ (endothermic by 1.6 eV). A 100 μsec electron beam pulse creates the ions and a subsequent 0.01 to 0.7 V/cm double resonance field is pulsed on for 100 μsec producing ion translational energies from thermal to about 20 eV. The normal reaction time is 3 msec. The product ions are detected by a fast response (0.1 msec) marginal oscillator and the signal is stored in a signal averager for signal accumulation.

A trapped-ion analyzer cell[10] (Figure 7) employs a number of significant advantages over cells used presently. The normal three region cell is replaced by a one region cell. The side plates provide trapping fields while the drift and end plates provide potentials slightly more negative than the trapping potential. Observing and heating rf fields are coupled in the normal way to the drift plates. Both the electron beam and the rf double resonance field are operated only in pulse mode. The electrons pass through a hole in the side plate and are collected on a collector plate after passing through a hole in the opposite side plate. Normal operation involves a 0.10 msec electron beam pulse controlled by pulsed grid modulation[6] followed by an rf pulse, if desired, of variable amplitude. After a desired reaction time the ion under study is detected by pulsing the off resonant magnetic field to a value satisfying the resonance condition. For example, at 153.57 kHz, N_2^+ is detected at a magnetic field at 2800 $\pm$ 5 gauss. If, during the reaction time period the base magnetic field is at 2780 gauss, no N_2^+ ions are detected and their translational energies are not perturbed by the observing oscillator. A pulse of 20 gauss caused by a pulse of current through helmholtz coils around the magnet pole caps increases the total applied field to 2800 gauss and the N_2^+ ion intensity can be monitored for a time τ_2 (Figure 8).

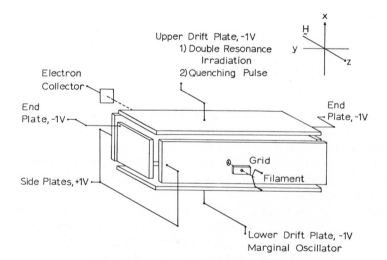

Fig. 7. The trapped-ion analyzer cell. The d-c voltages applied to each plate are suitable for trapping positive ions (McIver[10]).

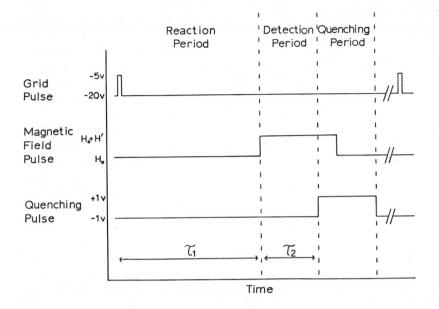

Fig. 8. The pulse sequence used in trapped-ion ICR. The reaction period may be varied from 0 to 80 msec, and the detection period may be varied from 1 to 6 msec (McIver[10]).

For experiments of this type to be made at a high repetition rate, no ions can be present before the initiation of the reaction period. Therefore, at the end of the detection period, the drift potential polarity is reversed and the ions rapidly drift to the walls of the cell and are neutralized. This also accurately limits the time of reaction and this time period can be accurately measured. The trapping time is increased over the normal continuous operation ICR cell from 3 to 100 msec. This greater time resolution also provides greater mass resolution. The observed N_2^+ linewidth is linear with $1/\tau_2$ and the pressure at constant emission current.

C. Scattering Experiments

A number of recent studies involving negative ions[11-14] point out the exciting possibilities that ICR provides using simple experimental techniques. Inelastic electron scattering from neutral molecules has been a subject of much interest during the recent past, and ICR has been applied to it.[11,12] O'Malley and Jennings[11] have used SF_6 as a scavenger gas in N_2, relying on the essentially zero appearance potential of SF_6^- as an indicator of low energy electrons. As the electron beam energy is swept, a number of inelastic nonreactive events may occur between the N_2 molecules and the high energy electrons and these result in energy transfer between the high energy electrons and internal modes of excitation of the N_2 molecules. The scavenger gas will attach the thermalized electrons and these negative ions can be observed by ICR. The number of thermalized electrons will be governed by the presence or absence at a particular scattering electron energy of a neutral molecule resonant excitation. Thus, peaks in the SF_6^- intensity as a function of scattering electron energy can point to internal N_2 excitation (below ionization) such as vibrational or triplet excitation. Ridge and Beauchamp[12] have done the same experiment on N_2 but with the use of CCl_4 as the scavenger and Cl^- as the detected ion. Substantially identical excitation curves were obtained in both cases. The prominent peaks were identified with the known N_2 excitation to $B^3\Pi g$ and $E^3\sum g^+$ states.[12] O'Malley and Jennings[11] have reported preliminary data on several ethylenes where evidence was found for vibrational excitation of ground electronic states on formation of transient negative ions and triplet excitation. Peaks in the 1,3-butadiene spectrum at 3.4 and 6.0 eV were assigned to the lowest triplet and singlet excited states respectively.[11] Since the normal optical selection rules do not apply here, a geat deal of activity can be expected in the evaluation of optically inaccessible excitations.

Henis[13] has related autoionization of negative ions observed in ICR to lower limits of autoionization lifetimes. SF_6^- was determined to have a lifetime of 1200 μsec in the ICR spectrometer. This result can be derived from the linewidth of the SF_6^- ion, and points out the practicality of determining autoionization lifetimes using ICR techniques.

Ionizing reactions of metastable molecules was the subject of a study by Huntress and Beauchamp.[14] They modified the ICR apparatus (Figure 9) so that a multichannel beam source produced a high intensity molecular beam of N_2 which was scattered off an energy selected electron beam at 24 eV. Benzene, admitted through a second inlet, was allowed to react with the excited N_2 molecules in the analyzer cell. $C_6H_6^{+\cdot}$ is observed only when the N_2 gas is present and the energy required is consistent with the excited resonant species, the $a^1\Pi g$ state of N_2.

D. Transient ICR

Dunbar[15] has developed a transient (or heterodyne) technique of Ion Cyclotron Resonance. It is found in a detailed analysis of the power absorption of an ion that in addition to a time independent absorption term, which is that utilized for detection in ordinary ICR, there appears

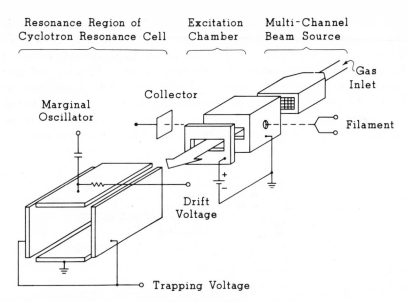

Fig. 9. A modified ICR analyzer cell for study of the ionizing reactions of metastable atoms and molecules. (Huntress and Beauchamp[14]).

a damped oscillation term with frequency $|\omega_c - \omega_{obs}|$. When ω_{obs} is close to the ω_c of a packet of ions, the marginal oscillator response is an oscillation which can be recorded using a fast response system and a signal enhancement device. In principle, it is also possible to separate the reactive and nonreactive collision events. The ions are produced by an electron beam pulse above the neutral's ionization potential. The oscillatory response is damped by ion-molecule collisions and their rates can be obtained without problems such as collisional scattering of ions into the cell walls, uncertainties regarding residence times, and nonhomogeneous fringing fields in the cell.

III. COLLISION-BROADENED ICR

A considerable amount of ICR research has been done in the area of small molecules. Apart from the interest in atmospheric studies, these molecules provide simple systems for detailed mechanistic investigation and also give clean, easily interpretable spectra.

The Cornell group has been interested in the determination of ion-molecule collision cross sections for a number of years.[2,3,16-19] Collision frequencies ν_0 were obtained for N_2^+ and Ar^+ in their parent gases from line width measurements using the relation $v_0 = \Delta\omega_{1/2}$ where $\Delta\omega_{1/2}$ is the frequency width of the absorption line at half-maximum.[2] The dependence of the $\Delta\omega_{1/2}$'s on E/P (observing rf electric field intensity/pressure) were detailed for N_2^+ and Ar^+ in their parent gases and the ion-molecule collision cross sections were found to be 125 and 140 $Å^2$, respectively, at high E/P, and 185 and 215 $Å^2$, respectively, at low E/P. The observed absorption and dispersion line shapes fit closely the Lorentz line shape derived theoretically with the assumption of constant mean free time.

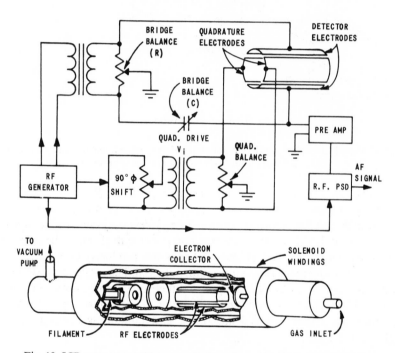

Fig. 10. ICR apparatus utilizing quadrature electrodes (Wobschall et al.[16]).

Wobschall, Graham and Malone[16] were the first to investigate negative ions in the ICR spectrometer. The cyclotron absorptions of O^-, O_2^-, H^-, OH^-, and H_2O^- in oxygen containing water vapor were observed. In order to observe negative ions unambiguously in their cylindrical analyzer, they developed a technique for applying two rf fields which could either cancel or reinforce one another, depending on the charge of the ion. The additional rf was placed on quadrature electrodes (Figure 10) perpendicular to the original plates, and was 90° out of phase. If the rf intensities are labeled by

E_D for detector and E_Q for quadrature fields, then the power absorption from the detector electrodes for a collision frequency of v_0 for ions of mass m and number n is

$$P_D = \frac{ne^2 E_D(E_D \pm E_Q)}{4mv_0 \left[1 + \frac{(\omega - \omega_c)^2}{v_0{}^2}\right]}$$

(3)

The positive sign holds for positive ions and the negative sign for negative

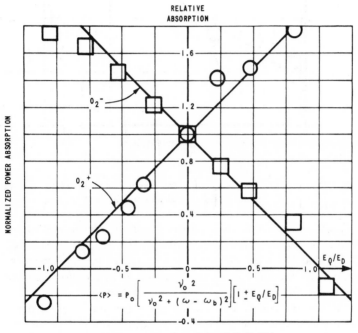

Fig. 11. Normalized power absorption as a function of quadrature voltage. (Wobschall et al.[16]).

ions. Thus, negative ions were observed without interference from positive ions when E_Q was set to match E_D (Figures 11 and 12).

The O_2^- production was dependent on the water vapor partial pressure as well as the O_2 pressure and, in addition, appeared at electron energies where dissociative attachment to water occurs. They suggest the mechanism $e^- + H_2O \rightarrow H^- + OH$ followed by $H^- + O_2 \rightarrow O_2^- + H$.

In a more recent study of collision cross sections, Wobschall, Fluegge

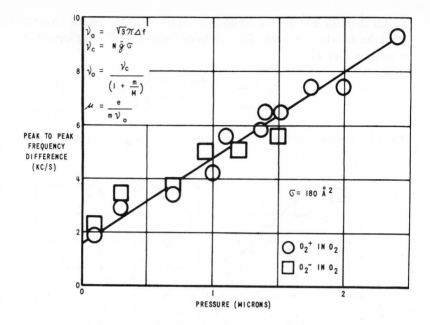

Fig. 12. Collision frequency for positive and negative molecular ions in oxygen for low values of E/P (Wobschall and Graham[19]).

and Graham[17] extended their measurements to higher E/P values. The ion-molecule collision cross section of H^+, H_2^+, H_3^+, He^+, Ar^+, N_2^+, N^+, O_2^+, O_2^-, and O^- were obtained in their parent gases. Agreement with d-c mobility data was found, and a minimum in collision frequency for H_3^+ in H_2 at $E/P = 60$ Vcm^{-1} $torr^{-1}$ was confirmed.

As before, the collision frequency was determined from a plot of the linewidth as a function of pressure at constant E/P. The slope of the resulting straight line is proportional to the reduced collision frequency. The cross section for H_2^+ in H_2, unobservable by d-c mobility methods, was measured as 275 and 68 $Å^2$ at low and high E/P, respectively.

The same investigators have examined the line-broadening at high E/P up to 400 Vcm^{-1} $torr^{-1}$ using rf detector heating, d-c heating, and cyclotron heating for varying ion energies.[18] The d-c heating method gave Lorentzian line shapes and permitted the determination of the collision cross section of He^+ in He as 30 $Å^2$ at 100 Vcm^{-1} $torr^{-1}$. Detailed analyses of the various ion heating methods were given.

Aside from these early investigations, the bulk of the applications of ICR to small molecule interaction with ions has appeared since 1967. The Stanford and Cal Tech—Jet Propulsion Laboratory groups in particular have generated an impressive attack on ion-molecule reactions of simple systems.

Beauchamp[20] has developed a theory of collision broadened ICR for resonant charge transfer. Equations are worked out for the velocity and energy of an average particle and the solutions of these equations are used to obtain the power absorption of the ions from the observing radio-frequency electric field. The formalism is then employed to calculate cyclotron resonance spectra in detail, as well as the effects of pressure and electric-field broadening. Beauchamp points out the applicability of the theory to inelastic collisions and ion-molecule reactions.

Recently Ridge and Beauchamp[21] have considered collision-broadened lineshapes of ions in methane. The ions CH_5^+, $C_2H_5^+$, $C_3H_7^+$, and $C_4H_9^+$ are unreactive toward methane neutrals. Experiments were performed to determine linewidths as a function of E_1/P to obtain reduced collision frequencies ξ_0. The reduced collision frequencies for all the above ions are found to be constant at a value of E_1/P of 80 Vcm^{-1} $torr^{-1}$ where losses to the trapping plates occur and have the respective values 7.13 $\pm$ 0.05, 4.32 $\pm$ 0.14, 3.30 $\pm$ 0.04, and 2.50 $\pm$ 0.07, all $\times 10^{-10}$ cc $molec^{-1}$ sec^{-1}. The reduced collision frequencies display the expected ($\mu^{1/2}/m$) dependence where μ is the collision pair reduced mass and m the mass of the reacting ion. The reduced collision frequencies are independent of ion energy up to $\gtrsim 0.2$ eV and this allows calculation of diffusion cross sections and reduced mobilities. The calculated and experimental diffusion cross sections agree very well and are independent of ion mass at a given temperature. The reduced mobilities are calculated directly from experimentally measured cyclotron resonance linewidths and agree well with those calculated on the basis of the assumption that the reduced collision frequencies are independent of ion energy. The authors point out the possibility of obtaining good values of angle averaged polarizabilities of nonpolar neutral molecules using these techniques. A useful discussion is also presented on the effect of ion neutral collisions on ion drift in the direction of the *drift* plates which becomes important at $\xi/\omega_c > a/2b$ where a is the cell height and b the distance from the electron beam to the end of the resonance region and ξ is the collision frequency. This can occur as an ion loss process at high pressures and would be most critical in long, rectangular cells.

IV. ENERGY DEPENDENCE OF ION-MOLECULE REACTIONS

A. Bimolecular Rate Constant

Beauchamp and Buttrill[22] have interpreted the double resonance phenomenon in terms of variations in the bimolecular rate constant with ion energy. The power absorption characteristic of an ion from an rf field in the case of a primary ion is given by[20]

$$A(\omega_1) = (n_p^+ e^2 E_1^2 / 4m_p)\{\xi_p / [(\omega_1 - \omega_p)^2 + \xi_p^2]\} \tag{4}$$

where n_p^+ is the total number of primary ions observed, ω_1 and E_1 are the observing oscillator frequency and electric field strength, ξ_p is the collision frequency for momentum transfer, and ω_p and m_p are the cyclotron frequency and mass of the observed primary ion. The same expression holds for the double resonance irradiating oscillator field E_2. This field contributes to the total energy of the ion which is given by

$$E_{ion} = \left(\frac{3}{2}\right)kT + \left[\frac{(m_p + M)A(\omega_2)}{2\xi m_p n_p^+}\right] \tag{5}$$

where M is the mass of the neutral molecule responsible for relaxing the reactant ion's momentum. At resonance,

$$E_{ion} = \left(\frac{3}{2}\right)kT + \left[\frac{e^2 E_2^2 (m_p + M)}{8\xi_p^2 m_p^2}\right] \tag{6}$$

In the case of molecular ions where $m_p = M$,

$$E_{ion} = \left(\frac{3}{2}\right)kT + \frac{e^2 E_2^2}{4\xi_p^2 m_p}. \tag{7}$$

The first term represents the thermal energy and the second the energy gained from the rf field. This holds in the limiting case where collisions are the processes responsible for momentum relaxation. Another process leading to interruption of power absorption is that of ions drifting out of the resonance region. In the limit of zero collisions the energy of the ion is

$$E_{ion} = \left(\frac{3}{2}\right)kT + \frac{e^2 E_2^2 \tau^2}{8m_p} \tag{8}$$

where τ is the time spent in the rf field.

As an approximation to the situation at low irradiating levels Beauchamp and Buttrill[22] expand the bimolecular rate constant k about the thermal energy value (k°) as

$$k = k^\circ + \left(\frac{dk}{dE_{ion}}\right)^\circ \left[E_{ion} - \left(\frac{3}{2}\right)kT\right] \cong k^\circ(1 + \alpha) \tag{9}$$

and derive an expression for the *pulsed* double resonance spectral intensity ΔI_s as

$$\Delta I_s = \frac{\left(\dfrac{dk}{dE_{\text{ion}}}\right)^{\circ} A(\omega_2)(M + m_p)e^2 E_1{}^2 P(0)}{8m_p\, m_s\, \xi_p\, \xi_s(nk^{\circ})^2}$$

$$\times \left[\exp\left(-nk^{\circ}\tau\right) - (1 + nk^{\circ}\tau' - nk^{\circ}\tau)\exp\left(-nk^{\circ}\tau'\right)\right] \quad (10)$$

where τ and τ' are the times at which the ion enters and leaves the cell. Thus, since all the factors are energy independent except the first, the size and magnitude of the pulsed double resonance spectrum should reflect the behavior of $(dk/dE_{\text{ion}})^{\circ}$ in the limit of vanishing E_2 field strength.

The bimolecular rate constants themselves can be arrived at by considering expressions for the intensities of the primary and secondary ions with time, the power absorption of the ions and good assumptions concerning the rates at which ions move through the cell.[23] The bimolecular rate constant has the form

$$k = 2m_p\, \xi_s\, I_s/n(m_p\, \xi_s\, I_s + m_s\, \xi_p\, I_p)(\tau'_p + \tau_p) \quad (11)$$

This can be generalized for more than one product ion (from the same reactants) to

$$k_j = 2m_p\, \xi_{sj}\, I_{sj}/nm_{sj}\left[\xi_p\, I_p + \sum_j (m_p/m_{sj})\xi_{sj}\, I_{sj}\right](\tau'_p + \tau_p) \quad (12)$$

A plot of $m_p I_{sj}/m_{sj}\left(I_p + \sum_j \dfrac{m_p}{m_{sj}} I_{sj}\right)$ versus $(\tau'_p + \tau_p)$ should give a straight line with a slope proportional to k_j. This can be accomplished by varying $\dfrac{1}{V}$ where V is the drift potential for both source and resonance regions. In this case, the system must be calibrated against a reaction of known rate constant and be corrected for any pressure change.

B. Interaction Forces

Dunbar[24] has considered theoretically the factors responsible for the change of k with ion energy. Taking a combination of inductive and dispersive potentials for a CH_4^+ ion reacting with a neutral methane molecule, an interaction potential (for small relative kinetic energies $\gtrsim 1$ eV) of the form $V(\mathbf{r}) = V_{\text{ind}} + V_{\text{disp}} = -18.7\mathbf{r}^{-4} - 54.2\mathbf{r}^{-6}$ eV was found. The orbiting cross section displayed an $E^{-0.45}$ dependence over the range 0.01 to

1.0 eV ion energy as compared to that which the $E^{-1/2}$ Langevin r^{-4} potential requires. The rate constant

$$k(T) = \int_0^\infty f(v)\sigma(v)v \, dv \tag{13}$$

was calculated assuming a Maxwellian velocity distribution for both reactants and was found to exhibit an $T/20$ dependence in contrast to the strict temperature independence of the Stevenson–Gioumousis model.[25] This residual temperature dependence gained from the r^{-6} potential, may be in part, that necessary to explain the double resonance response.

Another factor that has been employed in the consideration of ion-molecule rates is the assumed charge-dipole orientation. This rests on the assumption that a neutral molecule possessing a permanent dipole moment $\mathbf{D}$ will orient itself in the position of lowest energy with regard to a charged ion. This would assume greatest importance at low relative kinetic energy E and the classical Stevenson–Gioumousis[25] rate constant is modified by a term involving the neutral's polarizability α,

$$k(E) = \frac{2\pi q}{\mu^{1/2}} \left[\frac{\mathbf{D}}{\sqrt{2E}} + \sqrt{\alpha} \right] \tag{14}$$

Various experimental studies have supported this view,[26,27] other investigations have questioned it.[28,29] Dunbar[29] has examined the reaction $CH_3OH^{+\cdot} + CH_3OH \rightarrow CH_3OH_2^+ + neutral(s)\cdot$ in view of determining the importance of the dipole orientation effect on the rate constant. Using pulsed electron energy and double resonance techniques, the fractional decrease of $CH_3OH_2^+$ intensity was studied as a function of ion energy. The experimental dropoff at low energies was much shallower than that predicted by the "locked in" dipole theory and, in fact, at low energies approached a more energy independent situation predicted by the more sophisticated theory of Dugan and Magee[30] where the dipole averaging effect of molecular rotation is included. ICR is particularly useful in examining factors such as these which may or may not influence reaction rates.

Schaefer and Henis[31] have predicted ion-molecule rate constants as large or small for low energy reactions based on the idea that the probability for a reaction is determined by the nature of the change of electron density around the nuclear centers involved. They predicted eleven low energy reactions and reached agreement with observed values. The exothermic reactions $A^{(\pm)} + BC \rightarrow AB^{(\pm)} + C$ have cross sections greater or less than 1 Å^2 for positive ion reactions depending on whether the electronegativity of B is greater or lesser than that of A. The reverse holds for negative ion reactions. Best agreement is reached when the reactions being compared were simple and similar. Large cross sections are predicted for situations involving little electron density rearrangement. The exothermic reactions

$$N^+ + N_2O \longrightarrow NO^+ + N_2$$
$$N^+ + NO_2 \longrightarrow NO^+ + NO$$

unobserved to date, were predicted to have cross sections greater than 1 Å^2, while the exothermic reaction, also unobserved to date,

$$O^+ + NO_2 \rightarrow O_2^+ + NO$$

was predicted to have a cross section less than 1 Å^2. Most of the thermal energy reactions employed in this study were analyzed by ICR techniques.

Schaefer and Henis[32] have recently extended this concept of the importance of electron density rearrangement to predict *changes* in reaction cross section with increase in vibrational energy of ion or neutral.

V. ICR OF SMALL MOLECULES

Bowers, Elleman and King[33] have studied the kinetics of the diatomic systems H_2, D_2, HD, and N_2 and used the results to test current theories of ion-molecule reactions. Double resonance experiments verified the existence of the following reactions and kinetic analysis led to the thermal energy rate constants.

$$N_2^+ + D_2 \longrightarrow N_2D^+ + D \qquad 1.15 \times 10^{-9} \text{ cc molec}^{-1} \text{ sec}^{-1}$$
$$N_2^+ + H_2 \longrightarrow N_2H^+ + H \qquad 1.41 \times 10^{-9} \text{ cc molec}^{-1} \text{ sec}^{-1}$$
$$N_2^+ + HD \longrightarrow N_2H^+ + D \qquad 0.65 \times 10^{-9} \text{ cc molec}^{-1} \text{ sec}^{-1}$$
$$N_2^+ + HD \longrightarrow N_2D^+ + H \qquad 0.69 \times 10^{-9} \text{ cc molec}^{-1} \text{ sec}^{-1}$$

The result for N_2^+ reacting with H_2 compares favorably with that calculated from the charge induced dipole theory of Gioumousis and Stevenson[25] of 1.12×10^{-9} cc molec^{-1} sec^{-1}. The small isotope effect of 0.94 in the reaction of N_2^+ and HD indicates that the reaction mechanism is stripping and this is consistent with the results of Wolfgang et al. and Mahan et al.[34,35]

The energy dependence of the reaction rates was also considered, in particular that of the reaction of D_2^+ with N_2. There are two possible products which result from charge exchange or deuteron transfer. The Gioumousis–Stevenson[25] theory predicts the rate constant of an ion-molecule reaction to be

$$k = 2\pi q \left(\frac{\alpha}{\mu}\right)^{1/2} \qquad (15)$$

where α is the polarizability of the neutral reactant and μ is the reduced mass of the collision pair. This formalism predicts no dependence of k on ion energy. However, as seen above, a double resonance response has been associated with a change of k with kinetic energy.[22,23,24] In this particular case there is no disagreement with the Gioumousis–Stevenson theory because the charge exchange reaction decreases in rate with increasing ion kinetic energy while the reverse occurs in the D^+ transfer. In fact, in a plot of double resonance intensity, versus the square of the double resonance field strength, the algebraic sum of the two curves corresponding to these two reactions gives a net amount which changes very slowly with ion energy. Similar results were obtained earlier in the $C_2H_4^{+\cdot}$ reactions with C_2H_4.[36] Findings of this nature are in keeping with the phase-space theory of chemical kinetics of Light[37] which, in the strong collision limit, keeps the total reaction probability constant while redistributing particular reaction probabilities as the reactant ion energy is changed.

In the H_2, D_2, and HD system,[38] the ion-molecule reactions were

$$H_2^+ + H_2 \xrightarrow{k_1} H_3^+ + H$$
$$D_2^+ + D_2 \xrightarrow{k_2} D_3^+ + D$$
$$HD^+ + HD \begin{array}{c} \xrightarrow{k_{3a}} H_2D^+ + D \\ \xrightarrow{k_{3b}} HD_2^+ + H \end{array}$$

verified (Figure 13), and the rate constants are shown in Table I. These

TABLE I

Rate Constants in the H_2–D_2–HD System

	$k \times 10^9$ cc molec^{-1} sec^{-1}		Theory[a]	$Q \times 10^{16}$ cm^2		
	ICR[b]	Literature[c]	$\dfrac{k}{v_T}$	$\left(\dfrac{I_s^+}{I_p^+}\right)\left(\dfrac{1}{nL}\right)$		Lit.
$H_2^+ \to H_3^+$	2.11	2.02	2.08	63.0	78.5	83.5,[d] 74.5[e]
$D_2^+ \to D_3^+$	1.60	1.44	1.45	67.5	90.0	79.3[f]
$HD^+ \to H_2D^+$	0.75	0.75 ⎫		27.0	27.0	—
$HD^+ \to HD_2^+$	1.05	0.91 ⎭	1.66	46.0	52.0	—

[a] Reference 25
[b] Reference 38
[c] Reference 42
[d] Reference 43 extrapolated
[e] Reference 40 extrapolated
[f] Reference 44

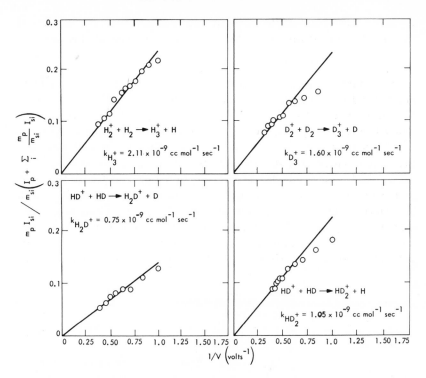

Fig. 13. Kinetic determination of rate constants in hydrogen. V is the drift voltage (Bowers et al.[38]).

reactions have been the subject of much study by a variety of techniques.[28,39-42] The isotopic rate constant ratio k_{3b}/k_{3a} of 1.40 compares favorably with the high pressure mass spectrometric ratio[42] of 1.21. Thermal energy reaction cross sections Q were estimated by two means. First, Q was approximated by $\dfrac{k}{v_T}$, where v_T is the average thermal velocity of the reactant ion, and secondly, Q was equated with $\left(\dfrac{I_s^+}{I_p^+}\right) \cdot \left(\dfrac{1}{nL}\right)$, a ratio of ion intensities modified by a factor containing the number of molecules and the path length of the primary ion. Details were given on the calculation of L and approximate expressions derived for single and double resonance. In addition to the quoted literature values[38] it appears that the linewidth derived ICR *collision* cross section obtained by Wobschall, Fluegge and Graham[17] of 68 Å^2 for H_2^+ in H_2 at high E/P compares favorably to the *reaction* cross section for the reaction $H_2^+ \rightarrow H_3^+$ obtained by Bowers, Elleman and King.[38] This is more meaningful in view of the

G. A. GRAY

finding[38] that charge exchange does not occur between H_2^+ and H_2 as evidenced by the absence of double resonance responses for the reactions

$$D_2^+ + H_2 \rightarrow (H_2^+ + D_2) \quad \text{or} \quad (HD^+ + HD)$$

The double resonance spectra in the H_2, D_2, HD system are complex and dependent on the strength of the irradiating field. In the pure systems H_2 and D_2 the double resonance response for H_3^+ or D_3^+ is energy dependent when H_2^+ or D_2^+ is irradiated. At low E_2 fields the response is positive, which, according to the interpretation of Beauchamp and Buttrill,[22] indicates a bimolecular rate constant which increases with energy. However, at higher field strengths an inversion of the signal is seen at the center of the resonance which may completely change the sign of the double resonance response as ω_2 is swept through the cyclotron frequency of the parent ion. Figure 14 shows this behavior. This could imply that the rate constant increases with ion energy to a maximum value and then decreases with added ion energy. However, Bowers, Elleman and King point out a

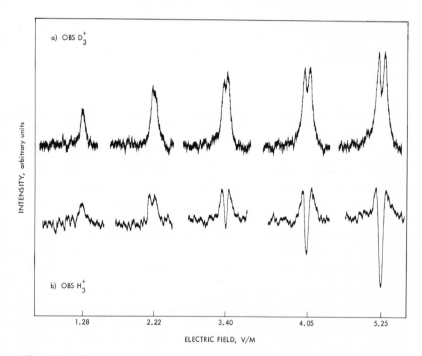

Fig. 14. Double resonance spectra in hydrogen: (a) D_3^+ observed and D_2^+ irradiated, (b) H_3^+ observed and H_2^+ irradiated. The spectra were taken in the pure gases at 3×10^{-6} torr and 31 eV electron energy (Bowers et al.[38]).

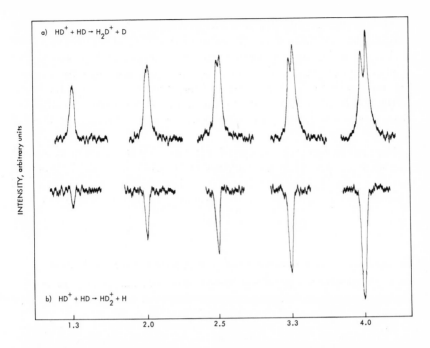

Fig. 15. Double resonance response as a function of irradiating field strength for reactions 3a and 3b. The product ion is observed while the reactant ion is irradiated. A positive response corresponds to an increase in the product ion population and *vice versa*. The spectra were taken at 31 eV electron energy and 2×10^{-6} torr (Bowers et al.[38]).

complex mechanism advocated by Chupka, Russell and Refaey[41] can also fit the experimental results. The kinetic scheme

$$H_2^+ + H_2 \xrightarrow{k_s} H_3^+ + H \quad \text{(stripping)}$$

$$H_2^+ + H_2 \underset{k_c^{-1}}{\overset{k_c}{\rightleftharpoons}} [H_4^+]^* \xrightarrow{k_c'} H_3^+ + H \quad \text{(complex)}$$

$$H_3^+ \xrightarrow{k_D} \text{products} \quad \text{(destruction)}$$

can account for the reversal of the sign of the double resonance response. However, when the HD system is considered, two products are possible for the stripping and complex mechanisms. The double resonance response for H_2D^+ (Figure 15) is analogous to H_3^+ or D_3^+ in H_2 or D_2, that is, going positive with sign reversal at higher E_2. Surprisingly, HD_2^+ has a different behavior—going negative and remaining so even at higher values of E_2. These experiments suggests that at low energies ($KE < 1$ eV) both complex formation and stripping occur, while at higher energies the stripping mechanism takes over. The ratio of $[H_2D^+]/[HD_2^+]$ increases from 0.73 at thermal energies to greater than one at higher HD^+ energies. This implies the dominance of the stripping mechanisms at high energies and the importance of the displacement of the center of mass from the center of the HD bond at HD rotational quantum numbers >0. Comparison of these results with those of Futrell and Abramson[45] allowed calibration of the HD^+ ion energy.

The ion ejection technique was further used by Bowers and Elleman[46] in a study of concurrent ion-molecule reactions in mixtures of argon and nitrogen with H_2, D_2, and HD. The following reactions were reported along with thermal energy rate constants (Table II) for the reactions

TABLE II

Thermal Energy Rate Constants Determined by Ion Ejection Methods

Reaction	$k \times 10^{10}$ cc molecule^{-1} sec^{-1} Bowers and Elleman[46]	Theory[a]	$k_{(exp)}/k_{(theory)}$
$Ar^+ + H_2 \rightarrow ArH^+ + H$	6.83	15.60	0.44
$Ar^+ + D_2 \rightarrow ArD^+ + D$	6.10	11.40	0.53
$Ar^+ + HD \rightarrow ArH^+ + D$	3.14	} 12.90	} 0.47
$Ar^+ + HD \rightarrow ArD^+ + H$	2.98		
$H_2^+ + Ar \rightarrow ArH^+ + H$	12.4	22.50	0.55
$D_2^+ + Ar \rightarrow ArD^+ + D$	9.91	16.3	0.60
$HD^+ + Ar \rightarrow ArH^+ + D$	6.11	} 18.6	} 0.67
$HD^+ + Ar \rightarrow ArD^+ + H$	6.28		
$H_3^+ + Ar \rightarrow ArH^+ + H_2$	3.65	19.1	0.19
$D_3^+ + Ar \rightarrow ArD^+ + D_2$	4.71	13.5	0.35
$N_2^+ + H_2 \rightarrow N_2H^+ + H$	14.10	15.40	0.92
$N_2^+ + D_2 \rightarrow N_2D^+ + D$	12.60	11.20	1.12
$N_2^+ + HD \rightarrow N_2H^+ + D$	5.63	} 13.5	} 0.82
$N_2^+ + HD \rightarrow N_2D^+ + H$	5.46		
$H_2^+ + N_2 \rightarrow N_2H^+ + H$	19.5	22.80	0.86
$D_2^+ + N_2 \rightarrow N_2D^+ + D$	16.10	16.60	0.97
$HD^+ + N_2 \rightarrow N_2H^+ + D$	8.05	} 20.00	} 0.82
$HD^+ + N_2 \rightarrow N_2D^+ + H$	8.24		
$H_3^+ + N_2 \rightarrow N_2H^+ + H_2$	10.30	19.00	0.54
$D_3^+ + N_2 \rightarrow N_2D^+ + D_2$	7.47	13.95	0.54

[a] Ref. 25.

below and other isotopic versions of the reactions:

$$Ar^+ + H_2 \longrightarrow ArH^+ + H$$
$$H_2^+ + Ar \longrightarrow ArH^+ + H$$
$$H_3^+ + Ar \longrightarrow ArH^+ + H_2$$
$$N_2^+ + H_2 \longrightarrow N_2H^+ + H$$
$$H_2^+ + N_2 \longrightarrow N_2H^+ + H$$
$$H_3^+ + N_2 \longrightarrow N_2H^+ + H_2$$

The proton transfer of H_2^+ to Ar displays a negative pulsed double resonance response while the charge exchange of H_2^+ to Ar gives a positive double resonance signal. In contrast to the results in the N_2—H_2^+ system[33] these do not balance each other out. If Light's theory[37] of reaction probabilities holds, then the results imply a significant amount of non-reactive scattering occurring in the H_2^+—Ar system and relatively little in the H_2^+—N_2 system. Deuteron transfer from D_2^+ and D_3^+ to N_2 give opposite ion energy dependence, the former's rate constant decreases with energy and latter's increases with energy. This was related[47] to the possibility that D_3^+ does not have the charge exchange channel available. In addition, nonreactive scattering is prevalent in the latter and absent in the former.

The hydrogen atom abstraction reactions of Ar^+ and N_2^+

$$Ar^+ + H_2 \longrightarrow ArH^+ + H$$
$$N_2^+ + H_2 \longrightarrow N_2H^+ + H$$

lead to strikingly different energy dependences. Ar^+ abstracts a hydrogen atom at a rate increasing with ion energy, in keeping with a stripping mechanism.[38] The absence of an isotope effect for this reaction at thermal energies also supports this thesis. However, the pulsed double resonance response for $N_2^+ \rightarrow N_2H^+$ goes positive then negative, with increasing energy. This suggests a mixed mechanism of possibly both complex formation and stripping as in the H_2^+—H_2 system.[38]

In general, the rate constants of the N_2—$(H, D)_2$ system agree with the charge induced dipole theory,[25] while those in the Ar—$(H, D)_2$ system are smaller than the theoretical value by approximately a factor of two. The reactions of H_3^+ and D_3^+ with Ar and N_2 led to rate constants less than theoretical by factors of two to five, and are consistent with kinetic data on reactions of H_3^+ with saturated hydrocarbons.[47]

Very recently, Bowers and Elleman[48] reexamined the Ar—D_2 system in an effort to understand the nature of the energetics and reactive species present. They find their data consistent with the reaction scheme

$$D_2^+ + D_2 \longrightarrow (D_3^+)^* + D \tag{A}$$

$$\downarrow \xrightarrow{\quad (D_2)^n \quad} (D_3^+)^\dagger + D_2 \tag{B}$$

$$(D_3^+)^* + Ar \longrightarrow ArD^+ + D_2 \tag{C}$$

$$(D_3^+)^\dagger + Ar \xrightarrow{\quad\times\quad} ArD^+ + D_2 \tag{D}$$

$$ArD^+ + D_2 \longrightarrow D_3^+ + Ar \tag{E}$$

$$Ar^+ + D_2 \longrightarrow ArD^+ + D \tag{F}$$

The existence of reaction (D) only above a threshold $(D_3^+)\dagger$ energy demonstrates that proton affinity of $(D_2)\dagger$ is greater than that of Ar. The pressure dependence studies of ArD^+ intensity shows that it increases as a function of D_2 pressure to a maximum and then falls off to zero at 4×10^{-4} torr, as indicated in (E). The fraction of ArD^+ reactant ions increases and levels off as a function of D_2 pressure, indicating a fraction of the ArD^+ is also being formed from D_3^+. The validity of the kinetic analysis based on the above mechanism is borne out by the pressure dependences of the various species involved. The number of collisions necessary to deexcite all the $(D_3^+)^*$ to $(D_3^+)\dagger$ was estimated to be two or three on the average, with ten as a maximum. By observing the threshold double resonance field strength necessary for initiating reaction (D) the estimate of $PA(D_2)\dagger = PA(Ar) + (7 \pm 5)$ kcal/mole was made.

Bowers and Elleman make the suggestion that $(D_3^+)\dagger$ is *not* ground state D_3^+, but an excited version. According to Leventhal and Friedman[49] $(H_3^+)^*$ is formed with approximately 2-eV excess energy from H_2^+ and H_2. They assumed that this was then deactivated to ground state H_3^+ under subsequent collisions. Bowers and Elleman argue that the primary mechanism of energy loss is proton transfer to H_2, leaving vibrationally excited H_2 behind. This process repeats itself until the residual excess energy (possibly up to 10 kcal/mole) is so small that it is insufficient for the product H_2 neutral to be in its first vibrational level after proton transfer (Figure 16).

The above conclusions depend heavily on the presumably small cross sections for vibrational to translational and vibrational to rotational energy transfer relative to vibrational to vibrational relaxation. Thus, the authors warn against the use of $(H_3^+)\dagger$ from reaction (B) for studies in determining the intrinsic proton affinity of H_2. These measurements do place bounds on the proton affinity of H_2 as $80 \pm 4 \leqslant PA(H_2)\dagger \leqslant 88$ kcal/mole when referenced to $(H_3^+)\dagger$ generated in (B). The intrinsic proton

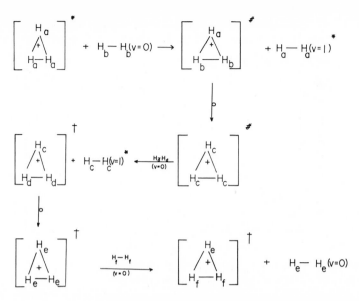

Fig. 16. Energy transfer mechanism for H_3^+ in H_2 (Bowers and Elleman[48]).

affinity has been calculated theoretically by Schwartz and Schaad[50] as > 96 kcal/mole and is not accessible experimentally with the above system. The $PA(D_2)$† of 110 kcal/mole of Leventhal and Friedman[49] does not fall within the error bounds in the ICR study, and Bowers and Elleman attribute this to the uncertainty (perhaps as much as one eV) in Leventhal and Friedman's threshold data.

Using ICR heterodyne experiments, Dunbar[15] has considered in detail the collisional events of N_2^+ in N_2 and $CH_4^{+\cdot}$ in CH_4. Literature values[17,51,52] of the momentum relaxation rate constant K_r for N_2^+ in N_2 are 0.79[17], 0.64,[52] and 0.71 × 10^{-9} cc molec^{-1} sec^{-1}[53] with significant decrease in N_2^+ mobility at E/P values corresponding to only hundredths of an electron volt. The heterodyne experiments give a K_r of

$$(0.90 \pm 0.10) \times 10^{-9} \text{ cc molec}^{-1} \text{ sec}^{-1}$$

(slope of Figure 17) which, when corrected for the average ion energy of 0.02 eV gives $K_r = (0.8 \pm 0.1) \times 10^{-9}$ cc molec^{-1} sec^{-1}. These results confirm that the collision rate is considerably larger than that predicted using only the charged induced dipole polarization[25] for momentum transfer $(0.43 \times 10^{-9}$ cc molec^{-1} sec$^{-1})$ and support a large charge transfer cross section. The rate of momentum relaxation of $CH_4^{+\cdot}$ in CH_4 was found to be 1.25 × 10^{-9} cc molec^{-1} sec^{-1} which, compared to the

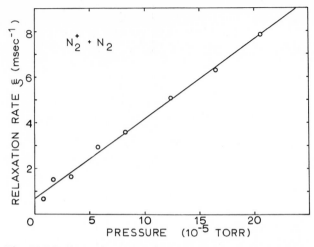

Fig. 17. ICR heterodyne relaxation data for N_2^+ in N_2 with least-squares fit (Dunbar[15]).

total rate of disappearance of $CH_4^{+\cdot}$ in CH_4 of 1.15×10^{-9} cc molec^{-1} sec^{-1} [28,53] indicates that the rate constant for nonorbiting charge transfer is $\leqslant 0.13 \times 10^{-9}$ cc molec^{-1} sec^{-1}. This is very small in comparison to the charge transfer rate constant in the $N_2^+ - N_2$ system and suggests that $CH_4^{+\cdot}$ may differ enough in geometry from CH_4 neutral so that a small Franck-Condon factor makes resonant charge transfer unlikely. The heterodyne technique can be easily extended so that energy dependence of the momentum relaxation rate can be investigated.

Charge exchange in xenon-methane mixtures has been studied by King and Elleman[54] using double resonance techniques. They observe the exothermic reactions.

$$Xe^+(^2P_{1/2}) + CH_4 \longrightarrow CH_4^{+\cdot} + Xe \qquad (\Delta H = -7 \text{ kcal/mole})$$
$$CH_4^{+\cdot} + Xe \longrightarrow Xe^+(^2P_{3/2}) + CH_4 \qquad (\Delta H = -23 \text{ kcal/mole})$$

They also reported a double resonance response for $CH_3^+ \rightarrow Xe^+$ which, at best, is $+15$ kcal/mole endothermic, and suggested the possibility of excited CH_3^+ being responsible. Clow and Futrell[55] reexamined this system and claimed that the $CH_3^+ \rightarrow Xe^+$ charge exchange did not occur, although they supported the reactions involving $CH_4^{+\cdot}$.

It should also be mentioned that Clow and Futrell[8] find no ion energy dependence of rate constants for the reactions

$$Ar^+ + D_2 \longrightarrow ArD^+ + D \qquad (k^\circ = 0.9 \pm 0.1 \times 10^{-9} \text{ cc molec}^{-1} \text{ sec}^{-1}) \quad \text{(A)}$$
$$D_2^+ + Ar \longrightarrow ArD^+ + D \qquad (k^\circ = 1.6 \pm 0.2 \times 10^{-9} \text{ cc molec}^{-1} \text{ sec}^{-1}) \quad \text{(B)}$$
$$D_2^+ + D_2 \longrightarrow D_3^+ + D \qquad \qquad \qquad \qquad \qquad \qquad \qquad \qquad \qquad \text{(C)}$$

from thermal energies to 20 eV using their modified analyzer cell construction. They do, however find energy dependences for the reactions

$$CH_4^{+\cdot} + CH_4 \longrightarrow CH_5^+ + CH_3\cdot \quad (k^\circ = 1.2 \pm 0.1 \times 10^{-9} \text{ cc molec}^{-1} \text{ sec}^{-1})$$
$$CH_5^+ \longrightarrow CH_3^+ + H_2$$
$$CH_3^+ + CH_4 \longrightarrow C_2H_5^+ + H_2 \quad (k^\circ = 1.0 \pm 0.1 \times 10^{-9} \text{ cc molec}^{-1} \text{ sec}^{-1})$$
$$C_2H_5^+ \longrightarrow C_2H_3^+ + H_2$$

Anders[9] has also shown similar rate constant energy dependence for the reactions

$$CH_3^+ + CH_4 \longrightarrow C_2H_5^+ + H_2 \quad (\Delta H = -0.9 \text{ eV})$$
$$CH_3^+ + CH_4 \longrightarrow C_3H_3^+ + 2H_2 \quad (\Delta H = +1.6 \text{ eV})$$

The usual explanations for a false double resonance response are coulombic (rather than chemical) coupling or that ions are lost to the plates because of their increased orbits. This mode of loss is most critical to light ions as their orbits are larger. Hence, for any *negative* double resonance response the latter might be an alternative explanation to a change in the bimolecular rate constant. This *could* then explain any observed double resonance for $D_2^+ + Ar \rightarrow ArD^+ + D$ as an artifact. However, the double resonance responses for the (A) and (C) reactions (Figure 13) obtained by Bowers and Elleman[46] are *positive* at lowest indicating levels (0.01 V/cm). It is difficult to rationalize an *increase* in a product ion by an ion loss mechanism.

Beauchamp[56] has independently obtained the same energy dependence results in the HD system as Bowers and Elleman.[46]

The disparity in these results certainly calls for additional work to clarify the energy dependences under question as well as extreme care in the execution and evaluation of double resonance experiments.

VI. SYSTEMS OF CHEMICAL INTEREST

A. Rate Constants

As has been the case in other techniques of chemical physics, Ion Cyclotron Resonance appears destined for application in more chemical areas involving analysis of complicated ion-molecule reaction schemes, ion structure, kinetics, mechanisms of reactions, and thermodynamics.

Buttrill[57] has developed a formalism for obtaining ion-molecule reaction rate constants and applied it to reactions in methane and acetylene. A complete expression was derived for the power absorption of an ion which undergoes no collisions. This power absorption at low pressure leads to a non-Lorentzian line shape. However, as collisions become important the line shape reverts to Lorentzian in form. A complete expression for a product ion intensity is also given for when ions undergo collision leading

to reaction. Reasonable approximations lead to an expression for the rate
constant.

$$k_i = \frac{m_p{}^2 I_s \left[3 - (\tau_p + 2\tau'_p)n \sum_{j \neq i} k_j \right]}{[nm_s{}^2 I_p(2\tau_p + \tau'_p) + nm_p{}^2 I_s(\tau_p + 2\tau'_p)]}$$

Which reduces in the special case of only one parent ion to

$$k = \frac{3m_p{}^2 I_s}{[nm_s{}^2 I_p(2\tau_p + \tau'_p) + nm_p{}^2 I_s(\tau_p + 2\tau'_p)]}$$

The following reactions were considered and their respective rate constants
evaluated by iterative computer analysis.

$CH_4^+\cdot + CH_4$	$\longrightarrow$	$CH_5^+ + CH_3\cdot$	9.5×10^{-10} cc molec^{-1} sec^{-1}
$CH_2D_2^+\cdot + CH_2D_2$	$\longrightarrow$	$CH_3D_2^+ + CHD_2\cdot$	5.1×10^{-10} cc molec^{-1} sec^{-1}
$CH_2D_2^+\cdot + CH_2D_2$	$\longrightarrow$	$CH_2D_3^+ + CH_2D\cdot$	4.1×10^{-10} cc molec^{-1} sec^{-1}
$C_2H_2^+\cdot + C_2H_2$	$\longrightarrow$	$C_4H_2^+\cdot + H_2$	3.9×10^{-10} cc molec^{-1} sec^{-1}
$C_2H_2^+\cdot + C_2H_2$	$\longrightarrow$	$C_4H_3^+ + H\cdot$	8.3×10^{-10} cc molec^{-1} sec^{-1}

   In a treatment to expand the theory to reaction schemes which allow
reactions of secondary ions to give tertiary and higher order product ions,
Marshall and Buttrill[58] have performed a very complete analysis in which
they determine the number of ions in a particular region of the analyzer
by calculating from the zero pressure ICR lineshape the power absorption
from primary, secondary, and tertiary ions in that region, and then summing such contributions over all regions of the analyzer. The ion populations are then compared to experimental single resonance intensities and
a computer controlled iterative analysis produces all bimolecular rate
constants for the system, assuming a given kinetic scheme. Methyl fluoride
was chosen to illustrate the method. The rate constants, independent of
pressure over the range $2-5 \times 10^{-5}$ torr, and their respective reactions
are

$CH_3F^+\cdot + CH_3F$	$\longrightarrow$	$CH_3FH^+ + CH_2F\cdot$	13.6×10^{-10} cc molec^{-1} sec
$CH_3F^+\cdot + CH_3F$	$\longrightarrow$	$C_2H_4F^+ + HF + H\cdot$	0.96×10^{-10} cc molec^{-1} sec
$CH_3FH^+ + CH_3F$	$\longrightarrow$	$C_2H_6F^+ + HF$	8.0×10^{-10} cc molec^{-1} sec

B. Methane-Hydrogen

   In an effort to explain the "Wilzbach"[59] labeling of methane by radioactive tritium, Inoue and Wexler[60] examined the mixtures CH_4-D_2 and
CD_4-H_2 at relatively high ICR pressures to estimate the contribution
and sequence of ionic reactions using ICR. The proposed reactions as

inferred by double resonance, are depicted in Figure 18. Although double resonance responses were observed for the reactions indicated by the dashed lines, the authors felt that it is more likely that the H_3^+ first reacts to give CD_4H^+, which then reacts to give the methyl species. The ethyl ions are inert to further reactivity with hydrogen, as opposed to the methyl ions. Xenon and nitric oxide were noted to inhibit the production of the tagged ethyl ions initiated by reactions of H_2^+ and H_3^+, but have little effect on their formation by chains of reactions beginning with CD_3^+ from methane. The rate constants for the reactions

$$CH_4^{+\cdot} + CH_4 \xrightarrow{k_1} CH_5^+ + CH_3 \cdot$$

$$CH_4^{+\cdot} + D_2 \xrightarrow{k_2} CH_4D^+ + D \cdot$$

were determined using the formalism developed by Beauchamp[36] extended to the binary system, and these were $k_1 = 3.1 \times 10^{-10}$ cc molec^{-1} sec^{-1} and $k_2 = 9 \times 10^{-12}$ cc molec^{-1} sec^{-1}. k_1 appears to be about a factor of three to four lower than that observed by most workers. The rate constant for the process

$$D_2^+ + D_2 \longrightarrow D_3^+ + D$$

was found to be 5.2×10^{-10} cc molec^{-1} sec^{-1} as compared to 16.0×10^{-10} cc molec^{-1} sec^{-1} obtained by Bowers, Elleman, and King.[38]

Bowers and Elleman[61] have also examined the methane-hydrogen system (Figure 19). Pulsed ion ejection studies showed that virtually all of the CD_4H^+ formed in the CD_4-H_2 system results from H_3^+. Minor amounts of CD_3^+ and CD_2H^+ are also formed from H_3^+. In order to accomodate the pressure dependence of the reactions, they found it necessary to postulate two forms of H_3^+, one an excited species $(H_3^+)^*$ and another, a deactivated form, H_3^+ (note that this interpretation preceded the $Ar-D_2^+$[48]

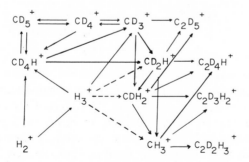

Fig. 18. Reaction scheme in CD_4–H_2 mixtures
(Inoue and Wexler[60]).

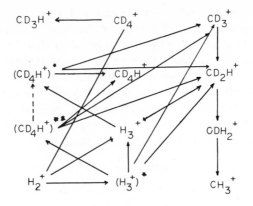

Fig. 19. Reaction scheme in CD_4-H_2 mixtures
(Bowers and Elleman[61]).

research where a slightly different mechanism is invoked). The CD_4H^+ is felt to arise primarily from the deactivated H_3^+, while the CD_3^+ and CD_2H^+ reactions involve some sort of direct displacement reaction from $(H_3^+)^*$. The CD_4H^+ formed from H_3^+ is either collisionally stabilized or decomposes to CD_3^+ or CD_2H^+. The lifetime of the $[CD_4H^+]^*$ ion was determined to be approximately 2×10^{-4} sec. The isotope exchange reactions of $C(D, H)_3^+$ and H_2 were reported although those reactions reported by Inoue and Wexler[60] incorporating *both* hydrogens of the neutral H_2 in the product methyl ions $(CD_3^+ \rightarrow CDH_2^+$ and $CD_2H^+ \rightarrow CH_3^+)$ were not reported. Whereas Inoue and Wexler[60] report a $H_2^+ \rightarrow CD_4H^+$ reaction (Table III) and no other reactions of H_2^+, Bowers and Elleman[61] (Table IV) do not report the $H_2^+ \rightarrow CD_4H^+$ coupling at low or high H_2 partial pressures. They do, however, observe the reactions

$$H_2^+ + CD_4 \longrightarrow CD_4^{+ \cdot} + H_2$$
$$\longrightarrow CD_3^+ + H_2 + D\cdot$$
$$\longrightarrow CD_2H^+ + H\cdot + D_2$$

and all have *positive* double resonance responses. This is also the case, in contrast to those of Inoue and Wexler,[60] for the isotope exchange reactions of the methyl ions and the reactions of H_3^+ to form CD_3^+ and CD_2H^+.

Considering the $H_3^+-CH_4$ reaction in terms of Light's[37] theory, Bowers and Elleman[61] examined the variation of reaction probability with ion energy. CH_5^+ decreased in intensity approximately 2.5 times faster than the CH_3^+ intensity rose, as the H_3^+ ion energy was increased. No significant charge transfer between H_3^+ and CH_4 was observed over the

ION CYCLOTRON RESONANCE

TABLE III

Double Resonance Spectra of CD_4-H_2 System
Inoue and Wexler[60]

Observed ion		Irradiated ion		Double resonance signal intensity (% of the single resonance intensity of observed ion)
m/e	ion	m/e	ion	
15	CH_3^+	3	H_3^+	
		16	CDH_2^+	−15.4
		17	CD_2H^+	−21.6
		18	CD_3^+	−36.2
		20	CD_4^+	+15.4
		21	CD_4H^+	+16.1
		22	CD_5^+	+16.9
16	CH_2D^+	17	CD_2H^+	−23.6
		18	CD_3^+	−40.5
		20	CD_4^+	−5.4
		21	CD_4H^+	−11.2
		22	CD_5^+	−11.8
17	CHD_2^+	18	CD_3^+	−26.7
		20	CD_4^+	−11.5
		21	CD_4H^+	+4.6
		22	CD_5^+	+4.9
18	CD_3^+	20	CD_4^+	−9.5
		21	CD_4H^+	−5.2
		22	CD_5^+	−9.5
20	CD_4^+	18	CD_3^+	−4.1
		21	CD_4H^+	+6.6
		22	CD_5^+	+5.7
21	CD_4H^+	20	CD_4^+	−49.5
		22	CD_5^+	−27.5
22	CD_5^+	20	CD_4^+	−66.1
		21	CD_4H^+	−8.7
34	$C_2D_5^+$	45	CH_3^+	+3.26
		16	CDH_2^+	+3.70
		17	CD_2H^+	−4.45
		18	CD_3^+	−15.0
		20	CD_4^+	+14.1
		21	CD_4H^+	+20.0
		22	CD_5^+	+42.2

TABLE III—(*Contd.*)

Observed ion		Irradiated ion		Double resonance signal intensity (% of the single resonance intensity of observed ion)
m/e	ion	m/e	ion	
33	$C_2D_4H^+$	16	CDH_2^+	-11.5
		17	CD_2H^+	-20.5
		18	CD_3^+	-11.1
		20	CD_4^+	$+13.7$
		21	CD_4H^+	$+20.8$
		22	CD_5^+	$+41.0$
32	$C_2D_3H_2^+$	15	CH_3^+	-15.8
		16	CH_2D^+	-24.8
		17	CHD_2^+	-10.8
		18	CD_3^+	-7.7
		20	CD_4^+	$+13.5$
		21	CD_4H^+	$+21.4$
		22	CD_5^+	$+43.5$
31	$C_2D_2H_3^+$	15	CH_3^+	-35.0
		16	CH_2D^+	-21.4
		17	CHD_2^+	-12.6
		18	CD_3^+	-8.7
		20	CD_4^+	$+15.5$
		21	CD_4H^+	$+21.8$
		22	CD_5^+	$+48.5$

same energy range. Aquilanti and Volpi[47] measured disappearance rate constants for H_3^+ impacted on CH_4, and found a rate constant approximately 50% of the classical orbiting rate constant (7.5×10^{-10} cc molec^{-1} sec^{-1}). This implies a significant number of nonreactive scattering collisions and supports the double resonance energy dependence.

Although Bowers and Elleman's pulsed ion ejection spectra show that essentially all the CD_4H^+ ions come from H_3^+, Inoue and Wexler's double resonance results for CD_4H^+ in a $CD_4–H_2$ mixture show that 49.5% of the CD_4H^+ ion single resonance intensity is eliminated upon irradiating $CD_4^{+\cdot}$, and 27.5% of the same intensity is lost when CD_5^+ is irradiated, apparently indicating that at least 77% of the CD_4H^+ intensity comes from reactions involving $CD_4^{+\cdot}$ and CD_5^+. Since supposedly there is little charge exchange between H_3^+ and CD_4 it is difficult to reconcile these viewpoints. Another factor that is not clear is the extent of the isotope exchange reaction

$$CD_5^+ + H_2 \longrightarrow CD_4H^+ + HD$$

TABLE IV

Double Resonance Results for Methane Hydrogen Mixtures:
Bowers and Elleman[61]

System	Observed		Irradiated	Neutral[a] Products	Response[b]
	amu	species			
$CD_4-H_2^{(c)}$	20	CD_4^+	H_2^+	H_2	+
	18	CD_3^+	H_2^+	$HD + H$	+
	17	CD_2H^+	H_2^+	$D_2 + H$	+
$CD_4-H_2^{(d)}$	21	CD_4H^+	H_3^+	H_2	−
	20	CD_4^+	H_3^+		+
	19	$\begin{cases} H_3O^{+(f)} \\ (CD_3H^+) \end{cases}$	H_3^+		−
	18	CD_3^+	H_3^+	$H_2 + HD$	+
	17	CD_2H^+	H_3^+	$H_2 + D_2$	+
	16	CDH_2^+	H_3^+	−	+
	15	CH_3^+	H_3^+	−	+
	15	CH_3^+	CH_2D^+	HD	+
$CH_3D-D_2^{(e)}$	19	$CH_3D_2^+$	D_3^+	D_2	−
	20	$\begin{cases} H_2DO^{+(f)} \\ (CH_2D_3^+) \end{cases}$	D_3^+		−

[a] Assumed. In all cases the neutral reactant is CD_4 or CH_3D except when CH_2D^+ is irradiated, where H_2 is the neutral reactant.

[b] A positive response indicates more product ion is formed when the reactant is irradiated and a minus sign indicates less.

[c] 8×10^{-6} torr CD_4 and 8×10^{-6} torr H_2.

[d] 8×10^{-6} torr CD_4 and 4.5×10^{-4} torr H_2.

[e] 8×10^{-6} torr CD_4 and 4×10^{-4} torr D_2.

[f] Most likely product is H_3O^+ or H_2DO^+.

Inoue and Wexler's[60] double resonance results point out the importance (27.5%) of this process in the production of CD_4H^+ and, indeed, use this as an explanation of $[T_2]^{3/2}$ dependence in the rate of tritiation of methane in "Wilzbach" labeling of methane. However, the pulsed ion ejection studies[61] of CD_4H^+ in CD_4-H_2 show no CD_5^+ responsible for the production of CD_4H^+. This was further apparent by the absence of an appropriate entry in the double resonance data.[61]

These conflicting reports on the ionic processes in methane-hydrogen systems point out the need for some reinvestigation and careful analysis.

C. Olefins

Olefins display an exceedingly rich ion-molecule chemistry, and have been the center of several ICR investigations. Bowers, Elleman and

Beauchamp[36] have examined the ion-molecule reactions in ethylene and display the reactive events as follows:

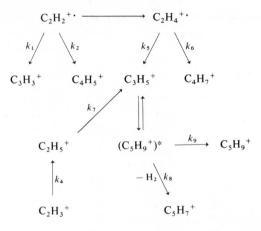

They studied in detail the mechanism responsible for formation of the $C_5H_7^+$ and $C_5H_9^+$ ions, and estimated lifetimes of the collision complex $(C_5H_9^+)^*$ from a kinetic analysis. The relative rates for the above reactions were measured and for $k_5 \equiv 1.00 : k_1 = 0.45$, $k_2 = 0.27$, $k_3 = 0.45$, $k_4 = 0.82$, and $k_6 = 0.10$. The complete randomization of hydrogen atoms in the $(C_4H_8^+)^*$ collision complex in reaction (5) of this display was indicated from the analysis of the isotopic distribution in a C_2H_4–C_2D_4 mixture. A specific isotope effect was observed in the reactions

$$C_2H_4^{+\cdot} + C_2D_4 \xrightarrow{\ k_{5b}\ } C_3H_4D^+ + CD_3\cdot$$

$$C_2D_4^{+\cdot} + C_2H_4 \xrightarrow{\ k_{5e}\ } C_3HD_4^+ + CH_3\cdot$$

where k_{5b} decreases faster than k_{5e} with increasing ion energy. It was felt that this may have resulted from a decreased lifetime of the complex formed at higher impacting ion energies leading to nonequilibration of the isotopic hydrogen.

The same condensation products as found in R2 and R3 of this display are formed in the reverse combination $C_2H_4^{+\cdot} + C_2H_2$. Since the internal energy of the complex is 20 kcal/mole greater, the relative rate k_2/k_3 may be different, and indeed the ratio is 4.4 instead of 1.55. The reaction of $C_2H_2 + C_2D_4^{+\cdot}$ gave good indication of H,D randomization in the intermediate complex for the C_3 products, but showed a significant isotope effect for the C_4 products, reflecting a favored tendency to lose H over D.

Through a consideration of double resonance intensities as a function of ion energy, Bowers, Elleman, and Beauchamp[36] showed, in certain cases where several product ions originate from a single reactant, that the partitioning of reaction products depends strongly on ion energy while the total number of reactive encounters remains essentially unchanged with ion energy. This was demonstrated in the case of $C_2H_4^{+\cdot}$ reacting with C_2H_4 and $C_2H_2^{+\cdot}$ reacting with C_2H_4.

Henis[62] extended the study of olefins from ethylene through hexene. Addition reactions giving higher molecular weight products were most generally noted, and reactivities and product distributions were found to be related to the molecular structures of the reactants. Reactants were of the form

$$P(i)^+ + M(j) \xrightarrow{\sigma_{ij}} \Sigma S(ij)^+ + \Sigma N$$

Here $P(i)^+$ is the i-olefin molecular ion and $M(j)$ is the j-olefin molecule. σ_{ij} is the total reactive cross section for $P(i)^+ + M(j)$ giving secondary ion(s) $S(ij)^+$ and the necessary neutrals for mass balance. The pure systems give primarily addition reactions where one, two, or three carbon atoms are added to the parent molecule. All 1-olefins add or transfer one carbon unit more efficiently than do 2- or 3-olefins. The $P(2)^+ + M(1)$ reactions are at least on order of magnitude more probable than the $P(1)^+ + M(2)$ reactions with respect to carbon unit addition in all of the mixed olefin systems. As a rule, 1-olefin neutrals are generally more reactive than the 2- and 3-olefin neutrals, regardless of the structure of the reacting ions.

Henis presented a mechanism for reactions of olefin parent ion with neutral olefins in which it is assumed that the positive charge is located at a secondary carbon site without rearrangement before reaction. This secondary carbon bonds to one member of the neutral's double bond, producing an excited dimer ion which can rearrange and fragment to products. Fragmentation occurs primarily at the tertiary carbon in the intermediate and does not involve more than one bond. The product distributions definitely point out that extensive randomization does not occur in the higher alkenes.

In an examination of isotope exchange, Henis[63] considered the products of ion-molecule reactions involving deuterated and undeuterated olefins. The results again indicated, especially in the larger olefins, that extensive exchange between H and D does not take place. Where exchange does occur, all products arising from a common intermediate show significantly different degrees of H—D exchange.

O'Malley and Jennings[64] investigated the ionic reactions in electron-impacted acetylene gas using ICR. The two major ionic processes in acetylene are

$$C_2H_2^{+\cdot} + C_2H_2 \longrightarrow C_4H_2^{+\cdot} + H_2$$

$$C_2H_2^{+\cdot} + C_2H_2 \longrightarrow C_4H_3^{+} + H\cdot$$

The ratio $I(C_4H_2^{+\cdot})/I(C_4H_3^{+})$ was determined to be 0.455 ± 0.020, in agreement with mass spectrometric studies. The ions C_3H^{+}, $C_3H_2^{+\cdot}$, and C_4H^{+} were found to be formed by reaction of $C^{+\cdot}$, CH^{+}, and $C_2^{+\cdot}$ respectively by double resonance studies. $C_4H_2^{+\cdot}$ was determined to originate from C_2H^{+}. The authors noted a reversal in sign of the normally negative double resonance signals for several reactions. They did not ascribe this to a reversal in the energy dependence of the rate constant or a mixture of mechanisms, but to a loss of ions to walls upon rf heating (the "flat" analyzer cell construction was used in this study).

Collisional stabilization of ion-molecule condensation products was the subject of an investigation by Buttrill[68], focusing on the reactions of secondary ions

$$C_4H_2^{+\cdot} + C_2H_2 \longrightarrow (C_6H_4^{+\cdot})^* \xrightarrow{C_2H_2} C_6H_4^{+\cdot} + C_2H_2^{*}$$

$$C_4H_3^{+} + C_2H_2 \longrightarrow (C_6H_5^{+})^* \xrightarrow{C_2H_2} C_6H_5^{+} + C_2H_2^{*}$$

in acetylene, and similar reactions in ethylene,

$$C_3H_5^{+} + C_2H_4 \longrightarrow (C_5H_9^{+})^* \Big\langle \begin{array}{l} \xrightarrow{C_2H_4} C_5H_9^{+} + C_2H_4^{*} \\ \\ \searrow C_5H_7^{+} + H_2 \end{array}$$

At a pressure of 5×10^{-5} torr, essentially all of the $C_2H_2^{+\cdot}$ ions are consumed with most of the intensity being in the $C_4H_2^{+\cdot}$, $C_4H_3^{+}$, $C_6H_4^{+\cdot}$, and $C_6H_5^{+}$ ions. There are minor amounts of $C_8H_6^{+\cdot}$ and $C_8H_7^{+}$ ions observed. Double resonance links $C_4H_2^{+\cdot}$ to $C_6H_4^{+\cdot}$. $C_4H_3^{+}$ can contribute to $C_6H_4^{+\cdot}$ through the reaction

$$C_4H_3^{+} + C_2H_2 \longrightarrow C_6H_4^{+\cdot} + H\cdot$$

but this apparently results only from excited $C_4H_3^{+}$, since at 12.5 eV there is no double resonance response. Double resonance further indicates that the reactions

$$C_6H_5^{+} + C_2H_2 \longrightarrow (C_8H_7^{+})^* \xrightarrow{C_2H_2} C_8H_7^{+}$$
$$\searrow^{-H\cdot}$$
$$C_6H_4^{+\cdot} + C_2H_2 \longrightarrow (C_8H_6^{+\cdot})^* \xrightarrow{C_2H_2} C_8H_6^{+\cdot}$$

occur. Mixing of hydrogens of ions and collisionally stabilizing neutrals proceeds to an appreciable extent. This was demonstrated in C_2H_2/C_2D_2 mixtures. For example, double resonance showed that the reaction

$$C_4D_2H^+ + C_2D_2 \longrightarrow (C_6D_4H^+)^* \xrightarrow{C_2D_2} C_6D_5^+ + C_2HD$$

occurs. Data were presented showing the contributions of the various isotopically labeled C_4 ions to each of the collision stabilized C_6 products in this mixture.

The reactions found in ethylene by Bowers, Elleman, and Beauchamp[36] were substantiated in this study. In addition, by consideration of the ratio $[C_5H_9^+]/[C_5H_7^+]$ as a function of pressure, Buttrill[65] was able to calculate a minimum lifetime of 3.1×10^{-4} sec for $(C_5H_9^+)^*$ and 1.4×10^{-3} sec for $(C_5D_6^+)^*$ with respect to dissociation into $C_5H_7^+ (C_5D_7^+)$ and $H_2(D_2)$. Hydrogen exchange also was found to occur in ethylene reactions, for example,

$$C_3H_3D_2^+ + C_2H_2D_2 \longrightarrow (C_5H_5D_4^+)^* \xrightarrow{C_2H_2D_2} C_5H_4D_5 + C_2H_3D^*$$

Complete scrambling does not occur on every collision, but evidence was presented for two types of stabilizing collisions, one in which the $(C_5(H, D)_9^+)^*$ complex is stabilized without any hydrogen mixing, and another in which an intimate $C_7(H, D)_{13}^+$ complex is formed, resulting in complete hydrogen scrambling.

As an extension of the work on unsaturated systems, Bowers et al.[66] detailed the ion-molecule reactions of allene ($H_2C=C=CH_2$) and propyne ($CH_3-C\equiv CH$). These systems have been studied using high pressure mass spectrometry by Myher and Harrison[67] and their results agree for the most part with the ICR studies. One exception is the role of the secondary $C_3H_5^+$ ions. Myher and Harrison state that $C_3H_5^+$ (3.4 eV ion exit energy) reacts primarily by hydride ion abstraction

$$C_3H_5^+ + C_3H_4 \longrightarrow C_3H_3^+ + C_3H_6 \qquad \Delta H \cong 0$$

while the ICR results (thermal ion energy) show that essentially the only reaction involving $C_3H_5^+$ is

$$C_3H_5^+ + C_3H_4 \longrightarrow C_6H_7^+ + H_2$$

Bowers et al. explain this discrepancy as an effect of $C_3H_5^+ \rightarrow C_3H_3^+$ being slightly endothermic and requiring translationally hot $C_3H_5^+$ ions to proceed. The difference in the $[C_3H_5^+]/[C_6H_7^+]$ ratio between the two studies was explained in terms of the energy dependences of the reactions

$$C_3H_4^{+\cdot} + C_3H_4 \xrightarrow{k_1} C_3H_5^+ + C_3H_3\cdot$$

$$C_3H_4^{+\cdot} + C_3H_4 \xrightarrow{k_2} C_6H_7^+ + H\cdot$$

Double resonance studies showed that k_1 increases and k_2 decreases with ion energy. This difference is responsible for the higher $[C_3H_5^+]/[C_6H_7^+]$ ratio in the study of Myher and Harrison.

The reaction of $C_3H_4^{+\cdot} + C_3H_4$ to give a $(C_6H_8^{+\cdot})^*$ complex ion was interpreted in terms of a four centered intermediate in which the four centers were the four unsaturated carbons. The isotopic data for ions of ethylene and acetylene showed that $C_2H_2D_2$ was lost approximately three times the random prediction in allene and one half the random prediction in propyne from the $(C_6H_4D_4^{+\cdot})^*$ complex ion.

The cross-reactions of ethylene and allene and propyne were reported as

$$
\begin{array}{ccc}
 & \text{Propyne} & \text{Allene} \\
\end{array}
$$

$$
C_2H_4^{+\cdot} + C_3H_4 \quad \begin{array}{c} \overset{0.98}{\nearrow} \\ \underset{1.00}{\searrow} \end{array} \quad \begin{array}{c} \overset{0.58}{\nearrow} \\ \underset{1.00}{\searrow} \end{array} \quad \begin{array}{c} C_4H_5^+ + CH_3\cdot \\ \\ C_5H_7^+ + H\cdot \end{array}
$$

$$
C_3H_4^{+\cdot} + C_2H_4 \quad \begin{array}{c} \overset{0.46}{\nearrow} \\ \underset{1.00}{\searrow} \end{array} \quad \begin{array}{c} \overset{0.21}{\nearrow} \\ \underset{1.00}{\searrow} \end{array} \quad \begin{array}{c} C_4H_5^+ + CH_3\cdot \\ \\ C_5H_7^+ + H\cdot \end{array}
$$

The relative rates stated are only valid in the particular pair cited. These results indicate that the $C_3H_4^{+\cdot}$ ions from propyne and allene are different in structure or internal energy. From detailed isotopic experiments it was evident that methyl groups play little part in the reaction dynamics and that the four center complex appeared to have general applicability in ion-molecule reaction in mixtures of 3-carbon unsaturates.

D. Reactions of Hydrocarbon Derivatives

Ion Cyclotron Resonance has served as the tool to unravel the ion-molecule reactions of H_2S with ethylene and acetylene.[68] In contrast to reactions of hydrocarbons, Buttrill[68] found that hydrogen rearrangement was relatively unimportant with respect to fragmentation of the ion-molecule complexes. In the $H_2S-C_2H_4$ system, protonated H_2S and $C_3H_5^+$ were observed. The cross reactions

$$H_2S^{+\cdot} + C_2H_4 \longrightarrow CH_2SH^+ + CH_3\cdot$$
$$C_2H_4^{+\cdot} + H_2S \longrightarrow CH_2SH^+ + CH_3\cdot$$

were observed and confirmed by double resonance, as well as the minor processes,

$$H_2S^{+\cdot} + C_2H_4 \longrightarrow CH_2S^{+\cdot} + CH_4$$
$$C_2H_4^{+\cdot} + H_2S \longrightarrow CH_2^{+\cdot} + CH_4$$

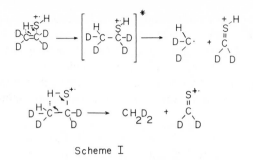

Scheme I

Fig. 20. Reaction scheme for $H_2S^{+\cdot} + C_2D_4$
$\rightarrow CD_2S^{+\cdot} + CH_2D_2$ (Buttrill[68]).

Small amounts of CHS^+ were presumably formed by $H_2S^{+\cdot}$ and $C_2H_4^{+\cdot}$. Labeled reactants showed that two of the three hydrogens in the CH_2SH^+ ion originate from the same carbon in the C_2H_4.

Reactions were reported in the $H_2S-C_2H_2$ system as

$$H_2S^{+\cdot} + C_2H_2 \longrightarrow HCS^+ + CH_3\cdot$$
$$H_2S^{+\cdot} + C_2H_2 \longrightarrow C_2H_3S^+ + H\cdot$$

whereas the analogous reactions where the charge is on the C_2H_2 do not proceed. These are examples of *exothermic* ion-molecule reactions which do *not* occur. The suggested mechanisms for reaction of $H_2S^{+\cdot}$ with C_2H_4 and C_2H_2 are presented in Figures 20 and 21. The lack of hydrogen scrambling was taken as indicative of the importance of delocalized positive

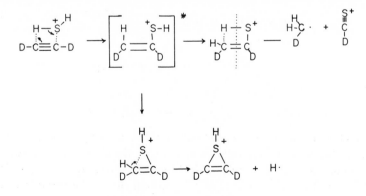

Scheme II

Fig. 21. Reaction scheme for $H_2S^{+\cdot} + C_2D_2 \rightarrow (DCS^+ + CH_3\cdot)$ and
$(D_2C_2SH^+ + H\cdot)$ (Buttrill[68]).

charge in promoting scrambling and the localization of a positive charge by a heteroatom in preventing scrambling.

The first ICR investigation of a "chemical" system was that of chloroethylene by Beauchamp, Anders, and Baldeschwieler.[69] The reaction scheme was analyzed by double resonance, and is illustrated in Figure 22. In addition, reactions in a mixture of methane and chloroethylene were reported as

$$CH_3^+ + C_2H_3Cl \longrightarrow C_3H_5^+ + HCl$$
$$C_3H_5^+ + C_2H_3Cl \longrightarrow C_5H_7^+ + HCl$$
$$C_2H_5^+ + C_2H_3Cl \longrightarrow C_4H_7^+ + HCl$$
$$C_2H_3^+ + CH_4 \longrightarrow C_3H_5^+ + H_2$$

The most general result is the predominance of reactions of the type $A^+ + C_2H_3Cl \rightarrow AC_2H_2^+ + HCl$ which can be thought of as a bimolecular electrophilic addition-elimination process. In the reaction of $C_2H_3Cl^{+\cdot}$ with C_2H_3Cl, the chlorine in any of the product ions comes with equal probability from the charged or neutral reactant.

Fluoroethylene was used by O'Malley and Jennings[70] in a recent ICR study for comparison to the chloroethylene system. At 3×10^{-6} torr $C_3H_5^+$, $C_2H_3F^{+\cdot}$, $C_3H_4F^+$, $CH_2CF_2^{+\cdot}$, $C_5H_7^+$, $C_4H_5^{+\cdot}$, $C_3H_3F_2^+$, and $C_5H_6F^+$ were observed. The $CH_2CF_2^{+\cdot}$ was attributed to a minor amount of CH_2CF_2 impurity. Double resonance results pointed out the reactions,

$$
C_2H_3F^{+\cdot} + C_2H_3F \begin{cases} C_3H_3F_2^+ + CH_3 \cdot \\ C_3H_4F^+ + CH_2F \cdot \\ C_3H_5^+ + CHF_2 \cdot \\ C_4H_5F^{+\cdot} + HF \end{cases}
$$

$$C_3H_5^+ + C_2H_3F \longrightarrow C_5H_7^+ + HF$$
$$C_3H_4F^+ + C_2H_3F \longrightarrow C_5H_6F^+ + HF$$

Although many of the reactions eliminate HF, the product ions from these reactions are weak in intensity, as opposed to the chloroethylene system. The major product ions at 1×10^{-4} torr are $C_3H_5^+$, $C_3H_4F^+$, $C_5H_7^+$, $C_3H_3F_2^+$, and $C_5H_6F^+$, with $C_3H_3F_2^+$ predominating. The general process $A^+ + C_2H_3F \rightarrow AC_2H_2^+ + HF$, however appeared to be the major source of the higher mass ions where A^+ is other than the molecular ion. In general, the reactions of fluoroethylene parallel ethylene more closely than chloroethylene.

Another fluorinated hydrocarbon, hexafluoroethane, was studied using

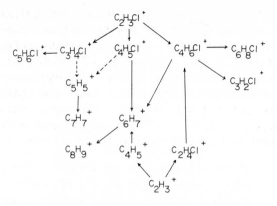

Fig. 22. Reaction scheme in chloroethylene (Beauchamp et al.[69]).

ICR techniques by King and Elleman[71] in order to explain the abnormally high $C_2F_5^+$ yield under electron impact. The reactions

$$CF_3^+ + C_2F_6 \longrightarrow C_2F_5^+ + CF_4$$
$$CF_2^+\cdot + C_2F_6 \longrightarrow C_2F_5^+ + CF_3\cdot$$
$$C_2F_5^+ + M \longrightarrow CF_3^+ + CF_2 + M$$
$$CF_2^+\cdot + C_2F_6 \longrightarrow CF_3^+ + C_2F_5\cdot$$
$$CF_3^+ + C_2F_6 \longrightarrow CF_2^+\cdot + neutrals$$

were observed to occur and were confirmed by double resonance. No collision induced decomposition of $C_2F_5^+ \rightarrow CF_2^+\cdot$ could be observed using double resonance.

Methanol was the subject of one of the earliest ICR investigations.[72] In this study, Henis documented the complex reactions shown in Figure 23.

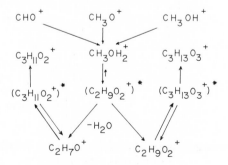

Fig. 23. Reaction scheme in methanol (Henis[73]).

The $C_2H_7O^+$ ion was shown to have the structure $(CH_3)_2OH^+$ by comparing the reactions of $C_2H_7O^+$ ions derived from dimethyl ether and ethanol with methanol. At low pressures, fragmentation of the $(C_2H_5O_2^+)^*$ ion predominates and the stable protonated dimethyl ether ion results. This can then react to produce $(C_3H_{11}O_2^+)^*$. At high pressures $(C_2H_9O_2^+)^*$ is collisionally stabilized and can react further to give $C_3H_{13}O_3^+$.

Gray has applied ICR techniques to a nitrogen-containing organic, acetonitrile (Figure 24).[73,74] The 23 ion-molecule reactions found illustrate the complexity of the reaction scheme. The most important reactions are

$$CH_3CN^{+\cdot} + CH_3CN \longrightarrow CH_3CNH^+ + CH_2CN\cdot$$
$$CH_2CN^+ + CH_3CN \longrightarrow C_3H_4N^+ + HCN$$
$$CH_3CNH^+ + CD_3CN \longrightarrow CD_3CNH^+ + CH_3CN$$

The first reaction has been a subject of prior investigation, in particular in regard to its mechanism. Martin and Melton[75] viewed it as hydrogen atom abstraction, while Moran and Hamill[76] considered it as an example of proton transfer. The data of Shannon and Harrison[77] indicated the possibility of both mechanisms occurring. The double resonance experiments, denoted schematically in Figure 25, prove the proton transfer mechanism.

The second reaction produces the second most intense ion-molecule reaction product ion. Studies involving mixtures of natural abundance,

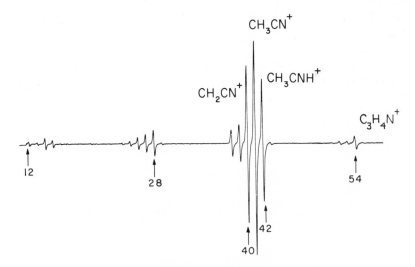

Fig. 24. Single resonance spectrum of acetonitrile at 2×10^{-6} torr using 60 eV ionizing voltage (Gray[73]).

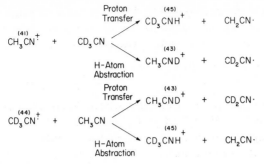

Predicted Double Resonance:

Proton Transfer	H-Atom Abstraction	Both	Experimental
$41^+ \longrightarrow 45^+$	$44^+ \longrightarrow 45^+$	$44^+ \longrightarrow 43^+, 45^+$	$41^+ \longrightarrow 45^+$
$44^+ \longrightarrow 43^+$	$41^+ \longrightarrow 43^+$	$41^+ \longrightarrow 43^+, 45^+$	$44^+ \longrightarrow 43^+$

Fig. 25. Possible reactions between the chemical species $CH_3CN^{+\cdot}$ and neutral acetonitrile in $1:1$ CH_3CN-CD_3CN, the predicted double resonance, and the observed double resonance (Gray[74]).

^{15}N-, and ^2H-labeled acetonitrile showed that the nitrogen and hydrogen eliminated in the (elements of) HCN come exclusively from the ionic and neutral reactants respectively. If the neutral fragments come off as an HCN molecule, the reaction must involve a means for the CN of the ionic reactant to approach the neutral molecule's methyl group after complex formation. This can be accomplished by a cyclic form (II) rather than the linear form (I) in Figure 26. Alternatively, if the process is two step in nature (I) can consecutively lose $H\cdot$ and $CN\cdot$. The labeling experiments certainly pointed out the high probability that the site of attack is the basic nitrogen atom of neutral acetonitrile. The collision complex has mass 81 amu, a value for which a resonance is found in the single resonance spectrum, so that collisional stabilization of the complex is possible.

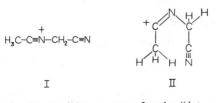

I   II

Fig. 26. Possible structures for the "intermediate" ion in the reaction $CH_2CN^+ + CH_3CN \rightarrow C_3H_4N^+ + HCN$ (Gray[78]).

$C_3H_4N^+$ is also formed by $CH_2^{+\cdot}$. Double resonance studies (Figure 27) of CD_3CN/CH_3CN analogous to the above showed that the hydrogen atom only comes from the neutral reactant and suggests the mechanism

$$CH_2^{+\cdot} + CH_3CN \longrightarrow [H_3C-C\equiv N^+-CH_2\cdot]^*$$
$$\longrightarrow H_2C=C=N^+=CH_2 + H\cdot$$

In addition to the proton transfer reactions described above, CH_2CN^+, $CHCN^{+\cdot}$, H_2CN^+, $HCN^{+\cdot}$, $C_2H_2^{+\cdot}$, and $CH_2^{+\cdot}$ all proton transfer to CH_3CN, in agreement with thermochemical predictions. Only CH_3^+ is predicted not to proton transfer to CH_3CN, as a 10 kcal/mole endothermicity is forecast. However, even at a 0.02 V/cm double resonance field strength, a negative response is evident. To back up this evidence a 50:1 CH_3I/CD_3CN mixture was examined, particularly with regard to the reactions

$$CH_3^+ + CD_3CN \longrightarrow CD_3CNH^+ + CH_2$$
$$CD_3^+ + CD_3CN \longrightarrow CD_3CND^+ + CD_2$$

Even though CH_3^+ outnumbered CD_3^+ by 50:1, the $CH_3^+ \rightarrow CD_3CNH^+$ double resonance signal was virtually negligible ($S/N \sim 2$) compared to the relatively strong ($S/N \sim 6$)$CD_3^+ \rightarrow CD_3CND^+$ double resonance signal. This points out the excited nature of the CH_3^+ and indicates that

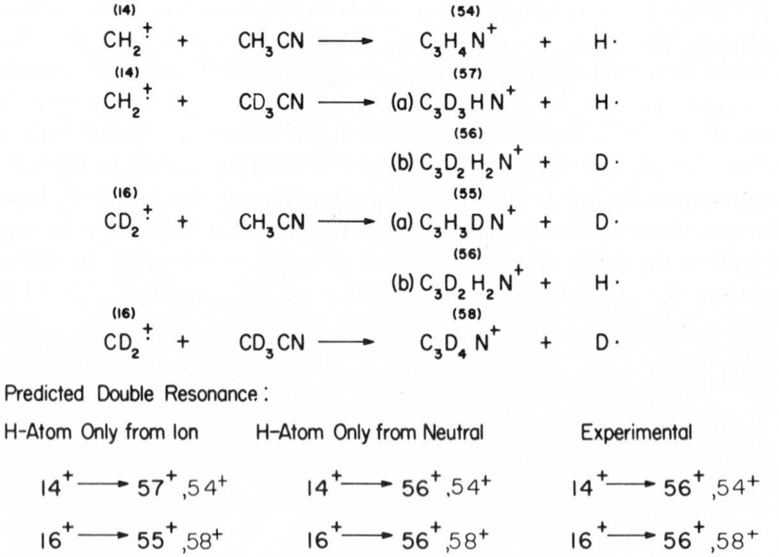

Fig. 27. Possible reactions between the chemical species $CH_2^{+\cdot}$ and neutral acetonitrile in 1:1 $CH_3CN\text{-}CD_3CN$, the predicted double resonance, and the observed double resonance (Gray[74]).

virtually all of the CH_3^+ ions resulting from electron impact decomposition of CH_3CN are excited.

The last of the above principal reactions, proved by considering CH_3CN/CD_3CN mixtures, points out the thermoneutral proton transfer of CH_3CNH^+ to CH_3CN. Similar experiments indicated that the proton transfer remains unique and no methyl protons are consumed in reactions of this type. This confirms the intuitive assignment of $CH_3—C≡N^+—H$ as that of protonated acetonitrile. The proton affinity of 186 kcal/mole as found by Moran and Hamill[76] was confirmed by running competing proton transfer reactions with acetone ($PA = 188$ kcal/mole) and acetaldehyde ($PA = 184$ kcal/mole).

Negative ion spectra of acetonitrile (Figure 28) showed a large CN^- resonance, with a sizable amount of C_3N^- formed by CN^- (in CH_3CN or CD_3CN) in the reaction

$$CN^- + CH_3CN \longrightarrow C_3N^- + NH_3\cdot$$

Mixtures of H_2O and CH_3CN lead to spectra in which CH_2CN^- appears as well as CNO^- (or OCN^-) presumably by the reactions

$$OH^- + CH_3CN \longrightarrow CH_2CN^- + H_2O$$
$$OH^- + CH_3CN \longrightarrow CNO^- \text{ (or } OCN^-) + CH_4\cdot$$

Gas phase ion-molecule reactions in HCN were investigated by Huntress, Baldeschwieler, and Ponnamperuma[78] in a recent study. The principal reactions are

$$HCN^{+\cdot} + HCN \longrightarrow H_2CN^+ + CN\cdot$$
$$HCN^{+\cdot} + HCN \longrightarrow HC_2N_2^+ + H\cdot$$

with $HCN^{+\cdot}$ and H_2CN^+ accounting for over 90% of the ionization. The hydrogen atom in $HC_2N_2^+$ comes equally well from ionic or neutral reactant, as shown in HCN/DCN mixtures. The reactions reported by Inoué and Cottin[79] include the above two as well as

$$CN^+ + HCN \longrightarrow C_2N_2^{+\cdot} + H\cdot$$
$$CH^+ + HCN \longrightarrow HC_2N^{+\cdot} + H\cdot$$
$$CH^+ + HCN \longrightarrow C_2N^+ + H_2$$
$$C^{+\cdot} + HCN \longrightarrow C_2N^+ + H\cdot$$

and were all observed by double resonance. Other reactions reported by Inoué and Cottin[79] are

$$C_2N^+ + HCN \longrightarrow C_3N_2^{+\cdot} + H\cdot$$
$$C_2NH^{+\cdot} + HCN \longrightarrow C_3N_2H^+ + H\cdot$$
$$C_2N_2^{+\cdot} + HCN \longrightarrow C_3N_3^+ + H\cdot$$

and were not observed by double resonance or single resonance.

NEGATIVE ION SINGLE RESONANCE SPECTRA

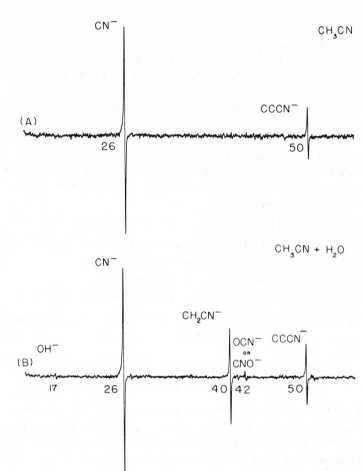

Fig. 28. (a) Single resonance negative ion spectrum of acetonitrile at 10^{-5} torr using 3 eV ionizing voltages; (b) Single resonance negative ion spectrum of an acetonitrile-water mixture at 10^{-5} torr using 8 eV ionizing voltage (Gray[74]).

Dunbar[80] has extended ICR measurements to the inorganic system B_2H_6. Both negative and positive ions were observed. BH_4^-, $B_2H_7^-$, $B_3H_8^-$, $B_4H_9^-$, $B_5H_{10}^-$, and $B_6H_9^-$ were noted in the negative ion spectra. Only the reaction

$$BH_4^- + B_2H_6 \longrightarrow B_2H_7^- + BH_3$$

gave strong double resonance signals. Detailed analysis showed that the three boron atoms in the reactants appear with equal probability in the neutral product.

Mixtures of H_2O and B_2H_6 showed no ion containing both boron and oxygen atoms, but a large $B_2H_5^-$ ion is produced. Double resonance shows that it is made by

$$OH^- + B_2H_6 \longrightarrow (BH_4^-)^* + \text{neutrals(s)}$$
$$(BH_4^-)^* + B_2H_6 \longrightarrow B_2H_5^- + BH_3 + H_2$$
$$\longrightarrow B_2H_7^- + BH_3$$

$(BH_4^-)^*$ is required since $B_2H_5^-$ is not observed in B_2H_6 alone. A single resonance spectrum of $H_2S—B_2H_6$ show BH_4^-, $B_2H_7^-$, $HSBH_3^-$, $B_2H_5S^-$, $B_2H_6S^-$, and BS_2^-, with BS_2^- predominating.

Positive ion spectra of B_2H_6 show B_2, B_3, B_4, B_5, and B_6 species involving from one to ten hydrogens in the various ions. The most significant of the exothermic ion-molecule reactions are of the form

$$A^+ + B_2H_6 \longrightarrow AB_2H_2^+ + 2H_2$$

where A is $B_2H_2^{+\cdot}$, $B_2H_3^+$, $B_2H_4^{+\cdot}$, $B_3H_5^{+\cdot}$, $B_4H_4^+$, and $B_4H_6^{+\cdot}$. Dunbar gives a useful treatment of double resonance intensities for a single reaction where the reactant and product ions are present in several isotopic forms.

E. Thermochemical Considerations

Beauchamp[23] has developed the useful technique of determining proton affinities of neutral molecules using ICR. The proton affinity of M is defined as the negative of the enthalpy change for the reaction

$$M + H^+ \longrightarrow MH^+$$
$$PA(M) = \Delta H_f(M) + \Delta H_f(H^+) - \Delta H_f(MH^+)$$

In an investigation by Buttrill and Beauchamp[22] values of 164 ± 4 kcal/mole and 178 ± 2 kcal/mole were determined for the proton affinities of H_2O and H_2S respectively. Limits on the proton affinities were realized by competition experiments such as

$$M_1H^+ + M_2 \rightleftharpoons M_1 + M_2H^+ \cdot$$

The reaction will proceed exothermically to produce the M_i of lower proton affinity. Hence by bracketing a given M_i by neutrals with corresponding higher and lower proton affinities it is possible to calculate a M_i proton affinity. All this is based on the assumption that the ions are in their ground states, but with enough cross checks reasonably reliable values can be obtained.

The reactions which best limit the H_2S proton affinity are

$$H_2S^{+\cdot} + CD_4 \longrightarrow H_2DS^+ + CD_3\cdot \quad \Delta H_f(H_3S^+) \leqslant 184 \text{ kcal/mole}$$
$$H_3S^+ + C_3H_6 \longrightarrow H_2S + C_3H_7^+ \quad \Delta H_f(H_3S^+) \geqslant 181 \text{ kcal/mole}.$$

Alternatively, for water the limiting reactions are

$$C_2H_5^+ + H_2O \longrightarrow C_2H_4 + H_3O^+ \quad \Delta H_f(H_3O^+) \leqslant 149 \text{ kcal/mole}$$
$$H_3O^+ + C_2H_4 \longrightarrow H_2O + C_2H_5^+ \quad \Delta H_f(H_3O^+) \geqslant 149 \text{ kcal/mole}.$$

Both reactions were observed with negative double resonance responses.

Brauman and Blair[81] utilized the above techniques but applied them to negative ion reactions of alkoxide ions.

$$ROH + R'O^- \longrightarrow RO^- + R'OH$$

The direction of the reaction depends on the relative acidities and double resonance techniques were used to test the exo- and endo-thermicity for the $R'O^- \rightarrow RO^-$ reactions. The following order of acidities was found: neopentyl alcohol > t-butyl > isopropyl > ethyl > methyl > water, and t-butyl $\approx$ n-pentyl $\approx$ n-butyl > n-propyl > ethyl. The reverse order holds true for basicity. The above order is reverse to that found in solution and brings into question the often invoked ability of a methyl group to "donate" electrons inductively. Brauman and Blair explain the results in terms of stabilization of charge by the greater polarizability of a methyl or methylene group over hydrogen. This stabilization is manifested thermodynamically in an increase in the electron affinity of the corresponding radical and will be more important in the gas phase than in solution where hydrogen bonding and specific solvation can occur. The acidity of toluene was found to be approximately equal to that of ethanol.

Using the same techniques, Brauman and Blair[82] determined that acetyl cyanide is a stronger acid than HCN and that acetylacetone is a stronger acid than acetyl cyanide. Toluene was found to be a stronger acid than water.

Gas phase acidities of primary and secondary amines were also investigated by Brauman and Blair in a recent study.[83] The order of acidities is diethyl > neo-pentyl $\geqslant$ t-butyl $\geqslant$ dimethyl $\geqslant$ isopropyl > n-propyl > ethyl > methylamine > ammonia. Under the assumption of constant RNH—H bond strength, the large alkyl groups increase acidity by increasing the electron affinity of the corresponding radical. The difference of 6 kcal/mole between the N—H bond strength of methyl- and dimethyl-amines, if assumed to hold for all primary and secondary alkyl amines, coupled with the approximately equal acidities of t–butylamine and dimethylamine, predicts that the electron affinity of the t–butylamino radical is about 6 kcal/mole greater than that of the dimethylamino radical.

Recently, Brauman and Smyth[84] utilized photodetachment techniques in ICR to obtain electron affinities of neutral radicals by using a conventional tungsten light source focused through a glass window in the flange positioned on the end of the analyzer cell vacuum container. The spectrometer was operated in the negative ion mode and the negative ion intensity monitored in the normal fashion. Since the analyzer cell was, in this case, open ended, the light beam path was parallel to the direction of ion drift and the decrease in negative ion intensity due to the process was easily

$$A^- + h\nu \longrightarrow A + e^-$$

monitored. Photodetachment of SH^- did not occur for wavelengths longer than 5840 Å, allowing an estimate of 2.28 ± 0.15 eV for the vertical detachment energy. No Cl^- detachment was observed, since the tungsten source did not produce enough quanta of energy, 3.6 eV ($Cl\cdot$ electron affinity). In most cases, the vertical detachment energy can be taken as an estimate of the neutral's electron affinity and in some cases should approach it very closely.

Holtz and Beauchamp[85] used competing reactions to establish a 185 ± 4 kcal/mole proton affinity for PH_3. This is 22 kcal/mole less than NH_3, and is comparable to the difference in solution basicities. Solvation thus appears to play a minor role in determining their relative basicities. The hydrogen affinities, or $M-H^+$ bond strengths, were calculated as 128 and 102 kcal/mole for NH_4^+ and PH_4^+ respectively.

The norbornyl cation has been the center of a long standing controversy in organic chemistry, in particular to its electronic structure. The central issue is whether the ion in Figure 29 is best considered as a transition state between two rapidly equilibrating classical ions or as an intermediate, lower in energy than either classical ion.

Kaplan, Cross, and Prinstein[86] have used ICR to study the relative stabilities of sets of bicyclic olefins and ketones, free of solvation effects, in the hope of determining an answer to the above question. Heats of formation of the above species were calculated from proton affinity measurements and the differences between these heats of formation noted (Figure 30).

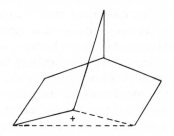

Fig. 29. Norbornyl cation (non-classical version).

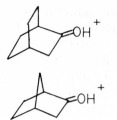

$$\Delta H_f(MH)^+ - \Delta H_f(M)$$

159 kcal/mole

161

171.5 ± 1.5

160 ± 1

199.5 ± 1.5

193 ± 1

Fig. 30. Protonated and neutral bicyclics and the differences in their heats of formation (Kaplan et al.[86]).

The differences show that the bicylo-(2·2·1)-heptyl cation compared to its saturated system is approximately 6 kcal/mole more stable than the bicyclo-(2·2·2)-octyl cation compared to its hydrocarbon. An even more pronounced effect is evident for the olefin systems. The keto systems show small difference between the (2·2·2) and (2·2·1) bicyclics, indicating the localization of the charge on the oxygen. The larger effects in the hydrocarbon ions were interpreted in terms of the importance of sigma delocalization of positive charge as in Figure 29. The ICR-derived heat of formation of 179 kcal/mole for the bicyclo-(2·2·1)-heptyl cation is approximately 20 kcal/mole lower in energy than that obtained by Klopman quantum-mechanically.[87]

F. Applications of ICR to Ionic Structure

A recurring problem in mass spectroscopy is the assignment of a structure to an observed ion of known composition from a variety of chemically possible structures. ICR has been shown by Beauchamp and Dunbar[88] to be helpful in this area through experiments aimed at comparing ion-molecule reactions of the unknown ion with reactions of ions of the same composition generated from known structures.

The $C_2H_5O^+$ (m/e 45) ion has been observed in many mass spectral decompositions and can have a variety of structures including

$$CH_3-\overset{+}{O}=CH_2 \qquad CH_3-\overset{\overset{+}{O}H}{\underset{H}{C}} \qquad H_2C-\overset{\overset{H}{\overset{|}{O^+}}}{CH_2} \quad CH_3CH_2O^+ \quad {}^+CH_2CH_2OH$$

	I	II	III	IV	V
ΔH_f(kcal/mole)[89]	170 ± 5	143 ± 2	170 ± 4	211 ± 2	—

For various reasons, previous investigators have felt I-III most probable.

Studies on isotopic variants of $CH_3OC_2H_5$ show that the m/e 45 ion is of structure I and undergoes hydride ion abstraction and methyl cation transfer, with $k_1/k_2 = 16/1$,

$$CH_3O^+=CH_2 + CH_3OC_2H_5 \xrightarrow{k_1} CH_3OCH_3 + CH_3O^+=CHCH_3$$

$$CH_3O^+=CH_2 + CH_3OC_2H_5 \xrightarrow{k_2} CH_2O + (CH_3)_2O^+-C_2H_5$$

Hydride abstraction occurs only from the ethyl group ($\Delta H_{rx} = -24$ kcal/mole) rather than the methyl group ($\Delta H_{rx} = -17$ kcal/mole). The protonated parent ion, formed only from the parent ion (at 13.0 eV), was found to be produced about equally by proton transfer and hydride abstraction.

Protonated acetaldehyde (II) was investigated directly and also as a decomposition product of 2-propanol and ethanol.

The main reaction involving $C_2H_5O^+$ in ethanol is the proton transfer reaction[23]

$$CH_3CHOH^+ + C_2H_5OH \longrightarrow CH_3CHO + C_2H_5OH_2^+$$

The electron impact ionization of 2-propanol leads to α-cleavage and the production of protonated acetaldehyde (II). Double resonance analysis detailed the ion-molecule chemistry of 2-propanol, and pressure dependence and electron energy dependence studies were performed. $C_2H_5O^+$ proton transfers to 2-propanol to give an excited $[(CH_3)_2CHOH_2^+]^*$ ion which then can decompose to $C_3H_7^+ + H_2O$ if the excitation is >25 kcal/mole. The $C_3H_7^+$ production is dependent on pressure and electron energy. The validity of structure II for protonated acetaldehyde is demonstrated by the various reactions it undergoes, which are characteristic of protonated carbonyl compounds.

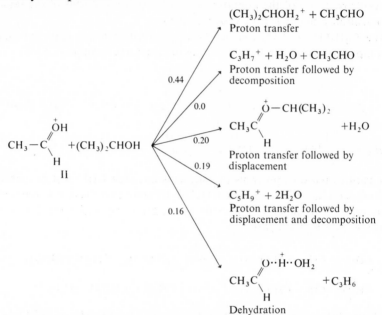

The protonated parent ion also undergoes the dehydration reaction and the proton transfer followed by displacement of H_2O. The product of the latter reaction can also decompose to $C_5H_9^+ + H_2O$. At high pressures (10^{-3} torr), the CH_3CHOH^+ ions lead predominately to proton-bound trimer of 2-propanol and the proton-bound dimer of isopropyl ether and 2-propanol. Labeled experiments again show the importance of the 1,2-elimination of H_2O in the dehydration reaction of CH_3CHOH^+ and 2-propanol. 2-butanol gives the same m/e 45^+ ion structure as shown by the similar product distributions, reactions, and variation of reactivity with electron energy.

Ions of structure (III) were observed to undergo reactions

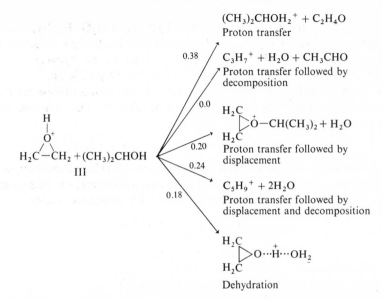

The similarity of the reaction schemes for III and II, coupled with the significantly higher heat of formation of III, indicates that III may rearrange to II after formation.

In any case where a $C_2H_5O^+$ ion is generated from a neutral M, reaction of the $C_2H_5O^+$ with the parent should easily determine between I and II or III. Observation of methyl cation transfer or hydride abstraction reactions would indicate I, while proton transfer reaction would point toward II or III. Beauchamp and Dunbar use these generalizations in examining the m/e 45 ion in dimethyl carbonate. High resolution mass spectra show that m/e 45 is 85% $C_2H_5O^+$ and 15% CO_2H^+. Double resonance experiments identify m/e 45 as producing one of the major product ions m/e 105. This product is assigned as trimethoxycarbonium ion and is produced by the reaction

$$CH_3\overset{+}{-}\overset{}{O}=CH_2 + \underset{CH_3O}{\overset{CH_3O}{\diagdown}}C=O \longrightarrow (CH_3O)_3C^+ + CH_2O$$
$$\text{I}$$

The reactant ion structure is assigned as I on the basis of this methyl cation transfer reaction. The small double resonance response linking m/e 45 to m/e 91 (protonated dimethyl carbonate) can be assigned to CO_2H^+. CH_3O^+ is the major precursor to m/e 91.

Applying these techniques, Beauchamp and Dunbar have found structure

I to occur for the $C_2H_5O^+$ ion in CH_3OCH_3, $CH_3OCH_2CH_3$, tetra-hydropyran, $CH_3OCH_2CH_2OCH_3$, and dimethylcarbonate; structure II (or III) in CH_3CH_2OH, $(CH_3)_2CHOH$, 2-butanol, diethylether, ethyl-isopropylether, diisopropylether, and α-methyl tetrahydrofuran.

Kaplan has utilized double resonance techniques to confirm mass spectral fragmentation pathways.[89] This was accomplished by observing collision induced fragmentation,

$$M^+ + \text{neutral} \longrightarrow [M-A]^+ + A\cdot + \text{neutral}$$

and relating this to unimolecular decomposition following electron impact. Using a $10:1$ mixture of nitrogen to p-chloroethylbenzene, the fragmentation scheme below was deduced from the double resonance studies.

$$C_8H_9{}^{37}Cl^{+\cdot} \searrow \quad \longrightarrow C_7H_6{}^{37}Cl^+$$
$$\searrow C_8H_9{}^+$$
$$C_8H_9{}^{35}Cl^{+\cdot} \nearrow \quad \longrightarrow C_7H_6{}^{35}Cl^+$$

Since there is a much higher probability for ion-molecule reactions than for the existence of the proper metastables in mass spectroscopy, the application of double studies to fragmentation patterns appears to be a promising approach.

Aliphatic alcohols undergo ionic dehydration in the gase phase.[72,90] In a recent study,[90] Beauchamp examined the ion-molecule reactions of t-butyl alcohol with emphasis on the reactions

$$RH^+ + R'OH \longrightarrow R\cdots H^+\cdots OH_2 + R''$$

Both $(CH_3)_3COH^{+\cdot}$ and $(CH_3)_3COH_2^+$ react with t-butyl alcohol to give ions which can be assigned the structure $R\cdots H^+\cdots OH_2$, and this suggested that the dehydration process is associated only with the presence of a labile proton on oxygen. The water molecule is displaced in a subsequent collision, for example,

$$(CH_3)_2CO\cdots\overset{+}{H}\cdots OH_2 + (CH_3)_3COH$$

$$\downarrow$$

$$(CH_3)_3C-\overset{\cdot}{\underset{\underset{H}{\diagdown}}{O}} \quad \overset{+}{H}\cdots OC(CH_3)_2 \quad + H_2O$$

Experiments with $(CD_3)_3CHOH$ show that it is HDO which is lost and a simple 1,2-elimination was suggested.

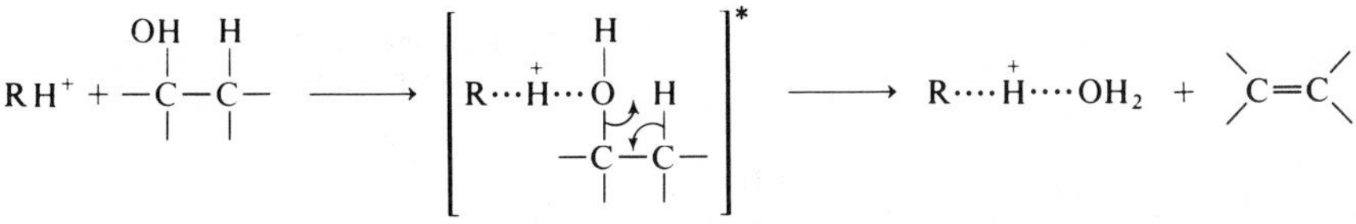

Caserio and Beauchamp[91] extended these ideas to 2-butanol, for which they used deuterium labeling in an attempt to determine the structures of the C_4H_8 hydrocarbon products and the reactant ions ROH^+. Pulsed double resonance confirmed that $CH_2=CHOH^+$, $CH_3CH=OH^+$, $CH_3CH_2CH=OH^+$, and $CH_3CH(O^+H_2)CH_2CH_3$ all participate in the dehydration reaction. They again show that dehydration occurs predominantly by 1,2- elimination of water. This result is required by the lack of observation of HDO elimination in $CH_3CD(OH)CH_2CH_3$ and the occurrence of elimination in $CD_3CH(OH)CD_2CH_3$. A small amount of HDO elimination occurs in $CD_3CH(OH)CH_2CH_3$ pointing out a $(1,2):(1,3)$ elimination ratio of $5:1$. Through the use of the two diastereomers of 2-butanol, $CH_3CH(OH)C(CH_3)HD$, the authors find equal distribution of isotopic products from all of the above reactant ions, but a $2:1$ preference for H_2O elimination over HDO. The cis-1,2 elimination of water results in equal amounts of *cis-* and *trans*-2-butene if the reactant ion, associated with the $-OH$ group of the alcohol, assumes a position *between* the two smallest substituents on the adjacent carbon. The unlabeled 2-butanol is then predicted to result in equal amounts of *cis-* and *trans-*2-butene. The observed $2:1$ preference for H_2O elimination reflects a significant isotope effect in the formation of 2-butene.

Diekman et al.[92] have also used ICR to observe reactions as a means of identifying a particular ionic structure from several possibilities. The $C_3H_6O^{+\cdot}$ ion arising from electron impact and electron impact decomposition of aliphatic ketones has been considered in detail. The central question is the structure of the $C_3H_6O^{+\cdot}$ ion from various reactions. The following scheme lists the several modes of preparation for this ion and the assumed structures.

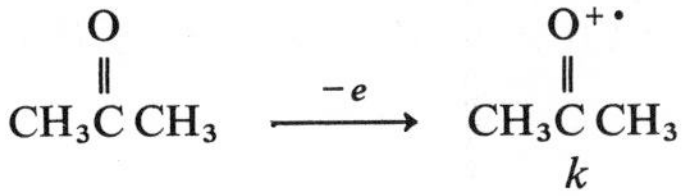

Keto species

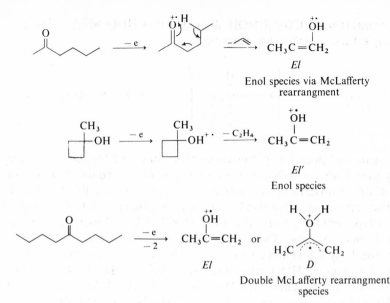

Enol species via McLafferty
rearrangment

El'
Enol species

El   D
Double McLafferty rearrangment
species

The authors compared reactivities by observing reactions of $C_3H_6O^{+\cdot}$ ions with parent neutrals and other neutrals, using labeled and natural abundance materials. The ion-molecule reactions they find which serve to distinguish between the enol and keto isomers of the $C_3H_6O^{+\cdot}$ ion are presented below

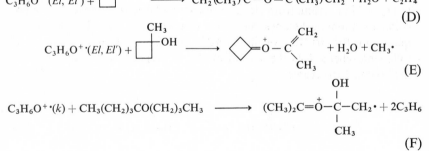

$$C_3H_6O^{+\cdot}(k) + CH_3COCH_3 \longrightarrow (CH_3)_2C{\overset{+}{=}}O{-}COCH_3 + CH_3\cdot \quad (A)$$

$$C_3H_6O^{+\cdot}(k) + CH_3CO(CH_2)_3CH_3 \longrightarrow CH_3C\overset{+}{O}(CH_2)_3CH_3 + CH_3COCH_3 \quad (B)$$

$$C_3H_6O^{+\cdot}(E1) + CH_3CO(CH_2)_3CH_3 \longrightarrow CH_3C\overset{+}{(OH)}(CH_2)_3CH_3 + CH_3COCH_2\cdot \quad (C)$$

$$C_3H_6O^{+\cdot}(El, El') + \longrightarrow CH_2(CH_3)C{-}\overset{+}{O}{=}C(CH_3)CH_2\cdot + H_2O + C_2H_4 \quad (D)$$

$$C_3H_6O^{+\cdot}(El, El') + \longrightarrow \quad + H_2O + CH_3\cdot \quad (E)$$

$$C_3H_6O^{+\cdot}(k) + CH_3(CH_2)_3CO(CH_2)_3CH_3 \longrightarrow (CH_3)_2C{\overset{+}{=}}\overset{OH}{\underset{CH_3}{\overset{|}{C}}}{-}CH_2\cdot + 2C_3H_6 \quad (F)$$

$C_3H_6O^{+\cdot}(E1, E1') + CH_3(CH_2)_3CO(CH_2)_3CH_3$

$$\longrightarrow CH_3(CH_2)_3C(\overset{+}{O}H)(CH_2)_3CH_3 + CH_3COCH_2\cdot$$

$$(G)$$

It can be seen that the keto species undergoes condensation followed by methyl elimination (A), charge transfer (B), and reaction (F), while the enol species can undergo proton transfer, as well as reactions (D) and (E). The ion resulting from the double McLafferty rearrangement from 5-nonanone did not participate in reaction of type (A) or (B) but does participate in reactions of type (C). In fact, this ion is indistinguishable in all reactive behavior from the species $E1$ and $E1'$ resulting from a single McLafferty rearrangement or electron impact decomposition of 1-methyl-cyclobutanol. The authors concluded that (1) the enol and double Mc-Lafferty ions have the same structure ($E1$); or (2) that this is the first instance of ions of different structure exhibiting identical reactivity; or (3) that the lifetimes in the ICR cell are so much longer than in conventional mass spectrometers that D isomerizes to $E1$ prior to undergoing ion-molecule reactions; or (4) this same long lifetime results in complete decomposition of D, if formed at all, prior to any further reaction.

In a subsequent study,[93] Eadon, Diekman, and Djerassi probed deeper into the double McLafferty rearrangement and the structure of the $C_3H_6O^{+\cdot}$ ion. Using 4-nonanone-1,1,1-d_3 (I) the following general scheme could apply.

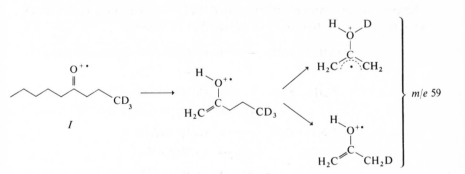

Double resonance of the (M + 2) ion showed no contribution from m/e 59, demonstrating absence of deuterium transfer. m/e 59 does proton transfer to the parent molecule, however. Studies on the neutral 4-non-anone-7,7-d_2 (II)

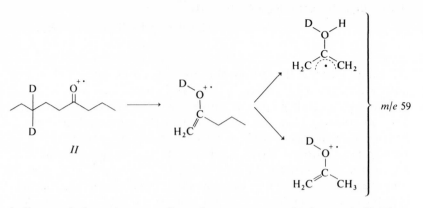

m/e 59

indicate *only* deuterium transfer and no proton transfer from *m/e* 59. Thus, no large isotope effects are important, and a substantial part, at least, of the double McLafferty ion must correspond to the enolic structure *E*1. Even if a symmetrical *D* structure is formed initially which then isomerizes *E*1, this must make it equally likely for H or D to be left bonded to oxygen (assuming no large isotope effect) so that proton and deuterium transfer should be observed.

Wilkins and Gross[94] used ICR to distinguish between the $C_8H_8^{+\cdot}$ ions from styrene and cyclooctatetraene (COT). The two single resonance spectra showed differences above *m/e* 104 with peaks appearing at *m/e* 208 and 130 for styrene. No similar peaks were observed in the spectrum of COT. Double resonance studies indicated that the following exothermic reactions take place in pure styrene at 20 eV and 1×10^{-5} torr.

$$C_8H_8^{+\cdot} + C_8H_8 \longrightarrow C_{16}H_{16}^{+\cdot}$$
$$\longrightarrow C_{10}H_{10}^{+\cdot} + C_6H_6$$
$$C_8H_7^+ + C_8H_8 \longrightarrow C_{10}H_9^+ + C_6H_6$$
$$\longrightarrow C_{10}H_8^{+\cdot} + C_6H_6 + H\cdot$$
$$C_6H_6^{+\cdot} + C_8H_8 \longrightarrow C_8H_8^{+\cdot} + C_6H_6$$
$$\longrightarrow C_{14}H_{12}^{+\cdot} + H_2$$
$$\longrightarrow C_{13}H_{11}^+ + CH_3\cdot$$
$$\longrightarrow C_{13}H_{10}^{+\cdot} + CH_4$$
$$\longrightarrow C_{12}H_{10}^{+\cdot} + C_2H_4$$

Positive double resonance responses (assumed to be indicative of endo-thermic reactions) were noted for

$$C_8H_8^{+\cdot} + C_8H_8 \longrightarrow C_{14}H_{12}^{+\cdot} + C_2H_4$$
$$\longrightarrow C_{13}H_{11}^+ + C_3H_5\cdot$$
$$\longrightarrow C_{13}H_{10}^{+\cdot} + C_3H_6$$
$$\longrightarrow C_{12}H_{10}^{+\cdot} + C_4H_6$$

Reactions observed in COT were identical except for the above mentioned production of m/e 208 and 130. Use of specifically labeled styrene alone and in mixtures of COT showed that the m/e 104 ions of styrene and COT are different. Further, styrene-α-$\underline{d}$ ion reacting with styrene-α-$\underline{d}$ gave 50% $C_{10}H_8D_2^{+\cdot}$ and 50% $C_{10}H_9\bar{D}^{+\cdot}$. Conversely, *trans*, -$\beta$-$\underline{d}$styrene gave 100% $C_{10}H_8D_2^{+\cdot}$ and no $C_{10}H_9D^{+\cdot}$. The complex ion $(C_{16}H_{14}D_2^{+\cdot})^*$ from which the above products result was suggested to have either the structure of a 1,2-diphenylcyclobutane or that of 1-phenyltetralin (Figure 31). Neither *cis*- or *trans*-1,2-diphenylcyclobutane give spectra which resemble the ICR data for the complex ion $(C_{16}H_{14}D_2^{+\cdot})^*$. No m/e 130 peak (corresponding to loss of C_6H_6) is found either. This ion is abundant in the spectrum of 1-phenyltetralin along with the occurrence of the proper (208 → 130, m^* 81.2) metastable in its mass spectrum. The authors conclude that styrene molecular ion can either isomerize to $COT^{+\cdot}$ (important at high ionizing voltages) or react to form the 1-phenyltetralin radical ion which can then, in part, eliminate benzene.

VII. CONCLUSION

It is apparent that ICR has become a valuable tool for the study of ion-molecule systems. The variety of systems and techniques spans virtually all of chemistry and much of physics. The detailed work on single systems allows many inferences to be drawn concerning reaction mechanism and

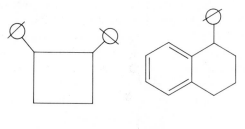

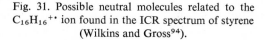

1,2-Diphenylcyclobutane   1-Phenyltetralin

Fig. 31. Possible neutral molecules related to the $C_{16}H_{16}^{+\cdot}$ ion found in the ICR spectrum of styrene (Wilkins and Gross[94]).

the factors responsible for their importance. Studies involving more complex organic molecules point out the utility in gaining an understanding of the sources and factors influencing chemical reactivity. As the store of rate data increases, conclusions concerning reactivity in terms of the molecular properties of the reactants will become more meaningful, and serve to assess the medium contribution in analogous solution studies.

The differences concerning ion-molecule reactions inferred from double resonance results should alert investigators to exercise extreme care in performing and intepreting double resonance experiments. Published double resonance data must include relevant experimental conditions such as double resonance field strength, electron beam emission current, cell configuration, pressure, and preferably drift and trapping potentials.

Although the number of laboratories participating in ICR research is still small, the achievements and promise of the technique certainly foretell an exciting and profitable future.

Note Added in Proof: In addition to the investigations described above, several other articles have appeared recently. Kriemler and Buttrill[95] have extensively explored the positive and negative ion-molecule reactions and proton affinity of ethyl nitrate. The positive ion chemistry of phosphine was studied by Eyler[96] who found P, P_2, P_3, and P_4 species along with 19 ion-molecule reactions. Huntress and Elleman have considered the ion-molecule chemistry of the methane-ammonia system.[97] Herod et al.[98] have compared the mass-spectrometric zero-field pulsing technique and various ICR techniques in determining product ion distributions of ion-molecule reactions.

Clow and Futrell have examined the kinetic energy dependence of ion-molecule reaction rates in methane, hydrogen, and rare gas-hydrogen systems.[99] The reaction of O^+ with CO_2 was studied by Mosesman and Huntress[100] while further investigations of gas-phase acidities of alcohols were reported by Brauman and Blair.[101] Phosphines remain interesting as a result of McDaniel, Coffman, and Strong's[102] report on the proton affinity of trimethylphosphine and Holtz'[103] coverage of a wide array of experiments on phosphine. Eadon et al.[104] applied ICR to elucidate the structure of the $C_3H_6O^{+\cdot}$ ion formed by the double McLafferty rearrangement. Rearrangements of molecular ions of dialkyl-N-nitrosamines were the subject of an ICR study by Billets, Jaffé, and Kaplan[105], while Jaffé and Billets also used ICR to examine the electrical effects of free-radical groups.[106] ICR was central in research on electrophilic aromatic substitution reactions, conducted by Benezra et al.[107], and on nucleophilic displacement reactions, conducted by Holtz et al.[108]

References

1. L. R. Anders, J. L. Beauchamp, R. C. Dunbar, and J. D. Baldeschwieler, *J. Chem. Phys.*, **45**, 1062 (1966); J. D. Baldeschwieler, *Science*, **159**, 263 (1968).
2. D. Wobschall, J. R. Graham, Jr., and D. P. Malone, *Phys. Rev.*, **131**, 1565 (1963).
3. D. Wobschall, *Rev. Sci. Instrum.*, **36**, 466 (1965).
4. Varian Associates, Palo Alto, California.
5. J. M. S. Henis and W. Frasure, *Rev. Sci. Instrum.*, **39**, 1772 (1968).
6. R. T. McIver, Jr., *Rev. Sci. Instrum.*, **41**, 146 (1970).
7. J. L. Beauchamp and J. T. Armstrong, *Rev. Sci. Instrum.*, **40**, 123 (1969).
8. R. P. Clow and J. H. Futrell, *Proc. Int. Conf. on Mass Spectroscopy*, Kyoto, Japan, Sept. 1969, in press.
9. L. R. Anders, *J. Phys. Chem.*, **73**, 469 (1969).
10. R. T. McIver, Jr., *Rev. Sci. Instrum.*, **41**, 555 (1970).
11. R. M. O'Malley and K. R. Jennings, *J. Mass. Spectr. and Ion Physics*, **2**, App. 1 (1969).
12. D. P. Ridge and J. L. Beauchamp, *J. Chem. Phys.*, **51**, 470 (1969).
13. J. M. S. Henis, *J. Chem. Phys.*, **53**, 2999 (1970).
14. W. T. Huntress, Jr., and J. L. Beauchamp, *Int. J. Mass. Spectr. and Ion Phys.*, **3**, 149 (1969).
15. R. C. Dunbar, to be published.
16. D. Wobschall, J. Graham, Jr., and D. Malone, *J. Chem. Phys.*, **42**, 3955 (1965).
17. D. Wobschall, R. A. Fluegge, and J. R. Graham, Jr., *J. Chem. Phys.*, **47**, 4091 (1967).
18. D. Wobschall, R. Fluegge, and J. R. Graham, Jr., *J. Appl. Phys.*, **38**, 3761 (1967).
19. R. A. Fluegge, *Technical Report, CAL Report #UA-1854-P-1*, Cornell Aeronautical Laboratory Inc. (1967).
20. J. L. Beauchamp, *J. Chem. Phys.*, **46**, 1231 (1967).
21. D. P. Ridge and J. L. Beauchamp, to be published.
22. J. L. Beauchamp and S. E. Buttrill, Jr., *J. Chem. Phys.*, **48**, 1783 (1968).
23. J. L. Beauchamp, *Ph.D. Thesis*, Havard University, Cambridge, Mass., 1967.
24. R. C. Dunbar, *J. Chem. Phys.*, **47**, 5445 (1967).
25. G. Gioumousis and D. P. Stevenson, *J. Chem. Phys.*, **29**, 294 (1958).
26. D. J. Hyatt, E. A. Dodman, and M. J. Henchman, *Adv. Chem. Ser.*, No. **32** (1961) Chapter 9.
27. K. R. Ryan and J. H. Futrell, *J. Chem. Phys.*, **43**, 3009 (1965).
28. S. K. Gupta, E. G. Jones, A. G. Harrison, and J. J. Myher, *Can. J. Chem.*, **45**, 3107 (1967).
29. R. C. Dunbar, *J. Chem. Phys.*, **52**, 2780 (1970).
30. J. V. Dugan and J. L. Magee, *J. Chem. Phys.*, **47**, 3103 (1967).
31. J. Schaefer and J. M. S. Henis, *J. Chem. Phys.*, **49**, 5377 (1968).
32. J. Schaefer and J. M. S. Henis, *J. Chem. Phys.*, **51**, 4671 (1969).
33. M. T. Bowers, D. D. Elleman, and J. King, Jr., *J. Chem. Phys.*, **50**, 1840 (1969).
34. Z. Herman, J. D. Kerstelter, T. L. Rose, and R. Wolfgang, *Discussions Faraday Soc.*, **44**, 123 (1967).
35. W. R. Gentry, E. A. Gislason, B. H. Mahan, and C. W. Tsao, *Discussions Faraday Soc.*, **44**, 137 (1967).

36. M. T. Bowers, D. D. Elleman, and J. L. Beauchamp, *J. Phys. Chem.*, **72**, 3599 (1968).
37. J. C. Light, *J. Chem. Phys.*, **40**, 3221 (1964).
38. M. T. Bowers, D. D. Elleman, and J. King, Jr., *J. Chem. Phys.*, **50**, 4787 (1969).
39. A. G. Harrison, J. J. Myher, and J. C. J. Thyne, *Advan. Chem. Ser.*, No. **58**, 150 (1966).
40. R. H. Neynaber and S. M. Trujillo, *Phys. Rev.*, **167**, 63 (1968).
41. W. A. Chupka, M. E. Russell, and K. Refaey, *J. Chem. Phys.*, **48**, 1518 (1968).
42. B. G. Reuben and L. Friedman, *J. Chem. Phys.*, **37**, 1636 (1962).
43. C. F. Giese and W. B. Maier II, *J. Chem. Phys.*, **39**, 739 (1963).
44. D. P. Stevenson and D. O. Schissler, *J. Chem. Phys.*, **29**, 282 (1958).
45. J. H. Futrell and F. P. Abramson, *Advan. Chem. Ser.*, No. **58**, 107 (1966).
46. M. T. Bowers and D. D. Elleman, *J. Chem. Phys.*, **51**, 4606 (1969).
47. V. Aquilanti and G. G. Volpi, *J. Chem. Phys.*, **44**, 2307 (1966).
48. M. T. Bowers and D. D. Elleman, *J. Am. Chem. Soc.*, **92**, 7258 (1970).
49. J. J. Leventhal and L. Friedman, *J. Chem. Phys.*, **49**, 1974 (1968).
50. M. E. Schwartz and L. J. Schaad, *J. Chem. Phys.*, **47**, 5325 (1967).
51. J. T. Moseley, R. M. Snuggs, D. W. Martin, and E. W. McDaniel, *Phys. Rev.*, **178**, 240 (1969).
52. M. Saporeschenko, *Phys. Rev.*, **139**, A352 (1965).
53. J. H. Futrell, T. O. Tiernan, F. P. Abramson, and C. D. Miller, *Rev. Sci. Instr.*, **39**, 340 (1968).
54. J. King, Jr., and D. D. Elleman, *J. Chem. Phys.*, **48**, 4803 (1968).
55. R. P. Clow and J. H. Futrell, *J. Chem. Phys.*, **50**, 5041 (1969).
56. J. L. Beauchamp, private communication.
57. S. E. Buttrill, Jr., *J. Chem. Phys.*, **50**, 4125 (1969).
58. A. G. Marshall and S. E. Buttrill, Jr., *J. Chem. Phys.*, **52**, 2752 (1970).
59. K. E. Wilzbach, *J. Am. Chem. Soc.*, **79**, 1013 (1957).
60. M. Inoue and S. Wexler, *J. Am. Chem. Soc.*, **91**, 5730 (1969). Although the $CD_4 - H_2$ system was primarily studied, for purposes of discussion the results were transformed into the reverse isotopic system where appropriate.
61. M. T. Bowers and D. D. Elleman, *J. Am. Chem. Soc.*, **92**, . 1847 (1970).
62. J. M. S. Henis, *J. Chem. Phys.*, **52**, 282 (1970).
63. J. M. S. Henis, *J. Chem. Phys.*, **52**, 292 (1970).
64. R. M. O'Malley and K. R. Jennings, *Int. J. Mass Spectr. and Ion Phys.*, **2**, 257 (1969).
65. S. E. Buttrill, Jr., to be published.
66. M. T. Bowers, D. D. Elleman, R. M. O'Malley, and K. R. Jennings, *J. Phys. Chem.*, **74**, 2583 (1970).
67. J. J. Myher and A. G. Harrison, *J. Phys. Chem.*, **77**, 1905 (1968).
68. S. E. Buttrill, Jr., *J. Am. Chem. Soc.*, **92**, 3560 (1970).
69. J. L. Beauchamp, L. R. Anders, and J. D. Baldeschwieler, *J. Am. Chem. Soc.*, **89**, 4569 (1967).
70. R. M. O'Malley and K. R. Jennings, *Int. J. Mass Spectr. and Ion Phys.*, **2**, 441 (1969).
71. J. King, Jr., and D. D. Elleman, *J. Chem. Phys.*, **48**, 412 (1968).
72. J. M. S. Henis, *J. Am. Chem. Soc.*, **90**, 844 (1968).
73. G. A. Gray, *J. Am. Chem. Soc.*, **90**, 2177 (1968).
74. G. A. Gray, *J. Am. Chem. Soc.*, **90**, 6002 (1968).
75. T. W. Martin and C. E. Melton, *J. Chem. Phys.*, **32**, 700 (1960).
76. T. F. Moran and W. H. Hamill, *J. Chem. Phys.*, **39**, 1413 (1963).

77. T. W. Shannon and A. G. Harrison, *J. Chem. Phys.*, **43**, 4206 (1965).
78. W. T. Huntress, Jr., J. D. Baldeschwieler, and C. Ponnamperuma, *Nature*, **223**, 468 (1969).
79. M. Inoué and M. Cottin, *Adv. in Mass Spectr.*, **3**, 339 (1966).
80. R. C. Dunbar, *J. Am. Chem. Soc.*, **90**, 5676 (1968).
81. J. I. Brauman and L. K. Blair, *J. Am. Chem. Soc.*, **90**, 6561 (1968).
82. J. I. Brauman and L. K. Blair, *J. Am. Chem. Soc.*, **90**, 5636 (1968).
83. J. I. Brauman and L. K. Blair, *J. Am. Chem. Soc.*, **91**, 2126 (1969).
84. J. I. Brauman and K. C. Smyth, *J. Am. Chem. Soc.*, **91**, 7778 (1969).
85. D. Holtz and J. L. Beauchamp, *J. Am. Chem. Soc.*, **91**, 5913 (1969).
86. F. Kaplan, P. Cross, and R. Prinstein, *J. Am. Chem. Soc.*, **92**, 1445 (1970).
87. G. Klopman, *J. Am. Chem. Soc.*, **91**, 89 (1969).
88. J. L. Beauchamp and R. C. Dunbar, *J. Am. Chem. Soc.*, **92**, 1447 (1970).
89. F. Kaplan, *J. Am. Chem. Soc.*, **90**, 4483 (1968).
90. J. L. Beauchamp, *J. Am. Chem. Soc.*, **91**, 5925 (1969).
91. M. C. Caserio and J. L. Beauchamp, *J. Am. Chem. Soc.*, to be published.
92. J. Diekman, J. K. MacLeod, C. Djerassi, and J. D. Baldeschwieler, *J. Am. Chem. Soc.*, **91**, 2069 (1969).
93. G. Eadon, J. Diekman, and C. Djerassi, *J. Am. Chem. Soc.*, **91**, 3986 (1969).
94 C. L. Wilkins and M. L. Gross, to be published.
95. P. Kriemler and S. E. Buttrill, Jr., *J. Am. Chem. Soc.*, **92**, 1123 (1970).
96. J. R. Eyler, *Inorg. Chem.*, **9**, 981 (1970).
97. W. T. Huntress, Jr., and D. D. Elleman, *J. Am. Chem. Soc.*, **92**, 3565 (1970).
98. A. A. Herod, A. G. Harrison, R. M. O'Malley, A. J. Ferrer-Correia, and K. R. Jennings, *J. Phys. Chem.*, **74**, 2720 (1970).
99. R. P. Clow and J. H. Futrell, *Int. J. Mass Spectr. and Ion Phys.*, **4**, 165 (1970).
100. M. Mosesman and W. T. Huntress, Jr., *J. Chem. Phys.*, **53**, 462 (1970).
101. J. I. Brauman and L. K. Blair, *J. Am. Chem. Soc.*, **92**, 5986 (1970).
102. D. H. McDaniel, N. B. Coffman, and J. M. Strong, *J. Am. Chem. Soc.*, **92**, 6697 (1970).
103. D. Holtz, J. L. Beauchamp, and J. R. Eyler, *J. Am. Chem. Soc.*, **92**, 7045 (1970).
104. G. Eadon, J. Diekman, and C. Djerassi, *J. Am. Chem. Soc.*, **92**, 6205 (1970).
105. S. Billets, H. H. Jaffé, and F. Kaplan, *J. Am. Chem. Soc.*, **92**, 6964 (1970).
106. H. H. Jaffé and S. Billets, *J. Am. Chem. Soc.*, **92**, 6965 (1970).
107. S. A. Benezra, M. K. Hoffman, and M. M. Bursey, *J. Am. Chem. Soc.*, **92**, 7501 (1970).
108. D. Holtz, J. L. Beauchamp, S. Woodgate, *J. Am. Chem. Soc.*, **92**, 7484 (1970).

STABILITY AND DISSIPATIVE STRUCTURES IN OPEN SYSTEMS FAR FROM EQUILIBRIUM

G. NICOLIS,

Faculté des Sciences de l'Université Libre de Bruxelles, Belgium

CONTENTS

210 G. NICOLIS

I. INTRODUCTION

Recent advances in nonequilibrium thermodynamics, especially in the nonlinear region, have brought up novel features of nonequilibrium states which have striking differences with close to equilibrium behavior.[1,2] Perhaps the most important concept introduced by this generalized thermodynamics has been that at a finite distance from equilibrium the stability of a macroscopic state of a system is not insured *a priori*. As a result, the system may quit such a state abruptly, through an unstable transition, and find itself in a new branch of states which is not a continuous extrapolation of the branch of near equilibrium states. Hydrodynamic instabilities arising from a coupling between internal convection and thermal transport processes are good examples of such type of behavior. More recently, purely dissipative systems, in particular open systems undergoing certain types of chemical reactions together with transport processes such as diffusion, have been shown, both theoretically and through computer and laboratory experiments, to give rise to similar effects.[2,3] Here again, the condition of large distance from equilibrium is quite essential. The analogy between the behavior of such systems past an unstable transition point and the structural and functional aspects of biological processes both at the metabolic and at the cellular level has incited Prigogine and coworkers to develop in a great detail the nonequilibrium aspects of open systems and to work out representative biological applications.

The object of this review is to present the theory developed by Prigogine and coworkers and to comment on the possible biological implications of the results. It should be mentioned that in a monograph in press by

Glansdorff and Prigogine, these points are also discussed in great detail. The presentation we follow herein will be oriented somewhat differently and also will focus special attention on a number of additional topics.

The presentation starts in Section II, with a brief outline of thermodynamics of irreversible processes especially in the nonlinear region. The far from equilibrium behavior of open systems is discussed more specifically in Section III. In this section the concept of "dissipative structure", that is, a spatial or temporal organization arising past a nonequilibrium unstable transition, is analyzed in some detail. The nonequilibrium phenomena arising in the nonliner domain are also classified in four broadly different categories:

(1) Oscillations of the same type as in conservative systems.

(2) Oscillations past an instability.

(3) Spatial structures past a symmetry-breaking instability and interference between space and time order.

(4) Transitions between multiple steady states.

Section IV is devoted to the study of oscillations in conservative systems. An example originally due to Lotka is studied in some detail, both from the kinetic and the thermodynamic viewpoint. In Section V, the properties of oscillations beyond a nonequilibrium instability are studied on the basis of Poincaré's theory of limit cycles and of the results of computer experiments. The symmetry-breaking instabilities leading to a spontaneous appearance of space order in a previously uniform system are considered in Section VI and illustrated by examples. Section VII deals with the problem of multiple steady states. In Section VIII, it is pointed out that the far-from-equilibrium behavior described in Sections II to VII is intimately connected to (and even dictated by) the behavior of fluctuations. A stochastic theory of fluctuations in nonequilibrium open systems is set up and illustrated by examples. Section IX deals with the biological implications of the theory. Some conclusions and a presentation of the main problems open to future research are discussed in Section X.

II. THERMODYNAMICS OF IRREVERSIBLE PROCESSES

A. General Theory and the Linear Region

The first attempts to an extension of classical thermodynamics to include nonequilibrium phenomena go back to the classical works of the French school and especially of P. Duhem.[4] However it is only during the last decades that we have witnessed the firm foundations and the rapid growth of thermodynamics of irreversible processes. We shall first define in this section, a set of conditions which will guarantee a first extension of thermodynamics to nonequilibrium situations. It is not claimed that these conditions apply to all irreversible changes, and it is quite possible that a consistent thermodynamic theory could be set up under less restrictive conditions.

Let a given thermodynamic system be divided into microscopically large but macroscopically small subsystems, each having a given volume $\mathcal{V}$. We also assume that it is meaningful to specify at a given moment in these subsystems the internal energy, E and the mole fractions, n_i of species i. At equilibrium all thermodynamic quantities such as temperature T, pressure p, chemical potential μ_i of component i, entropy S are well-defined quantities depending on E, $\mathcal{V}$, and n_i. If now equilibrium does not hold, it is necessary to redefine all these quantities. We assume that T, p, μ_i, and S for each subsystem of a globally nonequilibrium system depend on E, $\mathcal{V}$, and n_i in exactly the same way as in equilibrium. In other terms one proceeds as if equilibrium prevailed in each subsystem separately. This is known as the assumption of *local equilibrium*. Analytically, it implies first that a local formulation of nonequilibrium thermodynamics is possible. And second, that in this formulation the local entropy will be expressed in terms of the same independent variables as if the system were at equilibrium. In other words, if ns is the entropy density, ne the energy density, and v the specific volume, the well known Gibbs relation[1] will hold locally:

$$s = s(e, v, n_i)$$

$$Tds = de + pdv - \sum_i \mu_i \, dn_i \tag{1}$$

The local formulation of irreversible thermodynamics based on Eq. (1) has been worked out systematically by Prigogine.[1,5] A few years later the same author established the domain of validity of this local equilibrium assumption by showing that it implies the dominance of dissipative processes over purely mechanical processes. In more specific terms, at a given point the molecular distribution functions of velocities and relative positions may only deviate slightly from their equilibrium forms.[6] Recently it has also been possible to generalize Prigogine's proof to include arbitrary strongly coupled or dense systems.[7]

Clearly this is a *sufficient* condition which guarantees the applicability of the thermodynamic methods. It should be pointed out that this condition still permits treating a great variety of problems corresponding to situations quite far from complete thermal equilibrium. For instance, very complicated chemical reaction schemes with large affinities may be treated adequately provided they are not too fast. Similarly, all effects described by the Stokes–Navier equations, including hydrodynamical instabilities, are within the domain of validity of the local equilibrium theory.

Let us now outline the development of nonequilibrium thermodynamics based on Eq. (1). The starting point is of course the second law, which deals with the entropy change, dS in a system. It is convenient to divide dS into two parts.[1,3] We denote by d_eS the flow of entropy due to interactions with the external world and by d_iS the change in S due to (irreversible) processes inside the system, or *entropy production*. We have

$$dS = d_eS + d_iS \qquad (2)$$

The second law refers to d_iS and reads

$$d_iS \geqslant 0 \qquad (3)$$

In the limit of an isolated system $d_eS = 0$ and Eq. (3) reduces to

$$(dS)_E \geqslant 0$$

where the subscript E indicates that energy E is being kept constant. The equality in Eq. (3) applies for reversible changes.

In the local formulation of irreversible thermodynamics, it is allowed to define an entropy production per unit time and volume, σ:

$$P = \frac{d_iS}{dt} = \int d\mathcal{V}\,\sigma \qquad (4)$$

The local equilibrium assumption implies then

$$\sigma \geqslant 0 \qquad (5)$$

This quantity, σ, which plays a fundamental role in the theory may be calculated as follows. We first expand Eq. (2) using the Gibbs equation (1). The latter introduces the time evolution of energy, of density $n = 1/v$, and composition which are all given by the balance equations of mass, momentum, and energy. These equations are given in Refs. 1 to 3, and also in all textbooks of hydrodynamics and are not reproduced here. We only give the final result of the calculation of σ:

$$\sigma = \sum_i J_i X_i \qquad (6)$$

σ appears as a bilinear form summed over all irreversible processes i, of suitably defined flows, J_i, associated with these irreversible processes, and of generalized forces X_i giving rise to these flows. Table I gives one of the most commonly used choices of forces and flows corresponding to the usual transport phenomena and to chemical reactions.

TABLE I

Choices of Forces and Flows*

Phenomenon	Flow	Force	Tensor character
Heat conduction	Heat flow j_q	$\mathrm{grad}\left(\dfrac{1}{T}\right)$	Vector
Diffusion	Mass flow $j_{d,i}$	$-\left[\mathrm{grad}\left(\dfrac{\mu_i}{T}\right) - F_i\right]$	Vector
Viscous flow	Pressure tensor, P	$\dfrac{1}{T}\nabla\cdot\mathbf{u}$	Second rank tensor
Chemical reaction	Reaction rate w_ρ	Affinity $\dfrac{\mathscr{A}_\rho}{T}$	Scalar

* Flows and forces corresponding to the usual irreversible phenomena. F_i = external force per unit mass. $\mathbf{u}$ = hydrodynamic velocity.

It is clear that, as long as the flows are parameters not related to the corresponding forces, the equations of thermodynamics do not permit the explicit study of the evolution of a system subject to well defined boundary conditions. It is therefore necessary to combine the general balance equations with additional, phenomenological laws relating Js and Xs. Experiment shows that at thermal equilibrium there is no macroscopic transport of mass, momentum, or energy; as a result all currents J_i vanish. On the other hand, the conditions of thermal equilibrium imply the absence of constraints such as systematic temperature gradients, etc. Therefore, the generalized forces X_i vanish at the same time as J_i.

It is thus quite natural to assume that, in the neighborhood of equilibrium, linear laws between flows and forces will constitute a good first approximation. The phenomenological laws will therefore take the form

$$J_i = \sum_j L_{ij} X_j \tag{7}$$

where the sum is over (coupled) irreversible processes and the phenomenological coefficients $\{L_{ij}\}$ are in general functions of the thermodynamic state variables T, p, and so on. Eq. (7) defines the linear domain of irreversible processes.[1,8]

The coefficients $\{L_{ij}\}$ cannot be arbitrary. By the second law the diagonal coefficients are necessarily nonnegative:

$$L_{ii} \geqslant 0 \qquad (8)$$

As for the nondiagonal coefficients, it was shown in 1931 by Onsager[9] that it is always possible to choose the flows and forces such that the matrix (L_{ij}) is symmetrical:

$$L_{ij} = L_{ji} \qquad (9)$$

These are the celebrated Onsager reciprocity relations which later have been generalized by Casimir to a more general class of irreversible phenomena.[10]

The phenomenological laws (7) together with relations (8) and (9) constitute a convenient framework within which one can study a great number of irreversible phenomena in the linear approximation.[8] However, the local formulation of irreversible thermodynamics has been developed in yet another direction, the search of variational principles. As this point of view will be the most interesting one for the problems discussed in this review we will discuss in some detail, in this section, the variational principles in irreversible thermodynamics. The question is whether there exists a general principle (other than the second law) characterizing nonequilibrium states independently of the details of phenomena occurring in the system. In order to formulate this question quantitatively, it is necessary to analyze in some detail the character of a nonequilibrium state in thermodynamics. In an isolated system one has $d_eS = 0$, and the second principle implies that entropy increases until it reaches its maximum value. The system thus tends more or less rapidly to a uniquely determined permanent state which is the state of thermodynamic equilibrium. Consider, now, instead of an isolated system, a closed system which can exchange energy with the external world, or an open system which can exchange both energy and matter. In this case, and provided the external reservoirs are sufficiently large to remain in a time independent state, the system may tend to a permanent regime other than equilibrium. This will be a *steady nonequilibrium state*. Now this regime is no longer characterized by a maximum of entropy ($d_eS \neq 0$) or by a minimum of free energy. In other terms, the variational principles valid at thermal equilibrium cannot be extended beyond this state. It is therefore necessary to look for new principles which generalize the concept of a thermodynamic potential to steady (or slowly varying in time) nonequilibrium states. To this end we subdivide the domain of nonequilibrium phenomena into two parts: (a) the region close to equilibrium and (b) the region of states arbitrarily far from equilibrium. We first deal, in this subsection, with the first region, that is, the linear domain of irreversible processes, only considering systems in mechanical equilibrium. In his classical work on the reciprocity relations[9] Onsager had already proposed a variational principle for such nonequilibrium states which he called the *principle of least dissipation of energy*. In this principle, it is understood that the thermodynamic forces remain fixed and only the macroscopic currents may vary.

For the study of problems discussed in this review it will be much more convenient to refer to another variational principle, which is due to Prigogine,[1,5] and deals with the extremal properties of entropy production. Consider Eq. (6) combined with Eqs. (7) to (9). σ then becomes:

$$\sigma = \sum_{ij} L_{ij} X_i X_j \tag{10}$$

The currents $\{J_i\}$ are chosen such that at the steady nonequilibrium state one has

$$J_i = \sum_j L_{ij} X_j = 0, \qquad i = 1, 2, \cdots, \gamma \tag{11}$$

Moreover, it is assumed that for this same choice the phenomenological coefficients L_{ij} become constants independent of the thermodynamic state variables. We now differentiate expression (10) with respect to X_i ($i = 1, \ldots, \gamma$), bearing in mind that the right hand side is a symmetric quadratic form. We obtain:

$$\left(\frac{\partial \sigma}{\partial X_i}\right)_{X_{j,\, j \neq i}} = 2 \sum_{i,j} L_{ij} X_j = 0 \tag{12a}$$

at the steady state, and

$$\left(\frac{\partial^2 \sigma}{\partial X_i^2}\right)_{X_{j,\, j \neq i}} = 2 L_{ii} \geqslant 0 \tag{12b}$$

In other terms, the entropy production per unit time and volume, considered as a function of the generalized forces, is minimum at the steady state. It should be noticed that in this *minimum entropy production theorem* the flows vary at the same time as the forces and are only subject to the boundary conditions imposed on the system.

Imagine now that, as a result of a small fluctuation, the steady state is perturbed through a variation δX_i of the X_is. According to (12) this implies a variation of σ equal to

$$\Delta \sigma = \tfrac{1}{2} \sum_{i,j} \frac{\partial^2 \sigma}{\partial X_i \partial X_j} \delta X_i \delta X_j = \sum_{i,j} L_{ij}\, \delta X_i\, \delta X_j = \sum_i \delta J_i\, \delta X_i \geq 0 \tag{13}$$

by virtue of the properties of L_{ij} which guarantee the positive definiteness of the quadratic form (10) as required by the second law.

Therefore, σ plays the same role as the thermodynamic potentials in equilibrium theory[11] and may be considered as a nonequilibrium state function. On the other hand, it is intuitively clear that Eq. (13) guarantees the stability of the steady state. Indeed, a small fluctuation around this state can only increase σ and therefore cannot lead the system to a new steady state. As a result, the system will develop a mechanism for the decay of this fluctuation and will come back to the initial state. Alternatively, the theorem of minimum entropy production provides an *evolution criterion*, as it implies that a physical system will necessarily evolve to the steady nonequilibrium state starting from an arbitrary state close to it. This intuitive picture of stability and evolution will be substantiated in the following sections where a more rigorous stability theory will be developed. It is necessary to emphasize the interest of casting the whole evolution of the system in a compact thermodynamic criterion such as the minimum entropy production principle. This point of view will become even more apparent in the discussion of the problems related to the non-linear domain of irreversible processes.

B. Nonlinear Thermodynamics

A large number of important and quite frequently occurring phenomena cannot be described by the methods of irreversible thermodynamics even in a first approximation. For instance in the case of chemical reactions, it is often necessary to adopt nonlinear phenomenological laws. Also, wherever a system is not at mechanical equilibrium the coupling between dissipative and convective processes leads to effects of a new type which cannot be treated by the methods of linear theory.

The extension of the local formulation of thermodynamics to the non-linear region has been achieved during the last fifteen years by Glansdorff and Prigogine.[1,2] In its present form, it comprises three essential aspects: (1) The derivation of general evolution criteria for steady states far from equilibrium, (2) the search of thermodynamic potentials characterizing these states, (3) the study of stability of these states. In this subsection we only comment briefly on each of these points separately. A more detailed study of these problems with special emphasis on stability will be presented in Section III for the particular case of open system at mechanical equilibrium.

(1) The problem of evolution criteria was solved in two steps, the first involving a discussion of purely dissipative processes,[12] the second pro-

viding an extension to systems involving mechanical motion.[13] The final
result is as follows; in the whole domain of phenomena which are adequately described by a local theory it is possible to construct a differential
expression $d\Phi$ depending on the state variables such that

$$\frac{d\Phi}{dt} \leq 0 \qquad (14)$$

the equality being applicable at the stationary state. $d\Phi$ is a combination
of dissipative and convective processes. In the absence of convective
motion it can be shown that

$$d\Phi = \int d\mathscr{V} \sum_i J_i \, dX_i \equiv d_X P \qquad (15)$$

that is, $d\Phi$ is the variation of entropy production per unit time due to
a change in the generalized forces. In the general case, the flows J_i are
complicated functions of X_is. It follows that in principle $d\Phi$ *is a nontotal
differential*, that is, it does not represent the variation of a thermodynamic
state function. In the limit of linear phenomena and of validity of Onsager's
relations however Eq. (15) becomes

$$d\Phi = \int d\mathscr{V} \sum_{ij} L_{ij} X_j \, dX_i = \tfrac{1}{2} d \int d\mathscr{V} \sum_{ij} L_{ij} X_i X_j = \tfrac{1}{2} dP \qquad (16)$$

$2d\Phi$ becomes therefore in this case the differential of a state function, the
entropy production, and the evolution criterion reduces to the theorem
of minimum entropy production.

(2) The fact that $d\Phi$ is not a total differential in the general case gives
rise to the problem of the search of a variational principle in nonlinear
thermodynamics. This is a very complicated problem which has only
recently been properly formulated. In Section III we discuss more in detail
the different aspects of this question in the limit of purely dissipative
systems. Here is presented only the main idea, which amounts to realization
that in the general case it is necessary to formulate an *extended* variational
principle. This novel point of view gives rise to a function Ψ, the *local
potential* according to the terminology introduced by Glansdorff and
Prigogine,[13,14] which shares some of the properties of the potentials of
classical thermodynamics. However, it is necessary to look at Ψ as a functional of two sets of functions, average ones corresponding to (quasi steady)
solutions of the macroscopic equations and fluctuating quantities. The
extended variational principle must therefore be understood in terms of
fluctuation theory.[14,15] The Euler–Lagrange equations corresponding to
Ψ reduce then, in the average, to the equations of macroscopic physics.

This extended variational procedure is supplemented by a minimum property expressing that the excess local potential is positive definite around the nonfluctuating state. This fundamental property, which is largely responsible for the physical significance of Ψ has permitted establishment of the convergence of the variational procedure.[16] It also made it possible to treat in a unified way many interesting nonlinear hydrodynamical and stability problems.[17,18]

(3) The property of $d\Phi$ not to be a total differential and the lack of a true variational principle also imply that steady states far from equilibrium are no longer characterized by the extremum of a thermodynamic potential. As a result, the stability of these states is not always insured. This separation between evolution and stability thus leads to the search for independent stability criteria for states far from equilibrium. Recently, a complete *infinitesimal* stability theory of nonequilibrium states has been worked out.[2,19] The main result is as follows. Within the domain of validity of the local formulation of thermodynamics it is possible to construct a negative definite quadratic form

$$\delta^2 z = \delta^2 s - \frac{(\delta \mathbf{v})^2}{T} < 0 \tag{17}$$

where s is the specific entropy, $\mathbf{v}$ the average hydrodynamic velocity and δ denotes the variation of the corresponding quantity as a result of a fluctuation. It can be shown[19] that whenever the equilibrium stability conditions are satisfied $\delta^2 s$ is itself a negative semidefinite quadratic form even around states far from thermodynamic equilibrium:

$$\delta^2 s \leqslant 0 \tag{18}$$

Moreover, in the limit of small fluctuations,

$$\frac{\partial}{\partial t} \delta^2 z > 0 \tag{19}$$

in all cases the nonequilibrium state is stable. Clearly, this formulation is closely related to the ideas underlying Lyapounov's stability theory. This point will be discussed in detail in Section III.

The important point is that Eqs. (17) to (19) constitute a compact criterion which permits analysis of stability phenomena from a thermodynamic viewpoint. In general, Eq. (19) contains a complicated interplay between purely dissipative and convective processes. In the neighborhood of equilibrium it can be shown that the stability criterion is trivially fulfilled once $\delta^2 s < 0$. Alternatively, the existence of thermodynamic potentials guarantees the stability of the equilibrium states except in the neighborhood of phase transition points. Far from equilibrium, however, relation (19) does not follow from (17) and (18). Therefore, the stability of the system may be compromised, even when the equilibrium state is perfectly stable. If an instability occurs, the system tends necessarily to a new regime which may correspond to a completely different state of organization of matter. Since equilibrium remains stable we may say that unlike what happens in the linear range, this new regime is not a continuous extrapolation of the equilibrium behavior.

III. OPEN SYSTEMS AT MECHANICAL EQUILIBRIUM

A. General Comments

We consider a multicomponent mixture of n reacting chemical species. The system is in contact with reservoirs of chemicals and an exchange is allowed between these chemicals and the reacting species inside the system. In practically all cases, we will assume that the reservoir chemicals are at uniform composition. We also allow for diffusion of all reacting species but not for convective transport. Moreover, external fields and temperature variations will be neglected unless it is otherwise stated. These assumptions exclude from the very start systems usually encountered in hydrodynamics. In this review, hydrodynamic effects will be mentioned only occasionally in order to illustrate by more familiar examples some of the effects which will be analyzed in the limit of purely dissipative systems as defined in this section.

Let $n = \{n_i\}$ $(i = 1, 2, \ldots, n)$ be the vector of molar concentrations or of mole fractions, j the vector of diffusion flow vectors in an n-dimensional phase space spanned by the dependent variables n_i. As convective motion is excluded j_i will take the simple form

$$j_i = n_i \Delta_i \tag{20}$$

where Δ_i is the diffusion velocity of species i.

The equations of mass balance of the n species within the system may be written in vector form as[1,2,8]

$$\frac{\partial n}{\partial t} = -\nabla \cdot j + \sigma(n) \tag{21}$$

where $\sigma(n)$ is a source term whose components equal the production rates of species i due to the chemical reactions. Let w_ρ be the velocity of ρth reaction ($\rho = 1, \ldots, s$), $v_{i\rho}$ the stoichiometric coefficients of species i in the ρth reaction. Then[1,2,8]

$$\sigma_i(n) = \sum_{\rho=1}^{s} v_{i\rho} w_\rho \tag{22}$$

With (20) and (22) Eq. (21) takes the form

$$\frac{\partial n}{\partial t} = -\nabla \cdot n \, \Delta + \sum_{\rho=1}^{s} v_\rho w_\rho \tag{23}$$

This is the type of equation we shall be dealing with in this review. In practically all cases of interest, diffusion will be represented in a very good approximation, by a linear law of the Fick type:

$$\mathbf{j} = -D\nabla n \tag{24}$$

where D is the diffusion coefficient matrix. In principle, we allow for coupled diffusion effects and do not require D to be diagonal.

It should be noticed that Eq. (24) is a particular case of a more general phenomenological law which reduces to (24) in the limit of ideal mixtures. On the contrary, the chemical reaction velocities cannot be approximated in Eq. (23) by linear relations in terms of the affinities, especially as we will be interested in far from equilibrium effects.[1,2] It follows that in the general case the system of equation (23) will be *nonlinear*. This nonlinearity will be responsible, precisely, for the new effects such as instabilities described in detail in the subsequent sections.

B. Evolution Criterion

Equation (23), supplemented by the phenomenological relation (24) and the laws of chemical kinetics, constitute a closed system of equations which completely describe the kinetic behavior of the system. Let us now perform a thermodynamic analysis of the same system. The entropy production per unit time is (cf. Eq. (6) and Table I).

$$P = \int d\mathscr{V} \sum_\rho \frac{\mathscr{A}_\rho w_\rho}{T} - \int d\mathscr{V} \sum_i \mathbf{j}_i \cdot \nabla \frac{\mu_i}{T} \geq 0 \tag{25}$$

We recall that the affinity $\mathscr{A}_\rho$ is defined by[1,2,8]

$$\mathscr{A}_\rho = -\sum_i v_{i\rho} \mu_i \tag{26}$$

In order to analyze the behavior of the entropy production along the motion described by the kinetic equations of the previous subsection we differentiate Eq. (25) with respect to time and divide the right hand side into two parts: A part due to the variation of the flows and a second part due to the variation of the generalized forces. We obtain:

$$\frac{dP}{dt} = \frac{d_J P}{dt} + \frac{d_X P}{dt} \tag{27a}$$

$$\frac{d_X P}{dt} = \int d\mathscr{V} \left[\sum_\rho \frac{\mathscr{A}_\rho}{T} \frac{dw_\rho}{dt} - \sum_i \nabla\left(\frac{\mu_i}{T}\right) \cdot \frac{d\mathbf{j}_i}{dt} \right] \tag{27b}$$

$$\frac{d_X P}{dt} = \int d\mathscr{V} \left[\sum_\rho w_\rho \frac{d\left(\frac{\mathscr{A}_\rho}{T}\right)}{dt} - \sum_i \mathbf{j}_i \cdot \frac{d}{dt} \nabla\left(\frac{\mu_i}{T}\right) \right] \tag{27c}$$

We allow for arbitrary deviations from equilibrium. In this case the theorem of minimum entropy production is no longer valid and we do not expect dP/dt to yield information about the approach to a steady state. Let us now study more closely the part $d_X P/dt$. Using Eq. (26), we obtain:

$$\frac{d_X P}{dt} = -\int d\mathscr{V} \sum_{i,\rho} \nu_{i\rho} w_\rho \frac{d\left(\frac{\mu_i}{T}\right)}{dt} - \int d\mathscr{V} \sum_i \mathbf{j}_i \cdot \nabla \frac{d}{dt}\left(\frac{\mu_i}{T}\right) \tag{28}$$

In the last term, we perform a partial integration and then apply Gauss' theorem; we obtain:

$$\int d\mathscr{V} \mathbf{j}_i \cdot \nabla \frac{d}{dt}\left(\frac{\mu_i}{T}\right) = \int d\mathbf{r} \cdot \mathbf{j}_i \frac{d}{dt}\left(\frac{\mu_i}{T}\right) - \int d\mathscr{V} \frac{d}{dt}\left(\frac{\mu_i}{T}\right) \nabla \cdot \mathbf{j}_i \tag{29}$$

where $d\mathbf{r}$ is the unit vector surface element. We now assume that the system is subject to *time independent boundary conditions*. It follows that

$$\int d\mathbf{r} \cdot \mathbf{j}_i \frac{d}{dt}\left(\frac{\mu_i}{T}\right) = 0 \tag{30}$$

Eq. (28) becomes

$$\frac{d_X P}{dt} = -\int d\mathscr{V} \left[\sum_i \left(\sum_\rho \nu_{i\rho} w_\rho - \nabla \cdot \mathbf{j}_i \right) \frac{d}{dt}\left(\frac{\mu_i}{T}\right) \right] \tag{31}$$

Or, using Eq. (23)

$$\frac{d_X P}{dt} = -\int d\mathscr{V} \sum_i \frac{dn_i}{dt} \frac{d}{dt}\left(\frac{\mu_i}{T}\right) \tag{32}$$

Now μ_i/T is a function only of the composition n_i. It follows

$$\frac{d}{dt}\left(\frac{\mu_i}{T}\right) = \frac{1}{T}\sum_j \frac{\partial\mu_i}{\partial n_j}\frac{dn_j}{dt} \tag{33}$$

Eq. (33) becomes

$$\frac{d_X P}{dt} = -\frac{1}{T}\int d\mathscr{V}\sum_{ij}\frac{\partial\mu_i}{\partial n_j}\frac{dn_i}{dt}\frac{dn_j}{dt} \tag{34}$$

The coefficients $\partial\mu_i/\partial n_j$ are functions of the composition variables n. By the local equilibrium assumption which is implied throughout in this review, the form of this functional relation is the same as at equilibrium. Now stability of the equilibrium state with respect to diffusion implies that the quadratic form[1,2,8]

$$\sum_{ij}\frac{\partial\mu_i}{\partial n_j}\delta n_i\,\delta n_j \geq 0 \tag{35}$$

It follows that, in so far as the possibility of a phase separation is excluded, the quadratic form in the right hand side of Eq. (34) is also positive definite and therefore

$$\frac{d_X P}{dt} \leq 0 \tag{36}$$

with the equality occurring at the steady state only ($dn/dt = 0$).

Inequality (36) is quite general, as it is valid in the whole range of phenomena which are adequately described by kinetic equations of the form (23).[1,2,12] Its main importance is in the fact that it provides a compact *evolution criterion* which is independent of the details of the kinetic properties of the system (cf. also comments in Sec. II-B). However, it would be erroneous to conclude that Eq. (36) implies that for sufficiently long times the system always tends to a steady state regime for which $d_X P/dt$ vanishes. Indeed, as it was already observed in Sec. II-B, expression (34) cannot be written in the general case as the time derivative of a state function. This is easily seen by considering the chemical part of the differential form (27c). Because of the nonlinear dependence of w_ρ upon $\mathscr{A}_\rho$ there is no reason for this part to be a total differential. We will soon verify this general feature on examples. Let us only observe that in the limit of linear irreversible processes and provided that Onsager's reciprocal relations are valid,

$$w_\rho = \sum_{\rho'} L_{\rho\rho'}\frac{\mathscr{A}_{\rho'}}{T} \qquad L_{\rho\rho'} = L_{\rho'\rho} \tag{37}$$

the chemical term in Eq. (27c) becomes

$$(d_X P)_{ch} = \int d\mathcal{V} \frac{1}{T^2} \sum_{\rho\rho'} L_{\rho\rho'} \mathcal{A}_\rho \, d\mathcal{A}_{\rho'} = \frac{1}{2T^2} d \int d\mathcal{V} \sum_{\rho\rho'} L_{\rho\rho'} \mathcal{A}_\rho \mathcal{A}_{\rho'} \quad (38)$$

In other words $d_X P$ now becomes half the differential of the entropy production, which is a nonequilibrium state function. It is straightforward to see that the diffusion terms give the same result. Therefore in this range

$$\frac{dP}{dt} = 2 \frac{d_X P}{dt} \leq 0 \quad (39)$$

and the evolution criterion leads to the theorem of minimum entropy production (cf. Eq. (13)) implying that the system will necessary evolve to the steady state.

The property of $d_X P$ not to be a total differential in the general case does not make it possible to extend this near equilibrium result to the nonlinear range. In the next subsections we study the consequences of this point on the stability properties of the system.

C. Stability Theory

Let us first recall some of the essential points of the stability theory of nonlinear differential equations.[20,21] Consider a system of differential equations of the form (23). Let $X(\mathbf{r}, t)$ be a solution of this system, that is, a set of functions $X_i(\mathbf{r}, t)$ depending on the space and time variables $\mathbf{r}$ and t satisfying identically (23) together with initial and boundary conditions. We assume that the motion is defined in the time interval $(0, \infty)$ and that $X_i(\mathbf{r}, t)$ exist in this interval.

(1) We say that $X_i(t)$ is (Lyapounov) *stable* if, given $\varepsilon > 0$ and $t = t_0$, there exists an $\eta = \eta(\varepsilon, t_0)$ such that any solution $Y_i(t)$ for which $|X_i(t_0) - Y_i(t_0)| < \eta$, satisfies also $|X_i(t) - Y_i(t)| < \varepsilon$ for $t \geq t_0$. If no such η exists, the solution $X_i(t)$ is *unstable*. We see that stability in the sense of Lyapounov is equivalent to uniform continuity of $X_i(t; t_0)$ with respect to the initial conditions.

(2) If $X_i(t)$ is stable, and

$$\lim_{t \to \infty} |X_i(t) - Y_i(t)| = 0 \quad (40)$$

we say that $X_i(t)$ is *asymptotically stable*.

In other terms a solution (or motion) is asymptotically stable if all solutions coming near it approach it asymptotically.

(3) Consider now the case where the system (23) is autonomous, that is the case where the right hand side does not depend on time explicitly.

Let $X(t)$ be a solution of (23). It is then clear that any function $X(t + t_0)$, where t_0 is an arbitrary constant (the phase), is still a solution of (23) (translational invariance of autonomous systems). These infinitely many solutions differing from each other by the phase define a *trajectory* (or orbit) of the system. Let C be such an orbit. We say that C is *orbitally stable* if, given $\varepsilon > 0$ there exists an $\eta > 0$ such that, if X is a representative point of another trajectory within a distance η from C at time t_0 then X remains within a distance ε from C for $t \geq t_0$. Otherwise C is *orbitally unstable*.

(4) If C is orbitally stable and the distance between X and C tends to 0 as $t \to \infty$, C is *asymptotically orbitally stable*.

Lyapounov stability and orbital stability should not therefore be confused. Let us also notice that the concept of orbit becomes more natural when diffusion is suppressed in Eq. (23). In this case time becomes the only independent variable.

For the analytic study of stability, it is interesting to consider the behavior of the system of equation (23) *around* a given regime. This regime may correspond to a steady uniform state, a steady inhomogeneous state, or even a time dependent state. Let $X_0 = X_0(\mathbf{r}, t)$ be a solution corresponding to this "reference" regime. We may consider X_0 as an unperturbed solution. Let now $X(\mathbf{r}, t)$ be a perturbed solution. We set

$$X(\mathbf{r}, t) = X_0(\mathbf{r}, t) + x(\mathbf{r}, t) \tag{41}$$

where $x(\mathbf{r}, t)$ represent the perturbations.

From Eq. (23) and (41) one deduces a set of equations of evolution for x. Let us assume for simplicity that diffusion is represented by law (24) and that the matrix D is diagonal. We obtain

$$\frac{\partial x}{\partial t} = -\nabla \cdot D\nabla x + f(x) \tag{42}$$

where

$$f(x) = \sum_{\rho} v_{\rho}[w_{\rho}(X) - w_{\rho}(X_0)] = F(X) - F(X_0) \tag{43}$$

In general (42) is also a nonlinear differential system. In certain problems the local behavior of the system around the unperturbed motion is of primary interest. One obtains then extremely significant information from the linearized version of (42), where it is assumed that $|x_i|$ is sufficiently small compared to X_{0i}. In this limit we obtain the so called *variational equations* of (23):

$$\frac{\partial x_i}{\partial t} = \sum_{j} \left(\frac{\partial F_i}{\partial X_j}\right)_0 x_j - \nabla \cdot D_i \nabla x_i \tag{44}$$

The study of stability properties of system (44) is of great value, as its behavior is equivalent, in some cases, to the stability behavior of the non-linear system (42). We shall come back to this point later on. Let us only observe here that, by equations (41) to (44), the stability of the unperturbed motion is translated in Eq. (44), to the stability of the trivial solution $\{x_i = 0\}$.

To date there exist a variety of powerful methods which permit an analytic study of system (44). Some of them will be applied to the examples discussed in the following sections. It should be mentioned however that, in addition to this approach to stability, the so called second Lyapounov method, which gives conditions of stability without requiring the integration of the variational equations, has also been developed. As this approach is closer to the spirit of a thermodynamic analysis, we will compile in this subsection, a few basic results related to this method.

Consider a function $V (x_1, \ldots, x_n)$ of n variables, which we call the *Lyapounov function*. V is called *definite* in a domain D of the variables x_i defined by $|x_i| < \eta$ (η constant) if it takes only values of one sign and vanishes only for $x_1 = x_2 = \cdots = x_n = 0$. V is called *semidefinite* if it keeps the same sign or is zero in the domain. In all other cases V is called *indefinite*.

Consider first a differential system of the form (42) wherein diffusion terms are suppressed:

$$\frac{dx_i}{dt} = f_i(x_1, \ldots, x_n) \tag{45}$$

We define the (Eulerian) derivative of V along a solution of (45) by

$$\frac{dV}{dt} = \sum_{i=1}^{n} \frac{\partial V}{\partial x_i} \frac{dx_i}{dt} \equiv \frac{\partial V}{\partial x} \cdot \frac{dx}{dt} = f \cdot \text{grad } V \tag{46}$$

The first theorem of Lyapounov asserts:
Theorem (1) The steady state $(x_1 = x_2 = \cdots = x_n = 0)$ is stable in a domain D if one can determine in D a definite function V whose Eulerian derivative is either semidefinite of sign opposite to V or vanishes identically in D.

We shall not prove this theorem here, but refer the interested reader to the abundant literature in this field. Let us also state without proof two more fundamental theorems by Lyapounov:
Theorem (2) The state $(x_1 = x_2 = \cdots = x_n = 0)$ is asymptotically stable if one can determine a definite function V whose Eulerian derivative is definite and has a sign opposite to that of V.

Theorem (3) The state $(x_1 = x_2 = \cdots x_n = 0)$ is unstable if one can determine a function V whose Eulerian derivative is definite and V assumes in D values such that $V \, dV/dt > 0$.

These Lyapounov theorems, which give *sufficient* conditions for stability, admit a very intuitive geometrical interpretation. Consider the case of two independent variables and let $V(x_1, x_2)$ be positive definite, while $dV/dt \leqslant 0$. Consider the family of curves

$$V(x_1, x_2) = c \tag{47}$$

For sufficiently small values of c it can be shown that these curves are closed and that if $c_1 < c_2$ the curve $V = c_1$ is inside $V = c_2$ (See Ref. 21, Chapter 6). (See Figure 1.) Consider a trajectory S of the system starting at t_0 from a point close to $c = 0$. For $t > t_0$ this trajectory can never intersect the curve $V = c_1$ from inside to outside, otherwise dV/dt would have to be positive. Thus if $x(t_0)$ was inside curve $V = c_1$ at t_0 it will continue to remain inside that curve. For values of c_1 sufficiently small this ensures the stability of the origin. This argument also shows that for a general form of V stability and asymptotic stability do not imply monotonic approach to the origin.

So far, Lyapounov's theorems have been given for the ordinary differential system (45). When diffusion is not neglected, the general equation (42) becomes a (nonlinear) partial differential equation. The extension of Lyapounov's theory in that case has been carried out by Zubov[22] (cf. also

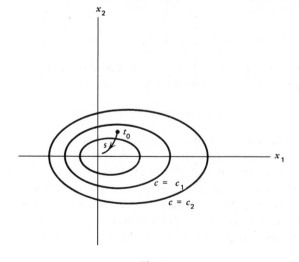

Fig. 1

Fowler, Ref. 23). Generally, the results by Lyapounov subsist provided the differential operators satisfy certain requirements (always met by the form (42)).

D. Stability Theory and Irreversible Thermodynamics

In this subsection the results of irreversible thermodynamics are combined with the ideas of stability theory in order to derive *thermodynamic stability criteria* for far from equilibrium states. This will provide a much deeper insight into the physical mechanism of stability.

In order to illustrate the nature of the problem let us first consider the simple case of states close to equilibrium and assume that Onsager's relations remain valid. In Sec. II-A and III-B, it was shown that the entropy production P satisfies in this case the inequality

$$\frac{dP}{dt} \leq 0$$

while

$$P > 0 \qquad \text{(Second law)} \qquad (48)$$

Or, in terms of the *excess entropy production*, ΔP introduced in Eq. (13):

$$\frac{d\,\Delta P}{dt} \leq 0$$

$$\Delta P \geq 0 \qquad\qquad\qquad (49)$$

where the equality applies only at the steady state, that is, at the state where the deviations of the thermodynamic variables

$$x_1 = x_2 = \cdots = x_n = 0$$

Clearly the function ΔP in (49) satisfies Lyapounov's second theorem (Sec. III-C) and therefore the steady state is asymptotically stable. We see therefore, in this limit, how a thermodynamic property (48) (essentially equivalent to the theorem of minimum entropy production) is sufficient to guarantee the asymptotic approach to the steady state. The situation is quite similar to equilibrium thermodynamics, where the extremal properties of the thermodynamic potentials guarantee the stability of the equilibrium state.

Let us now consider the general case of states arbitrarily far from equilibrium. A question which arises quite naturally is under what conditions the general evolution criterion (36):

$$\frac{d_X P}{dt} \leq 0 \tag{36}$$

may be transformed to a form

$$\frac{d\Phi}{dt} \leq 0, \qquad \Phi \geq 0 \tag{50}$$

where Φ is a state function? If this were possible, the stability of the steady state would be insured. Whenever Fick's law (24) is valid and the diffusion coefficient matrix is diagonal, it is not difficult to cast the part of (36) associated with diffusion in the form (50). The difficulty, if any, comes from terms due to chemical reactions. For the purpose of our discussion it is therefore sufficient to restrict coverage to the chemical kinetic equations

$$\frac{dn_i}{dt} = \sum_{\rho=1}^{s} v_{i\rho} w_\rho \tag{51a}$$

$$T \, d_X P = \sum_\rho w_\rho \, d\mathscr{A}_\rho \tag{51b}$$

It will be convenient to deduce first from (51b) the equations for the affinities of the independent reactions. By (26):

$$\frac{d\mathscr{A}_\rho}{dt} = -\sum_i v_{i\rho} \frac{d\mu_i}{dt} = -\sum_{ij} v_{i\rho} \left(\frac{\partial \mu_i}{\partial n_j}\right) \frac{dn_j}{dt}$$

$$\frac{d\mathscr{A}_\rho}{dt} = -\sum_{ij\rho'} v_{i\rho} v_{j\rho'} \left(\frac{\partial \mu_i}{\partial n_j}\right) w_{\rho'} = f_\rho(\mathscr{A}_1, \ldots, \mathscr{A}_s) \tag{52}$$

f_ρ may be thought of as a velocity vector field associated with the independent chemical affinities. We now introduce the divergence, $\tau(\mathscr{A}_\rho)$ and the curl, $\mathbf{c}$ of f_ρ in affinity space:

$$\tau(\mathscr{A}_\rho) = \mathrm{div}\, f = \sum_\rho \frac{\partial f_\rho}{\partial \mathscr{A}_\rho}$$

$$\mathbf{c}(\mathscr{A}_\rho) = \mathrm{curl}\, f \tag{53}$$

It can then be shown[24] that

$$f = -\,\mathrm{grad}\,\Phi + \mathrm{curl}\,\mathbf{A} \tag{54}$$

where Φ is a well defined scalar function ("scalar potential") and $\mathbf{A}$ is an n-dimensional vector ("vector potential") which depend both on τ and $\mathbf{c}$.

Suppose now curl $\mathbf{A} = 0$. It follows that

$$\frac{d\mathscr{A}_\rho}{dt} = -\operatorname{grad}_\rho \Phi = -\frac{\partial \Phi}{\partial \mathscr{A}_\rho} \tag{55}$$

and

$$\frac{d\Phi}{dt} = \sum_\rho \left(\frac{\partial \Phi}{\partial \mathscr{A}_\rho}\right) \frac{d\mathscr{A}_\rho}{dt} = -\sum_\rho f_\rho \frac{d\mathscr{A}_\rho}{dt}$$

$$= -\sum_\rho \left(\frac{d\mathscr{A}_\rho}{dt}\right)^2 \leq 0 \tag{56}$$

which is Eq. (51b) transformed to a form involving the total derivative of a state function, the "scalar potential". If in addition, Φ happens to be positive definite, the steady state will be asymptotically stable by Lyapounov's second theorem.

In the general case, curl $\mathbf{A}$ is nonzero. This is seen most easily by considering the example

$$A \rightarrow X$$
$$X + Y \rightarrow 2Y \tag{57}$$
$$Y \rightarrow E$$

We have

$$w_1 = A, \qquad w_2 = XY, \qquad w_3 = Y$$

$$\mathscr{A}_1 = \ln \frac{A}{X}, \qquad \mathscr{A}_2 = \ln \frac{X}{Y}, \qquad \mathscr{A}_3 = \ln \frac{Y}{E}$$

The independent affinities may be chosen as

$$\mathscr{A}_1 = \ln X, \qquad \mathscr{A}_2 = \ln Y \tag{58}$$

The kinetic equations read

$$\frac{dX}{dt} = A - XY$$

$$\frac{dY}{dt} = XY - Y \tag{59a}$$

Or, in terms of the affinities:

$$\frac{d\mathscr{A}_1}{dt} = Ae^{-\mathscr{A}_1} - e^{\mathscr{A}_2} = f_1$$

$$\frac{d\mathscr{A}_2}{dt} = e^{\mathscr{A}_1} - 1 \qquad = f_2 \tag{59b}$$

We verify that

$$(\text{curl } f)_{\substack{\text{perpend.} \\ \text{plane}}} = \frac{\partial f_1}{\partial \mathscr{A}_2} - \frac{\partial f_2}{\partial \mathscr{A}_1} = -e^{\mathscr{A}_2} - e^{\mathscr{A}_1} = -(X + Y) \neq 0 \qquad (60)$$

As a result, the vector potential is nonzero.

Eq. (51b) splits into a total differential of the form (56) plus an antisymmetric form associated with curl **A**, and the reduction to the differential of a state function is no longer possible. The addition of this antisymmetric form is indicative of a rotatory motion around the steady state point (in affinity space). This can be seen most easily in the case of 2 independent variables and provided Eq. (52) is truncated to its linear terms. We obtain after introduction of excess variables $\delta\mathscr{A}_i = \mathscr{A}_i - \mathscr{A}_{i0}$ the variational system of (52):

$$\frac{d \, \delta\mathscr{A}_1}{dt} = l_{11} \, \delta\mathscr{A}_1 + l_{12} \, \delta\mathscr{A}_2$$

$$\frac{d \, \delta\mathscr{A}_2}{dt} = l_{21} \, \delta\mathscr{A}_1 + l_{22} \, \delta\mathscr{A}_2 \qquad (61)$$

We write

$$l_{12} = \frac{l_{12} + l_{21}}{2} + \frac{l_{12} - l_{21}}{2} = l^s + l^a \qquad (62)$$

Eq. (61) becomes

$$\frac{d \, \delta\mathscr{A}_1}{dt} = l_{11} \, \delta\mathscr{A}_1 + l^s \, \delta\mathscr{A}_2 + l^a \, \delta\mathscr{A}_2 = f_1 = -\text{grad}_1 \, \Phi + \text{curl}_1 \, \mathbf{A}$$

$$\frac{d \, \delta\mathscr{A}_2}{dt} = l^s \, \delta\mathscr{A}_1 + l_{22} \, \delta\mathscr{A}_2 - l^a \, \delta\mathscr{A}_1 = f_2 = -\text{grad}_2 \, \Phi + \text{curl}_2 \, \mathbf{A} \qquad (63a)$$

where

$$\Phi = -l_{11} \frac{\delta\mathscr{A}_1^2}{2} - l_{22} \frac{\delta\mathscr{A}_2^2}{2} + l^s \, \delta\mathscr{A}_1 \, \delta\mathscr{A}_2 \qquad (63b)$$

$$(\mathbf{A}) = -l^a(\delta\mathscr{A}_1 \, \delta\mathscr{A}_2, \, -\delta\mathscr{A}_1 \, \delta\mathscr{A}_2)$$

We set

$$f_{i1} = -\text{grad}_i \, \Phi$$

$$f_{i2} = \text{curl}_i \, \mathbf{A}$$

Consider the distance vector $\delta\mathscr{A} \equiv (\delta\mathscr{A}_1, \delta\mathscr{A}_2)$ from the origin. We obtain:

$$\delta\mathscr{A} \cdot f_2 = \delta\mathscr{A}_1 \, \text{curl}_1 \, \mathbf{A} + \delta\mathscr{A}_2 \, \text{curl}_2 \, \mathbf{A} = 0 \qquad (64)$$

The displacement due to curl $\mathbf{A}$ is therefore orthogonal to the distance vector. The motion is divided into two parts: One "contraction" associated with Φ and one "rotation" associated with $\mathbf{A}$. The stability properties are contained in the contraction part only.

We see that in the general case the situation is far more complicated than in the domain of validity of the minimum entropy production theorem. In the subsequent sections, this situation will be illustrated by more examples. It is only indicated here that in some exceptional cases the Pfaffian form (51b) may be converted to a total differential through a suitable integrating factor. This point is not developed here but the reader is referred to the monograph by Glansdorff and Prigogine, where a much more thorough discussion of the different aspects of the evolution criterion is given.

In conclusion, we may say that the general evolution criterion does not imply always the stability of the steady state as it does not always give rise to the differential of a state function. However, as was shown by Glansdorff and Prigogine[2,19] it is still possible to set up a stability theory of states far from equilibrium provided one is limited to *infinitesimal stability*, that is, to stability with respect to small fluctuations. Let us outline here these ideas in the particular case of purely dissipative systems considered in this review.

Let S, P be the entropy and entropy production at a state of the system. We define the excess quantity

$$\Delta S = S - S_0 \qquad (65)$$

around some reference state (usually a steady state) whose stability is sought. In general

$$\Delta S = \delta S + \tfrac{1}{2}\delta^2 S + \cdots \qquad (66)$$

The deviation ΔS will be assumed small; higher order terms will therefore be neglected in Eq. (66). In general, $\delta S \neq 0$ for a state far from equilibrium. On the other hand, the fundamental local equilibrium assumption permits establishment of the following remarkable relation for $\delta^2 S$. (For a proof, see monograph by Glansdorff and Prigogine, Ref. 2):

$$\delta^2 S = -\int d\mathscr{V} \sum_{ij} \left(\frac{\partial \mu_i}{\partial n_j}\right)_0 \delta n_i \, \delta n_j \leq 0 \qquad (67)$$

where the subscript "0" implies evaluation at the steady state and $\{\delta n_i\}$ are the fluctuations of the state variables (molar concentrations etc.). The inequality sign in (67) is implied by the equilibrium stability conditions (see also discussion following Eq. (34)). The equality is applicable only for $\delta n_i = 0$. We will assume, universally, that equilibrium does remain stable and therefore that inequality (67) is always secured.

This result suggests an approach to stability wherein $\delta^2 S$ would play the role of a Lyapounov function.* In this way, whenever

$$\frac{\partial}{\partial t}(\delta^2 S) > 0$$

the steady state will be asymptotically stable. Equations (67) and (68) provide *sufficient* criteria for infinitesimal stability.

It is very instructive to study the connection between these stability criteria and the general evolution criterion.[2,3] Let us consider the case of chemical reactions only. The evolution criterion (36) becomes

$$T\,d_X P = \int d\mathscr{V} \sum_\rho w_\rho\,d\mathscr{A}_\rho \le 0 \qquad (69)$$

Let $\delta\mathscr{A}_\rho$ be an arbitrary variation of $\mathscr{A}_\rho$ (e.g., due to a fluctuation). The system will be in a stable steady state if (69) is violated for all possible variations $\delta\mathscr{A}_\rho$:

$$T\,\delta_X P = \int d\mathscr{V} \sum_\rho w_\rho\,\delta\mathscr{A}_\rho \ge 0 \qquad (70)$$

At the steady state, the equality sign applies in (70). This implies that the coefficients of the independent affinities vanish†:

$$w_\rho{}^0 = 0$$

and thus

$$w_\rho = \delta w_\rho + \tfrac{1}{2}\delta^2 w_{\rho+} \ldots \qquad (71)$$

It follows that a criterion for infinitesimal local stability is:

$$\sum_\rho \delta w_\rho\,\delta\mathscr{A}_\rho > 0 \qquad (72)$$

---

* This approach can also be considered as a slight generalization of the classical Lyapounov theory. Indeed, Lyapounov's theorems insure local stability, that is, for a given space point. For a nonuniform system in a certain region of space of volume $\mathscr{V}$, one can construct a stability theory using Lyapounov functionals which are integrals of state functions over the volume $\mathscr{V}$.

† Note that the $w_\rho s$ in the subsequent equations are *global* reaction velocities, that is, linear combinations of the velocities of the individual reactions.

where the variations are now assumed to be small around the steady state. Let us now evaluate (68) taking into account Eqs. (67) and (51a)

$$
\begin{aligned}
\frac{\partial}{\partial t}(\delta^2 S) &= -\int d\mathscr{V} \sum_{ij} \left(\frac{\partial \mu_i}{\partial n_j}\right)_0 \delta n_i \frac{\partial}{\partial t}(\delta n_j) \\
&= -\int d\mathscr{V} \sum_{ij\rho} \left(\frac{\partial \mu_i}{\partial n_j}\right)_0 v_{j\rho} \, \delta n_i \, \delta w_\rho \\
&= -\int d\mathscr{V} \sum_{j\rho} \delta \mu_j v_{j\rho} \, \delta w_\rho \\
&= +\int d\mathscr{V} \sum_{\rho} \delta w_\rho \, \delta \mathscr{A}_\rho
\end{aligned}
\tag{73}
$$

We obtain, therefore, an expression which is identical to (72) and in addition, an explicit form for the stability condition. Stability will be compromised as soon as negative terms appear which tend to change the sign of $\sum_\rho \delta w_\rho \, \delta \mathscr{A}_\rho$. Examples of this behavior will be given in the subsequent sections.

In conclusion, the stability properties of open systems arbitrarily far from equilibrium may be formulated in terms of thermodynamic quantities, the excess entropy $\delta^2 S$ and the excess entropy production $\partial/\partial t \, (\delta^2 S)$. This important result will enable us to interpret the models discussed in the subsequent sections in a much more intuitive and fundamental way which would not be possible by a purely kinetic stability analysis.

E. Dissipative Structures

In equilibrium thermodynamics, instabilities only occur at phase transition points. The new phase beyond instability has a markedly different structure; in particular it may correspond to a more ordered state. For instance, at the para-ferromagnetic transition at the Curie point, a system exhibiting spherical symmetry is replaced by a new one having a lesser cylindrical symmetry. Consequently, the ferromagnet which is being formed has a much higher degree of "organization". However, such structures are completely independent of the external world. Once they are formed they are selfmaintained and do not require an exchange of energy or matter with the environment.

In systems far from equilibrium, a new type of instability appears due to the existence of constraints which are responsible for the maintenance of a steady nonequilibrium state. Can one associate with these instabilities the formation of ordered structures of a new type? Such nonequilibrium structures would differ from equilibrium types in that their maintenance would necessitate the continuous exchange of energy and matter with the outside world. For this reason, Prigogine, who first suggested the existence of these states, has called them *dissipative structures*.[1-3]

Let us formulate the problem in thermodynamic terms in its most general form. Consider a nonisolated system (closed or open), subject to constraints which give rise to a steady nonequilibrium state. In this state, the values of different thermodynamic variables such as flows, etc., depend parametrically on a number of quantities, $\{A\}$ measuring the deviation of the system from equilibrium. For instance, A may be a gradient of composition, the overall affinity of a set of coupled chemical reactions, and so on. Let us adopt the convention that the state $\{A = 0\}$ is the state of thermodynamic equilibrium. For $\{A \neq 0\}$, but small, the equilibrium regime is continued by the steady states close to equilibrium, for which stability is guaranteed once equilibrium is stable (minimum entropy production theorem). On the contrary, for $\{A \neq 0\}$ and arbitrarily large, although it is always possible to define a continuous extrapolation of the equilibrium regime, the stability of the states belonging to this branch, which will be referred to as *thermodynamic branch*, is no longer insured automatically. In addition, the uniqueness property of the equilibrium state is not applicable in this case and the system may present multiple stationary states, provided it obeys nonlinear laws. One of these stationary states belongs to the thermodynamic branch but is not necessarily stable. It is therefore possible, *a priori*, to have a number of new effects, for instance; the system may not decay monotonously to the steady state belonging to the thermodynamic branch, once it is perturbed from it; in the limit, it may even never return to this state but evolve to a time dependent regime; under similar conditions it may finally deviate and evolve to a new stationary regime corresponding to a branch different from the thermodynamic one. This transition will be manifested abruptly as an *instability*, that is, as a fundamentally discontinuous process.

This situation is frequently met in hydrodynamics. In this domain, the problem of instabilities is a classical one which has been studied thoroughly since the early years of the present century. Recently, Glansdorff and Prigogine have shown that the general formulation of nonlinear thermodynamics provides a framework which permits a thermodynamic analysis of such instability phenomena. They have also been able to formulate their conclusions in terms very similar to phase transitions. Among the different problems treated are: the onset of thermal convection in a horizontal fluid layer heated from below (Bénard problem); stability of waves and the formation of shocks and detonations; stability of parallel flows; and others.[2]

The occurrence of instabilities is a much less obvious effect for purely dissipative systems, that is, systems without a mechanical motion, discussed in this review. In fact it is only during recent years that this possibility has been studied systematically and a theory of dissipative structures has been set up for such systems.

This problem presents a special interest because of the possible implications of the results in the understanding of biological phenomena. Indeed, typical biological phenomena inside a cell appear to occur under the same conditions as in open systems (exchange of ions through membrane processes, ADP $\rightleftarrows$ ATP transformations inside the cell, and so on). Therefore, it is tempting to associate certain types of biological structures with dissipative structures arising beyond a chemical instability. This point will be discussed in detail in Sec. IX. We shall first analyze in the subsequent sections IV to VII, the type of nonequilibrium phenomena which may arise in open systems in the neighborhood of instabilities. As the monotonic approach to a unique steady state is no longer insured far from equilibrium, we may distinguish between three possible situations; oscillations around steady states; symmetry-breaking instabilities; and multiple steady states.

Problems related to oscillations are particularly interesting in the limit of negligible damping, that is, of sustained oscillations. In this limit, the concentrations of chemicals are periodic in time and the steady state is never attained. In Secs. IV and V, we discuss this phenomenon for two widely different types of systems; systems characterized by some conservation property, and systems without such a property. In the latter case, sustained oscillations present certain peculiar features as they share the properties of *limit cycles* first discussed by Poincaré in nonlinear mechanics.

Symmetry-breaking instabilities refer to the spontaneous appearance of spatial structure in a previously homogeneous system. This spontaneous selforganization has interesting implications from the point of view of both the space order and the function of the system, which are discussed in Sec. VI.

A change in functional behavior may also arise in systems which keep their macroscopic structure unchanged but which may switch between several simultaneously stable steady states; the latter may only differ by the levels of concentration of the different constituents. This problem will be discussed in Sec. VII.

Summarizing, we may say that instabilities in the thermodynamic branch of solutions can lead to time or space organization and to a change in functional behavior in open systems undergoing chemical reactions. These instabilities can only arise at a finite distance from thermodynamic equilibrium, that is, their occurrence necessitates a minimum level of dissipation. *Structure and dissipation appear, therefore, to be intimately connected far from equilibrium.* This point will be repeatedly stressed in the subsequent sections.

IV. CHEMICAL OSCILLATIONS—GENERAL THEORY AND CONSERVATIVE SYSTEMS

A. Introduction

In this and the following section, the properties of open systems which are maintained spatially uniform are considered. In this limit, the general kinetic equation (23) does not involve diffusion terms and therefore becomes an ordinary differential equation:

$$\frac{dn_i}{dt} = \sum_{\rho=1}^{s} v_{i\rho} w_\rho = v_i(n_j) \tag{73}$$

where v_i denotes "global velocities" of reactions involving chemical i.

In this section we study, more specifically, systems described by Eq. (73) which possess, in addition, in the limit of very large deviations from equilibrium, a function of the dependent variables which is *constant of motion* and therefore plays a role similar to energy in Mechanics. We shall see that the behavior of such systems is very different from the behavior of systems operating beyond instability (cf. Sec. V).

In general, a system of the form (73) whose coefficients are time independent, that is, an autonomous system, admits a number of steady state solutions and, in addition, a number of oscillatory solutions which are sustained (or weakly damped). In Secs. IV and V we deal only with the oscillatory solutions; the problem of multisteady states will be discussed in Sec. VII.

By a (first) integral of system (73), we will understand a (nonconstant) differentiable function $V = V(n_1, n_2, \ldots, n_n)$ defined in a domain D of the phase space such that $V(X_1, X_2, \ldots, X_n) = $ constant, when X_i are solutions of (73). By definition, a system is conservative in D if it admits a *single valued* integral in this domain and if D is such that every trajectory having one point in D lies entirely in D for all times.

Let us now consider the kinetics of system (73) in phase space. We may think of (73) as defining a flow in this space. A point lying at $\{n_{i0}\}$ at $t = t_0$ evolves at time t to another point $\{n_i\}$ of the same space. The system (73) therefore defines a "mapping" between the totality of initial conditions (at t_0) in a domain D_{t_0} to another domain D_t at $t > t_0$. We define the (Lebesgue) measure $\mathscr{V}_t$ of D_t as the phase space "volume" at time t:

$$\mathscr{V}_t = \int_{D_t} dn_1 \ldots dn_n \tag{74}$$

The following remarkable result can now be established (for a proof see Cesari, Ref. 20):

Theorem: If div. $v = 0$, then $\mathscr{V}_t = \mathscr{V}_{t_0}$ for every t that is, Eq. (73) defines a measure preserving transformation. Alternatively (74) is an *integral invariant* of the system.

This theorem is particularly important, since for $n = 2m$, the system of Hamiltonian equations describing the motion of a mechanical conservative system with constraints independent of t, does verify the condition div $v = 0$ and possesses in addition, a single valued first integral, the total energy. We shall verify in examples, later on in this section that, whenever a constant of motion V exists, it is often possible to transform to new phase variables such that div $v = 0$ even if this was not true in the initial variables. There is therefore a certain analogy between mechanical and nonlinear conservative systems, even though the latter may be purely dissipative.

The evolution of a conservative system also presents certain peculiar stability properties. Let us introduce the concept of an invariant set D in phase space. A closed bounded set D is said to be invariant for a system (73) if every trajectory having one point in D lies entirely in D for all times. Clearly, the existence of a constant of motion guarantees the existence of infinitely many invariant sets. The following properties may then be proved (see Cesari, Ref. 20, Chapter 3, Sec. 6.8).

Property (1): A steady state solution $\{X_{0i}\}$ of Eq. (73) is (Lyapounov) stable if and only if there exists a sequence of invariant sets $D_n (n = 1, 2, \ldots)$ closing down to X_0. Clearly, this condition is satisfied by the existence of single valued first integrals. Notice that asymptotic stability is not guaranteed by this theorem.

Property (2): Let D_0 be a neighborhood of an initial state $X(t = t_0)$. Then according to Poincaré, if div $v = 0$, and an invariant set exists, an arbitrary motion returns infinitely often to the neighborhood of its initial state. Poincaré called this property "stability in the sense of Poisson". We see that this concept is closely related to the orbital stability discussed in Sec. III-C. Again, asymptotic orbital stability is not insured.

Further properties of systems of the form (73) will be discussed in the next subsection in the more easily visualized case of two independent variables.

B. Two Independent Variables—Singular Points and Variational Equations

We consider a set of chemical reactions involving only two intermediates of instantaneous molar concentrations X and Y. Equation (73) then reads:

$$\frac{dX}{dt} = v_X(X, Y)$$
$$\frac{dY}{dt} = v_Y(X, Y) \tag{75}$$

The system is assumed to be autonomous, which amounts to imposing time independent chemical constraints.

We assume that v_X, v_Y are Lipschitz-continuous in a domain of phase space. The Cauchy–Lipschitz theorem then asserts the existence and uniqueness of solutions of (75). This theorem applied to the autonomous system (75) implies that *through every point of the phase plane there passes one and only one integral curve.*[20,21]

As v_X and v_Y do not depend on time explicitly, one can eliminate t from (75). The result is:

$$\frac{dY}{dX} = \frac{v_Y(X, Y)}{v_X(X, Y)} \tag{76}$$

The solution of this equation defines a *trajectory* of the system. Using the same arguments as in point (3) of Sec. III-C, we can then ascertain that *to a given trajectory correspond infinitely many solutions of system (75) differing from each other by the phase.*

A closed trajectory in phase plane necessarily represents periodic solutions of the differential system.[20,21]

A phase space point (X_0, Y_0) for which v_X and v_Y do not vanish simultaneously is called an *ordinary point*, otherwise it is called a *singular point* of the differential system (75). We see that a singular point necessarily corresponds to a steady state solution of the system ($dX/dt = dY/dt = 0$). The study of the behavior of the system close to the steady state is performed most conveniently by means of a stability analysis of the latter. For this purpose, we consider the variational system of (75) (cf. also Eq. (44)).

We set

$$X = X_0 + x(t)$$
$$Y = y_0 + y(t) \tag{77}$$

and obtain

$$\frac{dx}{dt} = v_X(x, y; X_0, Y_0)$$

$$\frac{dy}{dt} = v_Y(x, y; X_0, Y_0)$$

(78)

The stability to be studied now is that of the zero solution of system (78). It is illuminating to write (78) in a form wherein linear terms in x and y are separated from the others:

$$\frac{dx}{dt} = a_{11}x + a_{12}y + v_x(x, y)$$

$$\frac{dy}{dt} = a_{21}x + a_{22}y + v_y(x, y)$$

(79)

where

$$a_{11} = \left(\frac{\partial v_X}{\partial X}\right)_{\substack{X=X_0 \\ Y=Y_0}}, \qquad v_x = 0(x^2, y^2, xy)$$

(80)

and so on.

Consider first the system (79) where nonlinear terms are neglected. Alternatively, we focus attention on the infinitesimal stability properties of the steady state. As (79) now becomes linear, it admits solutions of the form[20]

$$x = x_0 e^{\omega t}$$

$$y = y_0 e^{\omega t}$$

(81)

A solution of the form (81) is called a *normal mode* of frequency ω.* Inserting (81) into (79), we obtain (exp $[\omega t]$ simplifies on both sides):

$$(\omega - a_{11})x_0 - a_{12}y_0 = 0$$

$$- a_{21}x_0 + (\omega - a_{22})y_0 = 0$$

(82)

This linear homogeneous algebraic system admits nontrivial solutions $x_0 y_0 \neq 0$ if and only if

$$\begin{vmatrix} a_{11} - \omega & a_{12} \\ a_{21} & a_{22} - \omega \end{vmatrix} = 0$$

(83)

---

* The concept of normal mode is straightforwardly generalized to linear systems involving an arbitrary number of variables.

This is the *characteristic*, or *secular equation* of system (79). In general, it admits two solutions ω_i which are called the eigenvalues (or characteristic exponents) of (79). The main point is that this equation guarantees the existence of nonvanishing solutions of the form (81). In a more explicit form, Eq. (83) yields:

$$\omega^2 - (a_{11} + a_{22})\omega + (a_{11}a_{22} - a_{12}a_{21}) = 0$$

or   (84)

$$\omega^2 - Tr\omega + \Delta = 0$$

where Tr, Δ are the trace and the determinant of the coefficient matrix of the linear system. We observe that by (80), Δ is also equal to the Jacobian of the original system, evaluated at the steady state point. The stability features of the steady state of the linearized system depend upon the value of ω calculated from (84). Indeed, we will have:

$$x(t) = x_{01}\, e^{\omega_1 t} + x_{02}\, e^{\omega_2 t}$$
$$y(t) = y_{01}\, e^{\omega_1 t} + y_{02}\, e^{\omega_2 t}$$
(85)

(1) If $\text{Re }\omega_i < 0$, $x(t) \to 0$ as $t \to \infty$, and the zero solution of (79) is asymptotically stable.

(2) If at least one of the ω_i has a positive real part then $x(t) \to \infty$ as $t \to \infty$ and the zero solution of (79) is unstable.

Moreover, if both ω_i are real and of the same sign, the approach to (or departure from) the steady state is monotonic; the singular point is a *node*. The conditions for a node are: $Tr \neq 0$, $Tr^2 - 4\Delta > 0$. If $\omega_i = 0$ around a node, the system is at a state of *marginal stability*.

If ω_i are real and of opposite sign, the (unstable) singular point is a *saddle point*.

If ω_i are complex conjugate with nonvanishing real parts, the motion around the steady state is spiral. The singular point is a *focus*.

Finally, if ω_i are purely imaginary, the system rotates around the steady state. We say that this state is *marginally stable*. The singular point is a *center*. The conditions for a center are: $Tr = 0$, $\Delta > 0$.

In more general problems involving more than two variables, the stability analysis proceeds along the same lines. However, the investigation of the character of the roots of the secular equation becomes more complicated and requires additional considerations. We only mention here the stability conditions given by the criteria of Hurwitz, and refer the reader to the abundant literature for details. A good survey is given in Cesari, Ref. 20, Chapter 1.

Up to now we have studied the (infinitesimal) stability around a steady state point. As a result, the coefficients of the linearized variational system (79) were constant. A very interesting type of problems however is described by time dependent, periodic solutions of Eq. (75). In this case, the investigation of the motion around these states requires a linear stability analysis where the coefficients are now time dependent:

$$\frac{dx}{dt} = a_{11}(t)x + a_{12}(t)y$$

$$\frac{dy}{dt} = a_{21}(t)x + a_{22}(t)y$$

(86)

with

$$a_{11} = \left(\frac{\partial v_X}{\partial X}\right)_{\substack{X = X_0(t) \\ Y = Y_0(t)}}$$

and so on.

It follows that a_{ij} are periodic with period T, inasmuch as the unperturbed motion (X_0, Y_0) is periodic with period T. The investigation of stability is now more complicated mainly because the problem is not reducible immediately to an algebraic system as in the case of constant coefficients.

The fundamental result regarding Eq. (86) is due to Floquet, and reads (for more details and proofs see Refs. 20 and 21):

Theorem: The system (86) admits solutions of the form

$$x(t) = \exp(h_i t)\phi_i(t)$$

$$y(t) = \exp(h_j t)\phi_j(t)$$

(87)

where $\phi_i(i = 1, 2)$ are periodic functions of period T. The numbers h_i are called the characteristic exponents of (86).

We immediately deduce the corollary: If the characteristic exponents of the variational equation around an unperturbed periodic solution have negative real parts, the solution is asymptotically stable. If one characteristic exponent has a positive real part, the solution is unstable.

Again, these results are straightforwardly generalized to problems involving more than two variables.

Let now $X_0(t)$, $Y_0(t)$ be a periodic solution of the autonomous system (75). We differentiate with respect to t the identity obtained by inserting these functions into (75). We obtain:

$$\frac{d}{dt}\left(\frac{dX_0}{dt}\right) = \left(\frac{\partial v_X}{\partial X}\right)_0 \frac{dX_0}{dt} + \left(\frac{\partial v_X}{\partial Y}\right)_0 \frac{dY_0}{dt}$$

$$\frac{d}{dt}\left(\frac{dY_0}{dt}\right) = \left(\frac{\partial v_Y}{\partial X}\right)_0 \frac{dX_0}{dt} + \left(\frac{\partial v_Y}{\partial Y}\right)_0 \frac{dY_0}{dt}$$

(88)

It follows that dX_0/dt, dY_0/dt satisfy an equation identical to the variational equation (86). Since dX_0/dt, dY_0/dt are also periodic, it follows that (86) always admits a periodic solution; hence (cf. (87)) one of the characteristic exponents is equal to zero. As a result, a periodic solution of an autonomous system can never be asymptotically stable in the sense of Lyapounov. Referring to Sec. III-C, we see that all one can guarantee in this case is asymptotic orbital stability, provided the remaining characteristic exponents have negative real parts. On the other hand, it can be shown that[20,21]

$$\sum_i h_i = \frac{1}{T}\int_0^T dt \, Tr \, a(t)$$

(89)

where $a(t)$ is the coefficient matrix of system (86).

Consider now the case where (75) has a constant of motion $V(X, Y)$. Clearly, the stability properties of (75) are then identical with those of the system

$$\frac{dX}{dt} = V(X, Y)v_X$$

$$\frac{dY}{dt} = V(X, Y)v_Y$$

(90)

From (90) and $dV/dt = 0$ we obtain:

$$\text{div } vV = 0$$

or using the definitions of a_{ij} in (86),

$$\sum_i h_i = 0$$

(91)

for any orbit of (90).

It follows that if an h_i has a negative real part there will be at least one other h_i with positive real part and the system will be orbitally unstable. Hence, a conservative system possesses orbital stability only if all its h_is are zero. Necessarily then orbital stability for such systems can never be asymptotic.

Before we go over to the thermodynamic analysis of conservative systems, we discuss briefly the connection between infinitesimal (linear) stability and global stability of the complete, nonlinear equation (79). Using Lyapounov's second method discussed in Sec. III-C, one can establish the following results (see Ref. 21 for details and proofs):

(1) If all roots of the characteristic equation (83) have negative real parts, the steady state is asymptotically stable for (79) whatever the nonlinear terms.

(2) If at least one characteristic value has a positive real part, the steady state is unstable whatever the nonlinear terms.

(3) If the characteristic equation has roots with zero real parts, then the system may be stable or unstable depending on the form of the nonlinear terms.

We see that the nonlinear character influences the stability of the steady state only in the "critical" case where the singular point is a center or a marginally stable mode.

C. Thermodynamic Interpretation[2]

In this section we combine the results of the thermodynamic analysis of Secs. III-B and III-D with the normal mode analysis outlined in the previous section and derive the behavior of a single normal mode around the steady state. In a linear problem, this would cover all relevant aspects of infinitesimal stability; indeed, as we have seen, the stability of the system is compromised as soon as a single normal mode has a characteristic exponent with a positive real part.

We first observe that for a single normal mode of complex frequency the fluctuation amplitudes appearing in the evolution criterion are complex. As on the other hand, $d_X P$ has to be real, one has to reformulate slightly the criterion in the form (see Eqs. (65) and (66)):

$$
T \frac{d_X P}{dt} = T \frac{d_X \delta P}{dt} = \frac{1}{2} \int d\mathscr{V} \sum_{ij} \left(\frac{\partial \mu_i}{\partial n_j} \right) \left[\frac{d \, \delta n_i^*}{dt} \frac{d \, \delta n_j}{dt} + \frac{d \, \delta n_i}{dt} + \frac{d \, \delta n_j^*}{dt} \right]
$$

$$
= \frac{1}{2} \int d\mathscr{V} \sum_{\rho} \left(\delta w_\rho^* \frac{d \, \delta \mathscr{A}_\rho}{dt} + \delta w_\rho \frac{d \, \delta \mathscr{A}_\rho^*}{dt} \right) \tag{92}
$$

Notice that we have written here the evolution criterion in terms of excess variables and limited the expressions to second order terms in the deviations from the steady state (see Eq. (67)).

For a single normal mode

$$
\delta w_\rho = \delta w_\rho{}^0 \exp[(\omega_1 + i\omega_2)t]
$$
$$
\delta \mathscr{A}_\rho = \delta \mathscr{A}_\rho{}^0 \exp[(\omega_1 + i\omega_2)t] \tag{93}
$$

Eq. (92) becomes

$$\frac{1}{2}\int d\mathscr{V}\left\{\omega_1 \sum_\rho (\delta w_\rho^* \,\delta\mathscr{A}_\rho + \delta w_\rho \,\delta\mathscr{A}_\rho^*)\right.$$

$$\left. + i\,\omega_2 \sum_\rho (\delta w_\rho^* \,\delta\mathscr{A}_\rho - \delta w_\rho \,\delta\mathscr{A}_\rho^*)\right\} \leq 0 \qquad (94)$$

The coefficient of ω_1 in the left hand side is just the excess entropy production, δP, associated with the normal mode. Using Eq. (69) we obtain:

$$\delta P = \frac{1}{2}\int d\mathscr{V} \sum_\rho (\delta w_\rho^* \,\delta\mathscr{A}_\rho + \delta w_\rho \,\delta\mathscr{A}_\rho^*)$$

$$= \frac{\partial}{\partial t}(\delta^2 S) \qquad (95)$$

where $\delta^2 S$ is also defined in terms of complex normal modes by analogy to (92).

The coefficient of ω_2

$$\delta\Pi = \frac{i}{2}\int d\mathscr{V} \sum_\rho (\delta w_\rho^* \,\delta\mathscr{A}_\rho - \delta w_\rho \,\delta\mathscr{A}_\rho^*) \qquad (96)$$

is a real quantity which presents also some interesting properties.

Now for a single normal mode

$$\omega_1\,\delta P = \omega_1 \frac{\partial}{\partial t}(\delta^2 S) = \omega_1{}^2\,\delta^2 S \leq 0 \qquad (97a)$$

$$\omega_2\,\delta\Pi = T\,\frac{d_x P}{dt} - \omega_1\,\delta P$$

$$= (\omega_1{}^2 + \omega_2{}^2)\,\delta^2 S - \omega_1{}^2\,\delta^2 S = \omega_2{}^2\,\delta^2 S \leq 0 \qquad (97b)$$

by Eq. (63).

Inequality (97a) gives the stability criterion for a single normal mode: indeed, for stability $\delta P > 0$ (see Eq. (69)) as $\omega_1 < 0$. The second inequality imposes that the sign of $\delta\Pi$ be opposite to that of the imaginary frequency ω_2. We may say that it "fixes" the direction of rotation around the steady state. From these considerations, we may therefore define two thermodynamic transition points; one satisfying the condition:

$$\delta P = 0 \qquad (98a)$$

implying that the system has attained a critical regime where there exist nontrivial normal modes with $\omega_1 = 0$, that is, the system is at (or around) a marginally stable state.

The second transition point, satisfying the condition

$$\delta\Pi = 0 \qquad (98b)$$

corresponds to a critical state where $\omega_2 = 0$ and the motion around the steady state becomes aperiodic.

The equations just derived may be given a more explicit form by expanding, as in Sec. III-D, the excess reaction velocities δw_ρ in terms of the excess forces. In the linear approximation considered in this section, this expansion reads

$$\delta w_\rho = \frac{1}{T} \sum_{\rho'} l_{\rho\rho'} \, \delta\mathscr{A}_{\rho'} \qquad (99)$$

The coefficients $l_{\rho\rho'}$ are evaluated at the (steady) state of reference. As the latter is a far from equilibrium state, the matrix $(l_{\rho\rho'})$ need not be symmetric; let it be divided into a symmetric and an antisymmetric part $l^s_{\rho\rho'}$ and $l^a_{\rho\rho'}$, respectively. As Eqs. (95) and (96) contain, respectively, a symmetric and an antisymmetric quadratic form we obtain:

$$\delta P = \frac{1}{T^2} \int d\mathscr{V} \sum_{\rho,\rho'} l^s_{\rho\rho'} \, \delta\mathscr{A}^*_\rho \, \delta\mathscr{A}_{\rho'} \qquad (100)$$

$$\delta\Pi = \frac{i}{T^2} \int d\mathscr{V} \sum_{\rho,\rho'} l^a_{\rho\rho'} \, \delta\mathscr{A}^*_\rho \, \delta\mathscr{A}_{\rho'} \qquad (101)$$

From this and Eq. (98), it follows that the thermodynamic criteria for a marginal state and for the appearance of oscillations may be related to the properties of the coefficients in the expansion (99), the latter being also related to the coefficients of the original equations of evolution for the molar concentrations of the chemicals. However, this relation is not quite explicit.

For instance, even when $l^a_{\rho\rho'}$ is nonvanishing, $\delta\Pi$ and ω_2 may be zero, as long as the excess forces $\delta\mathscr{A}_\rho$ may be represented by real normal modes. Alternatively, one may say (see Eqs. (97), (100), and (101)) that ω_1 and ω_2 depend *explicitly* on l^s and l^a, respectively, but *implicitly* upon both sets of coefficients through the excess affinities $\delta\mathscr{A}_\rho$. The latter should satisfy the linearized equations of evolution (transformed to affinity space, as in Sec. III-D).

We see therefore that the properties of singular points discussed in the previous subsection may be cast into properties of well defined thermodynamic quantities. The important point now is to realize that, according to the results of this section and of Section III-D, expression (101) (and therefore also ω_2) always vanishes close to equilibrium, whereas δP can never vanish in this region by virtue of the minimum entropy production theorem. The appearance of oscillations and of instabilities are therefore typical far from equilibrium effects. Oscillations not involving instabilities occur on the thermodynamic branch. This is necessarily the case for oscillations in conservative systems involving two variables (see Eq. (91)). Examples of this behavior will be given in this section. On the other hand, all effects beyond instabilities, including oscillations, occur on a new, non-thermodynamic branch. Examples will be given in Secs. V, VI, and VII.

D. The Volterra–Lotka model

In this section we illustrate the results derived previously on a simple model involving two variables. We consider the following scheme of homogeneous chemical reactions

$$
\begin{array}{lll}
A + X \;\rightleftharpoons\; 2X & \text{(a)} & \\
X + Y \;\rightleftharpoons\; 2Y & \text{(b)} & \quad (102) \\
Y + B \;\rightleftharpoons\; E & \text{(c)} &
\end{array}
$$

We call k_i the forward, and k_{-i} the reverse, reaction constants ($i = 1, 2, 3$). The system is open, in contact with reservoirs of A, B, E which are assumed to be such that the concentrations of A, B, and E remain constant (or slowly vary) in time. The system is therefore characterized by two dependent variables, the molar concentrations X, Y of the intermediates.* We observe that reactions (a) and (b) are autocatalytic, whereas (c) describes a simple decomposition of Y. The thermodynamic state of this system is determined by the values of chemical constraints. It is easily seen that there is a single overall reaction

$$
A + B \;\rightleftharpoons\; E \qquad\qquad (103)
$$

In general, the affinity of this reaction is a complicated function of the molecular interactions through the chemical potentials (see Eq. (26)). In order to avoid unimportant complications, we consider in this review only *ideal systems* behaving as mixtures of perfect gases. In this case, elementary statistical mechanical calculations lead to the following expression of chemical potential:[11]

$$
\mu_i = \mu_i^{\,0} + RT \ln n_i \qquad\qquad (104)
$$

---

* Throughout this review we use the same symbols to denote the various chemical substances and their concentrations.

where μ_i^0 is independent of the composition, R is the gas constant and T is the temperature. It follows that for reaction (103), the affinity reads

$$\mathscr{A} = \mathscr{A}_1 + \mathscr{A}_2 + \mathscr{A}_3 = RT \ln \left(\frac{k_1 k_2 k_3 AB}{k_{-1} k_{-2} k_{-3} E} \right) \qquad (105)$$

When the assumption of ideal mixtures breaks down, corrections may be introduced through the activity coefficients.

We shall first state without proof a number of results which have been established elsewhere for system (102) (see Refs. 2, 25).

(1) In the linear range of thermodynamics around equilibrium

$$\delta P > 0, \quad \delta \Pi = 0 \qquad (106)$$

The system therefore is stable and cannot undergo oscillations.

(2) As the affinity (105) grows, the steady state of the system never becomes unstable: $\omega_1 \leq 0$. However, there exists a critical value $\mathscr{A}_c$ such that for $|\mathscr{A}| > |\mathscr{A}_c|$, the motion around the steady state becomes spiral. At this point $\delta \Pi$ switches to nonzero values as $\mathscr{A}$ increases to infinity.

We see that the general conclusions of the previous section are verified in this example. We shall now be especially interested in the limit of large affinity. It is convenient to express this limit by neglecting all back reactions in scheme (102) setting $k_{-i} = 0$. The kinetic equations giving the evolution of X and Y become (assuming the system to be ideal):

$$\frac{dX}{dt} = k_1 AX - k_2 XY \qquad \text{(a)}$$

$$\frac{dY}{dt} = k_2 XY - k_3 BY \qquad \text{(b)}$$

$$\qquad (107)$$

From these equations we see that scheme (102) becomes isomorphic to a model originally introduced by Lotka[26] and studied in detail later on by Volterra[27] to describe the behavior of a number of predator-prey biological species in interaction. More recently, the model (suitably extrapolated) has been used to discuss quite fundamental biological phenomena such as biological clocks, cellular control processes[28] or time dependent properties of neural networks.[29]

Let us now investigate the properties of the system described by (107). We immediately see that it admits a single nontrivial steady state solution:

$$X_0 = \frac{k_3 B}{k_2}, \qquad Y_0 = \frac{k_1 A}{k_2} \qquad (108)$$

and the trival solution

$$X_0 = Y_0 = 0$$

The properties of the perturbed motion of the system around the steady state will be investigated by the normal mode analysis outlined in Sec. IV-B. We immediately obtain the linearized equations corresponding to the nontrivial state

$$\omega x + k_3 By = 0$$
$$k_1 Ax + \omega y = 0 \tag{109}$$

and the secular equation

$$\omega^2 + k_1 k_3 AB = 0 \tag{110}$$

We see that small fluctuations around the state (108) are *periodic*, with a universal frequency depending only on the parameters descriptive of the system:

$$\omega = \pm i(k_1 k_3 AB)^{1/2} \tag{111}$$

The singular point is therefore a center. A similar analysis shows that the trivial steady state is always a saddle point.

Therefore, the Volterra–Lotka scheme provides a model for sustained oscillations in a chemical system. It should be realized that this result is a consequence of the limit of infinite affinity taken above. Certainly this limit is only an idealization of a physical process wherein back reactions subsist but are small. The evolution of a system described by (102) should therefore be considered as being separated into two stages, corresponding to two largely different time scales; one *fast* scale corresponding to the angular frequency (111) and one *slow* scale corresponding to the (small) inverse reaction constants k_{-i}. As a result the system, after spending a more or less long time in the regime described by Eq. (107), will finally decay slowly to the steady state (108). Here we are essentially dealing with the first stage of evolution corresponding to Eqs. (107).

Let us now investigate the evolution of perturbations at finite distance from the steady state. Let us introduce, instead of X and Y, new variables corresponding to the two independent affinities of reactions (102). We choose

$$\mathscr{A}_{\mathrm{I}} = -\mathscr{A}_1 = \ln \frac{k_{-1}X}{k_1 A} = \ln \frac{k_{-1}}{k_1 A} + \ln X$$
$$\mathscr{A}_{\mathrm{II}} = \mathscr{A}_3 = \ln \frac{k_3 YB}{k_{-3}E} = \ln \frac{k_3 B}{k_{-3}E} + \ln Y \tag{112}$$

$\mathscr{A}_\mathrm{I}$ and $\mathscr{A}_\mathrm{II}$ involve two constants added to $\ln X$, $\ln Y$. The study of the tine evolution is not affected by these constants. We therefore take as new variables

$$u_1 = \mathscr{A}_\mathrm{I} - \ln \frac{k_{-1}}{k_1 A} = \ln X$$

$$u_2 = \mathscr{A}_\mathrm{II} - \ln \frac{k_3 B}{k_{-3} E} = \ln Y \tag{113}$$

Dividing through Eqs. (107) by X and Y, we obtain (for simplicity we set all $k_i = 1$):

$$\frac{du_1}{dt} = A - e^{u_2} = v_1$$

$$\frac{du_2}{dt} = e^{u_1} - B = v_2 \tag{114}$$

We multiply both sides by $(e^{u_1} - B)$, $(e^{u_2} - A)$, respectively and add. We obtain:

$$(e^{u_1} - B)\frac{du_1}{dt} + (e^{u_2} - A)\frac{du_2}{dt} = 0$$

or

$$\frac{d}{dt}[e^{u_1} + e^{u_2} - Bu_1 - Au_2] = 0 \tag{115a}$$

or, in terms of X and Y:

$$X + Y - B \ln X - A \ln Y = \text{Constant} = V \tag{115b}$$

The system is therefore *conservative* as it admits a constant of motion.* The latter is additive over the different constituents. Clearly, (115b) defines an infinity of trajectories in phase space corresponding to different initial conditions. For X and Y close to the steady state (case of small perturbations) these trajectories become concentric ellipses. Indeed, expanding (115b) around X_0, Y_0, and retaining terms to first nontrivial order we obtain:

$$\frac{(X - X_0)^2}{2X_0} + \frac{(Y - Y_0)^2}{2Y_0} = \text{Constant} = -(V - V_0) \tag{116}$$

* See also Ref. 27, 33. A systematic study of the Volterra–Lotka model in the variables u_1 and u_2 has been performed by Kerner, Ref. 30–32. Here we see that this description amounts to switching to the space of generalized thermodynamic forces, that is, the affinities.

For finite $X - X_0$, $Y - Y_0$, the trajectories are deformed but remain *closed* provided V does not exceed some critical value.[27,33] According to Sec. IV-B, this implies that finite perturbations around the steady state *are also periodic*. It is easy to calculate the periods of these trajectories. We multiply the nonlinear equations (107) by $(Y - Y_0)$, $(X - X_0)$ respectively, and subtract. We obtain:

$$(X - X_0)\frac{dY}{dt} - (Y - Y_0)\frac{dX}{dt} = Y(X - X_0)^2 + X(Y - Y_0)^2 \quad (117)$$

We introduce polar coordinates in phase space around the point (X_0, Y_0):

$$X = X_0 + \rho \cos \phi, \qquad Y = Y_0 + \rho \sin \phi \quad (118)$$

We obtain:

$$\frac{d\phi}{dt} = (A + \rho \sin \phi) \cos^2 \phi + (B + \rho \sin \phi) \sin^2 \phi \quad (119)$$

This equation implies a periodic motion with period[27]

$$T = \int_0^{2\pi} d\phi[(A + \rho \sin \phi) \cos^2 \phi + (B + \rho \cos \phi) \sin^2 \phi]^{-1} \quad (120)$$

where $\rho = \rho(\phi)$ is the equation of the trajectory obtained by eliminating t from Eqs. (107). We see that each trajectory with finite ρ has a different period depending on the value of the constant V, that is, on initial conditions. The Volterra–Lotka model has therefore a continuous spectrum of frequencies associated with the existence of infinitely many periodic trajectories. This is a very important point, as it implies the lack of asymptotic orbital stability (see also Eq. (91)), that is, the lack of decay of fluctuations. As a result of infinitesimal fluctuations, the system will continuously switch to orbits with different frequencies and there will be no average, "preferred" orbit. Intuitively, oscillations of this type (in fact, oscillations for all conservative systems) may only describe noise type of effects rather than periodic effects characterized by sharply defined amplitudes and frequencies. We shall come back to this point in Sec. V.

Let us also mention another basic property of the model. Integrating both sides of Eq. (114) over a period and taking into account definitions (113) we obtain:

$$0 = \frac{1}{T} \int_0^T dt \frac{d \ln X}{dt} = A - \frac{1}{T} \int_0^T dt\, Y(t) \quad (121)$$

or

$$(\overline{Y})_T = \frac{1}{T} \int_0^T dt \, Y(t) = A = Y_0 \tag{122a}$$

Similarly

$$(\overline{X})_T = B = B_0 \tag{122b}$$

In other terms, the average concentration of X and Y over an arbitrary trajectory are equal to the steady state values.[27]

The properties of the Volterra–Lotka model outlined in this subsection are also shared by other chemical systems. Perhaps the most important class of such systems is that describing cellular control processes. The simplest set of equations describing such processes has been studied by Goodwin.[28] Let Y_i be the concentration of a protein whose synthesis is controlled by the corresponding species of mRNA of concentration X_i. Goodwin's equations read

$$\frac{dX_i}{dt} = \frac{a_i}{A_i + k_i Y_i} - b_i$$

$$\frac{dY_i}{dt} = \alpha_i X_i - \beta_i \tag{123}$$

where a_i, b_i, α_i, β_i, A_i, k_i are parameters maintained constant. Goodwin shows that system (123) is conservative and it therefore exhibits Volterra–Lotka type of oscillations. Equations of the same form as (123) may also be used to describe enzyme controlled reactions involving various types of negative feedback.

E. Thermodynamic properties of the Volterra–Lotka system

We first observe that in the space of affinities the Volterra–Lotka model is described by a divergence-free velocity field (see Eq. (114)). As a result, the conditions of the theorem of Sec. IV-A are satisfied and the system admits the phase space volume as an integral invariant. On the other hand, it is intuitively appealing to define the entropy of the system in terms of the phase space volume available by the classical relationship:[11]

$$S = k \ln \mathscr{V} + \text{Constant}$$

It follows that

$$\frac{dS}{dt} = 0 \quad \text{or} \quad S = \text{Constant} \tag{124}$$

This conjecture is substantiated by an explicit calculation in the limit of small fluctuations around the steady state. Using the notations of Sec. IV-C, we obtain (per unit volume):

$$T \frac{\partial\, \delta^2(ns)}{\partial t} = T\delta\sigma = \sum_\rho \delta w_\rho\, \delta \mathscr{A}_\rho \tag{125}$$

Now

$$\delta w_1 = \delta(AX) = A\delta X \qquad\qquad \delta \mathscr{A}_1 = \delta \ln \frac{AX}{X^2} = -\frac{\delta X}{X_0}$$

$$\delta w_2 = \delta(XY) = X_0\, \delta Y + Y_0\, \delta X \qquad \delta \mathscr{A}_2 = \delta \ln \frac{X}{Y} = \frac{\delta X}{X_0} - \frac{\delta Y}{Y_0}$$

$$\delta w_3 = \delta(YB) = B\delta Y \qquad\qquad \delta \mathscr{A}_3 = \delta \ln \frac{YB}{E} = \frac{\delta Y}{Y_0} \tag{126}$$

It follows that:

$$T \frac{d}{dt} \delta^2(ns) = \frac{Y_0 - A}{X_0} (\delta X)^2 + \frac{B - X_0}{Y_0} (\delta Y)^2 = 0 \tag{127}$$

$\delta^2(ns)$ is therefore a constant of motion. In fact, by applying definitions (63) and (104) one can show that $\delta^2(ns)$ is proportional to the excess constant of motion $V - V_0$ derived in Eq. (116). In terms of stability theory (see Sec. III-C and III-D) this result implies that the conditions of the first Lyapounov theorem are satisfied $\left(\delta^2(ns) < 0, \frac{d}{dt} \delta^2(ns) = 0 \right)$, and therefore the system is stable (around the steady state) but *not* asymptotically stable. We can call this type of stability "indifferent" or "marginal" stability. Notice that this result is in agreement with the normal mode analysis of the previous section where it was shown that the real part of the frequency vanishes.

An essential point to keep in mind is that this peculiar behavior is a consequence of the nonlinearity of the equation for X and Y which in turn is a consequence of the autocatalytic character of the original reaction scheme (102). This general feature will be repeatedly stressed in the following sections.

Another related property of the system is that the average entropy production over an arbitrary trajectory remains equal to the value of entropy production at the steady state. Indeed

$$T\sigma = \sum_\rho w_\rho \mathscr{A}_\rho = AX \ln A - Y \ln E$$

$$+ (XY - AX) \ln X + (BY - XY) \ln Y \tag{128}$$

Averaging over a period and taking into account Eq. (122) we obtain:

$$T(\overline{\Delta\sigma})_T = T(\overline{(\bar{\sigma})_T - \sigma_0}) = -\overline{\left(\frac{dX}{dt}\ln X + \frac{dY}{dt}\ln Y\right)}_T$$

$$= -\frac{d}{dt}\overline{(X\ln X + Y\ln Y)}_T + \left[\frac{d}{dt}(X+Y)\right]_T = 0 \quad (129)$$

The properties of the motion along a *given* trajectory are conveniently described (for small fluctuations) by the quantity $\delta\Pi$ (per unit volume), Eq. (96). We use the linearized equations (97) for one normal mode in order to eliminate δY in terms of δX. We obtain:

$$\delta\Pi = i\left[\frac{\omega}{B}|\delta X|^2 + \frac{A}{\omega^*}|\delta X|^2\right]$$

Let $\omega = \pm i(AB)^{1/2}$. It follows:

$$\delta\Pi = -(\delta X|^2\left[\pm\left(\frac{A}{B}\right)^{1/2} \pm \left(\frac{A}{B}\right)^{1/2}\right] = \mp 2\left(\frac{A}{B}\right)^{1/2}|\delta X|^2 \quad (130)$$

This has a definite sign once a particular ω is chosen as solution of the secular equation. Alternatively we can say that, since $\delta P = 0$,

$$T\frac{d_X\sigma}{dt} = \omega_2\delta\Pi = \pm 2(AB)^{1/2}(\mp)\left(\frac{A}{B}\right)^{1/2}|\delta X|^2 \le 0 \quad (131)$$

This inequality is independent of the sign of ω. It is equivalent to (117) in the limit of small deviations from the steady state and can be considered as "fixing" the direction of rotation around this state.

An additional interesting property of the system may be expressed in terms of the function V and the thermodynamic variables u_1 and u_2 introduced in Eq. (113). Indeed, from (114) and (115b) we obtain:

$$\frac{\partial V}{\partial u_2} = e^{u_2} - A = -v_1$$

$$\frac{\partial V}{\partial u_1} = e^{u_1} - B = v_2$$

$$(132)$$

It follows that

$$\frac{du_1}{dt} = -\frac{\partial V}{\partial u_2}$$

$$\frac{du_2}{dt} = \frac{\partial V}{\partial u_1}$$

$$(133)$$

and

$$\frac{\partial^2 V}{\partial u_2\, \partial u_1} = \frac{\partial v_2}{\partial u_2} = -\frac{\partial v_1}{\partial u_1} = \frac{\partial^2 V}{\partial u_1\, \partial u_2} \tag{134}$$

The existence conditions for V are thus satisfied, and V may be considered as a state function. The equations of motion therefore take the same form as the Hamiltonian equations of classical mechanics: V plays the same role as energy does in mechanics.[30,32] Referring to the discussion of Sec. III-D, we can also see that V may be considered as the component of a vector potential along an axis perpendicular to the phase plane (u_1, u_2).

Clearly the existence of an "energy like" constant of motion is very exceptional and is peculiar to Volterra–Lotka type of systems. In addition, V does not admit a straightforward thermodynamic interpretation in the general case, except in the limit of small fluctuations where it reduces to the excess entropy.

Further consequences of this "Hamiltonian" formulation will be discussed in Sec. VIII.

V. LIMIT CYCLES

A. Introduction

In this section, we investigate the behavior of chemical systems which may attain a point of instability of the steady state and then go to a new type or regime. In this investigation, we will be greatly helped by the fundamental discovery of Poincaré who has shown that the differential equations of the form (75):

$$\frac{dX}{dt} = v_X(X, Y)$$

$$\frac{dY}{dt} = v_Y(X, Y) \tag{135}$$

may admit solutions represented by closed curves in the phase plane (hence periodic solutions) which are such that another trajectory which is also closed is necessarily at a finite distance from the former. He called these trajectories *limit cycles*.[20,21] If all neighboring trajectories approach a limit cycle as $t \to \infty$ we say that it is asymptotically orbitally stable. If they go away from it, it is orbitally unstable. Finally, if the trajectories approach the limit cycle from the one side and depart from it from the other side, the limit cycle is semistable.

The fundamental importance of limit cycles is that they represent self-sustained oscillations *in nonlinear, nonconservative* systems. Such oscillatory states have been known for a long time to arise in electron tube circuits or even in mechanical systems. By the definition of the limit cycle, we see that oscillations of this kind do not depend on the initial conditions (in contrast to what happens in the Volterra–Lotka model) but are determined by the differential equation itself, that is, by the parameters descriptive of the system. This also explains why nonlinearity is necessary to obtain such a behavior. Similarly, the condition of lack of a constant of motion guarantees the absence of closed trajectories infinitely near the limit cycle.

The following important results may now be established for limit cycles (see Ref. 20, 21 for proofs and details):

(1) A closed trajectory surrounds at least one singular point. In others terms (see Fig. 2) inside a limit cycle there exists a point representing a steady state of a system. The proof of this property is based on the concept of the so called Poincaré topological index and is not reproduced here.

(2) Negative criterion of Bendixson: consider a system of the form (135). If the expression div *v* does not change sign (or vanish identically) in a domain *D* of phase space, there can be no closed trajectory in *D*.

(3) If a trajectory remains in a finite domain *D* of phase space at a finite distance from any singularity then it is either a closed trajectory or it approaches such a trajectory.

More explicit criteria for the existence of limit cycles have been worked out primarily by Liénard[34] and later on by N. Levinson, O. K. Smith et al.[20,21] These authors consider second order nonlinear differential

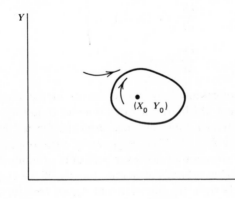

Fig. 2

equations (which are equivalent to a nonlinear first order system (135)) of the form:

$$\frac{d^2X}{dt^2} + f\left(X, \frac{dX}{dt}\right)\frac{dX}{dt} + g(X) = 0 \tag{136a}$$

and establish the existence of periodic solutions provided f and g satisfy certain conditions. Eq. (136) contains as a special case the famous van der Pol equation:[20,21,33]

$$\frac{d^2X}{dt^2} + \mu(X^2 - 1)\frac{dX}{dt} + X = 0 \tag{136b}$$

which describes the selfoscillations in a diode and which has been largely responsible for the early developments of Lienard's theory. We do not give here the various theorems of existence of periodic solutions for Eq. (136). We only consider the case of autonomous systems and state, without any proof, a theorem due to Levinson and Smith which we shall use in Sec. V-B. For details we refer to Ref. 20, Chapter III, Section 9.5.

Theorem: If there are positive constants $a, b, M, a < b$ such that $f(X, X') \geq 0$ for $|X| \geq a$ and all X', $f(X, X') \geq -M$ for all X, X', $f(0, 0) < 0$, $Xg(X) > 0$ for all $X \neq 0$, $G(X) = \int_0^X dXg(X) \to \infty$ as $X \to \pm\infty$, $\int_a^b f(X, X')\, dX \geq 10Ma$ for every decreasing positive function $X'(x)$, then (136) has at least one nonconstant periodic solution.

There exist additional, analytic and approximate methods which permit to predict the existence and calculate the characteristics of limit cycles. We will refer to some of these methods later on in this section.

Usually the way to arrive at a regime described by a limit cycle is the following. A physical system (for example, a set of chemical reactions) described by equations of the form (135) contains a set of parameters $\{A\}$. The solutions of the differential equations thus become functions of $\{A\}$. If for some changes of $\{A\}$, the solutions vary smoothly without any qualitative modifications in the topology of the trajectories, these values of $\{A\}$ will be called *ordinary* values. For instance, referring to Fig. 2, suppose that for A between 0 and A_c, the steady state (X_0, Y_0) is stable. The trajectories in phase space will then be curves which will coalesce to (X_0, Y_0) in a more or less identical way independently of the explicit value of A. However, if for some value $A = A_c$ the topological structure of the trajectories changes qualitatively, we will say that A_c is a *critical*, or *bifurcation* value. For instance, for $A \geq A_c$ the steady state point may become unstable, and for $A > A_c$ a self sustained oscillation may become possible. We say that we will have, in this case, the bifurcation of a stable limit cycle from a singular point. Additional examples of bifurcation may include; the appearance of multiple limit cycles, the appearance of additional stable singular points and so on. Examples will be given in this section and in Section VII.

B. Chemical Instabilities—A Simple Example

We shall now show that it is possible to conceive chemical reaction schemes which exhibit the type of behavior outlined in the previous subsection. We know from this analysis that nonlinearity is a necessary premise. For this reason we choose as a first example, an autocatalytic scheme. The necessity for nonlinearity and the role of autocatalysis will also be clarified in the thermodynamic analysis of Sec. V-C.

The set of reactions considered is

$$
\begin{array}{rcl}
A &\rightleftharpoons& X \\
B + X &\rightleftharpoons& Y + D \\
2X + Y &\rightleftharpoons& 3X \\
X &\rightleftharpoons& E
\end{array}
\tag{137}
$$

It is assumed that the concentrations of "initial" substances A, B and of "final" products D and E are maintained constant.

The autocatalytic step involving a trimolecular reaction is a convenient way to introduce nonlinearity. Mechanism (137) therefore has to be considered as a model rather than as a representation of a particular chemical process.

The scheme contains two distinct reactions

$$
\begin{array}{rcl}
A &\rightleftharpoons& E \\
B &\rightleftharpoons& D
\end{array}
\tag{138a}
$$

the overall reaction being

$$
A + B \rightleftharpoons E + D \tag{138b}
$$

The properties of the system as a function of the affinities of reactions (138) have been investigated in detail elsewhere.[2] It has been shown that beyond a critical affinity corresponding to a far from equilibrium state, the system exhibits damped oscillations. Here we are interested in the stability properties of the system. For this reason we take the extreme case of an infinite overall affinity by setting all inverse kinetic constants equal to zero. The chemical kinetic equations describing (137) become then, in the limit of an ideal mixture (all forward kinetic constants are put equal to one):

$$\frac{dX}{dt} = A + X^2Y - (B + 1)X = v_X$$

$$\frac{dY}{dt} = BX - X^2Y = v_Y$$

(139)

In this limit, the system admits a single steady state solution:

$$X_0 = A, \qquad Y_0 = \frac{B}{A}$$

(140)

which is the continuation of the thermodynamic branch of solutions in the limit as the affinity tends to infinity.

Let us perform a normal mode analysis of the system. We consider infinitesimal perturbations in the form

$$X = X_0 + xe^{\omega t}$$

$$Y = Y_0 + ye^{\omega t}, \qquad \left|\frac{x}{X_0}\right|, \left|\frac{y}{Y_0}\right| \ll 1$$

(141)

Linearizing the system (139) with respect to x and y we get:

$$\omega x = (B - 1)x + A^2y$$

$$\omega y = -Bx - A^2y$$

(142)

The secular equation of (142) is:

$$\omega^2 + (A^2 + 1 - B)\omega + A^2 = 0$$

(143)

We see that if

$$B > B_c, \qquad B_c = A^2 + 1$$

(144)

the real parts of the roots of (143) become positive and the steady state (140) becomes unstable. For $B \neq B_c$, the system is aperiodic as long as

$$B < (A - 1)^2, \qquad \text{or} \qquad B > (A + 1)^2$$

$$(B < B_c) \qquad\qquad\qquad (B > B_c)$$

(145)

Therefore at the bifurcation point, $B = B_c$, the system admits two modes with purely imaginary frequencies

$$\omega = \pm iA$$

(146)

In order to see what happens beyond the bifurcation value of B we set:

$$X(t) = A + x(t), \qquad Y(t) = \frac{B}{A} + y(t)$$

(147)

and study the exact, *nonlinear* system (139). We convert this system to a single second order differential equation for x by differentiating both sides of the equation for dX/dt and eliminating Y and dY/dt using the original equations. We obtain the following equation for x:

$$\frac{d^2x}{dt^2} + \frac{1}{A+x}\left[x^3 + 3Ax^2 + (3A^2 - B - 1)x + A(A^2 - B - 1) - 2\frac{dx}{dt}\right]\frac{dx}{dt}$$
$$+ x(x + A)^2 = 0 \quad (148)$$

This nonlinear equation is of the Liénard type (136) but does not satisfy in this form the conditions of the theorem of Sec. V-A. For this reason we introduce a new variable u through:

$$x = \frac{1}{1/A + u} - A \quad \text{with } u > -\frac{1}{A} \quad (149)$$

We obtain

$$\frac{d^2u}{dt^2} + \left[\frac{A^2}{(Au+1)^2} + 2Au - (B - 1)\right]\frac{du}{dt} + \frac{A^2u}{1+Au} = 0 \quad (150)$$

In this form, the conditions of the Levinson-Smith theorem given in Sec. V-A are satisfied as long as $B > B_c$. This is easily seen for the particular condition $f(0, 0) < 0$ (see Eq. (136)). We have:

$$f(0, 0) = A^2 - (B - 1) = B_c - B$$

that is, $f(0, 0) < 0$ for B beyond the bifurcation value.

The system therefore admits a periodic solution of the limit cycle type. The latter is established abruptly, beyond a point of instability of the thermodynamic state (140). *A limit cycle is therefore a dissipative structure.*

It is interesting to observe that this oscillatory behavior is a purely nonlinear phenomenon which occurs necessarily at a finite distance from the steady state (for $B > B_c$). Indeed, referring to the negative criterion of Bendixson (see Sec. V-A) we see that the condition for the change of sign of div v is:

$$Y = \frac{X}{2} + \frac{B+1}{2X} \quad (151)$$

Now it can be seen that curve (151) cannot be in the neighborhood of the steady state unless B is very close to the bifurcation value B_c. Indeed, referring to Eqs. (80), (84) and (143) we see that, in the neighborhood of the steady state,

$$(\text{div } v)_0 = Tr = \omega_1 + \omega_2 = B - B_c \tag{152}$$

At the marginal state, $B = B_c$ and curve (151) passes through the steady state point: the system admits an infinity of oscillatory modes close to the steady state (see normal mode analysis of this section). For $B > B_c$, however, the curve is at a finite distance from the steady state. As a result, the limit cycle will not be in the immediate neighborhood of this state. In fact we see from Eqs. (80), (84) and (152) that this result can be generalized to all cases involving two dependent variables and a point of oscillatory marginal stability.

We therefore verify on our example that the limit cycle is a purely non-linear phenomenon whose characteristics, that is, period and amplitude, are uniquely determined by the parameters describing the system (initial and final product concentrations, etc.) independently of the initial conditions.

Additional insight about the location and the general characteristics of the limit cycle may be obtained by using approximate methods of analysis of Eqs. (139). We briefly outline here the application of the strobo-scopic method, referring to the monograph by Minorsky[21] for a detailed presentation of the method.

We first write system (139) in the new variables

$$X + Y = \frac{Z}{A} \tag{153a}$$

$$X = W \tag{153b}$$

Introducing excess quantities x, z as in Eq. (147) and taking into account that

$$Z_0 = A^2 + B \tag{153c}$$

we obtain:

$$\frac{dz}{dt} = -Ax$$

$$\frac{dx}{dt} = (B - B_c)x + Az + \left(\frac{B}{A} - 2A\right)x^2 + 2xz + \frac{zx^2}{A} - x^3 \tag{154}$$

We introduce planar polar coordinates by

$$x = r \cos \theta$$
$$y = r \sin \theta \tag{155}$$

and deduce from (154) equations for $\rho = r^2$ and θ. After a few algebraic manipulations we find:

$$\frac{1}{2}\frac{d\rho}{dt} = (B - B_c)\rho \cos^2 \theta + \left[\left(\frac{B}{A} - 2A\right)\cos^3 \theta + 2 \sin \theta \cos^2 \theta\right]\rho^{3/2}$$

$$+ \left(\frac{1}{A}\cos^3 \theta \sin \theta - \cos^4 \theta\right)\rho^2 \quad (156a)$$

$$\frac{d\theta}{dt} = -A - (B - B_c)\sin \theta \cos \theta$$

$$- \left[\left(\frac{B}{A} - 2A\right)\cos^2 \theta \sin \theta + 2 \cos \theta \sin^2 \theta\right]\rho^{1/2}$$

$$+ \left(\cos^3 \theta \sin \theta - \frac{1}{A}\cos^2 \theta \sin^2 \theta\right)\rho \quad (156b)$$

We observe that for $B = B_c$, and when ρ/A, is small, these equations yield:

$$\rho = \rho_0 = \text{constant}, \qquad \theta = \theta_0 - At, \qquad \theta_0 = \text{constant} \quad (157)$$

This behavior is identical to the motion predicted by the normal mode analysis. The system exhibits an infinity of oscillatory modes around the steady state and the trajectories are swept counterclockwise at a constant angular velocity $|\omega_0| = A$ (see Eq. (146)) in the suitably chosen phase plane of z and x. Consider now the case

$$\frac{|B - B_c|}{B} \ll 1 \qquad \text{but finite}$$

$$\quad (158)$$

$$\frac{\rho}{A} \ll 1 \qquad \text{but finite}$$

We seek for solutions of (156) around the harmonic motion (157). For simplicity we set $A = 1$. Let

$$\rho = \rho_0 + \rho_1, \qquad \theta = \theta_0 + \theta_1 \quad (159)$$

where the deviations ρ_1, θ_1, are of the order $(B - B_c)/B$. Linearizing in ρ_1, θ_1 we obtain:

$$\frac{1}{2}\frac{d\rho_1}{dt} = (B - B_c)\rho_0 \cos^2 \theta_0 + \left[\left(\frac{B}{A} - 2A\right)\cos^3 \theta_0 + 2 \sin \theta_0 \cos^2 \theta_0\right]\rho_0^{3/2}$$

$$+ \left(\frac{1}{A}\cos^3 \theta_0 \sin \theta_0 - \cos^4 \theta_0\right)\rho_0^2 \quad (160a)$$

$$\frac{d\theta_1}{dt} = -(B - B_c) \sin \theta_0 \cos \theta_0$$

$$- \left[\left(\frac{B}{A} - 2A \right) \cos^2 \theta_0 \sin \theta_0 + 2 \cos \theta_0 \sin^2 \theta_0 \right] \rho_0^{1/2}$$

$$+ \left(\cos^3 \theta_0 \sin \theta_0 - \frac{1}{A} \cos^2 \theta_0 \sin^2 \theta_0 \right) \rho_0 \qquad (160b)$$

We first deal with Eq. (160). We integrate over a period, recalling that $|\omega_0| = 1$. Taking into account that odd powers in $\sin \theta_0$ or $\cos \theta_0$ vanish by this integration, we obtain

$$\rho_1(2\pi) = 2\pi\rho_0[(B - B_c) - \tfrac{3}{4}\rho_0] \qquad (161a)$$

This equation gives the evolution of ρ in the "stroboscopic time scale" $2\pi(B - B_c)$. Passing to the limit of a continuous time dependence we obtain instead of (161a) the equation

$$\frac{d\rho}{dt} = \rho[(B - B_c) - \tfrac{3}{4}\rho] \qquad (161b)$$

This equation admits a single stationary state solution,

$$\rho^{(0)} = \tfrac{4}{3}(B - B_c), \qquad \text{that is, } r^{(0)} = \frac{2}{\sqrt{3}} (B - B_c)^{1/2} \qquad (162)$$

For $B > B_c$, this solution is physical and is at a finite distance from the steady state. Clearly, it represents a limit cycle with an "average" radius given by (162). It also shows that no other limit cycles exist for this problem. The stability of this limit cycle is investigated by considering the variational equation of (161b):

$$\frac{d \, \delta\rho}{dt} = [(B - B_c) \, \delta\rho - \tfrac{3}{2}\rho^{(0)} \, \delta\rho] = -(B - B_c) \, \delta\rho \qquad (163)$$

We see that, beyond the bifurcation point, the limit cycle is always stable with respect to arbitrary perturbations.

Using Eq. (160b), one can also calculate the modification of the frequency of the harmonic regime as one goes to a limit cycle. We obtain:

$$\theta_1(2\pi) = -\frac{\pi}{4} \rho^{(0)} = -\frac{\pi}{3} (B - B_c) \qquad (164a)$$

In the continuous limit:

$$\frac{d\theta_1}{dt} = -\frac{(B - B_c)}{6} \tag{164b}$$

It follows that the average frequency along the limit cycle will be (for $A = 1$ and $\rho \ll 1$)

$$\omega = \pm i\left(1 + \frac{B - B_c}{6}\right) \tag{165}$$

A more refined description would require higher order stroboscopic equations and is not discussed here. The main conclusion to be drawn from this quantitative analysis is again that the characteristics of the limit cycle depend only on the parameters descriptive of the system and are independent of the initial conditions.

We finally report, in this subsection, the results of a numerical analysis of Eqs. (139) which confirm the theoretical predictions given above. We consider a typical case beyond instability corresponding to the numerical values ($A = 1$, $B = 3$). In Fig. 3, the trajectories of the system obtained by numerical integration are plotted. It is observed that starting from the steady state ($X_0 = 1$, $Y_0 = 3$), as initial condition, the system attains asymptotically, a closed orbit in (X, Y) space, that is, the system exhibits an oscillatory behavior. It can be verified that the characteristics of these oscillations, including their period, are independent of the initial conditions. Indeed, starting with initial states as different as ($X = Y = 0$),

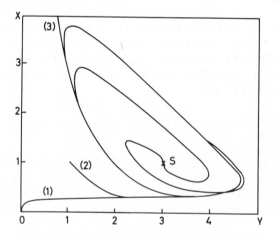

Fig. 3. Approach to a limit cycle in the model (139) for $A = 1$, $B = 3$ and for various initial conditions.

(X = Y = 1), (X = 10, Y = 0) it is seen on Fig. 3, that for long times the system approaches the same asymptotic trajectory as above. It is also interesting to observe that the system approaches the limit cycle faster, the further it is in the unstable region.

The results of Fig. 3 also imply the stability of the limit cycle with respect to *arbitrary* perturbations. Indeed, the fact that the trajectories with initial steady state condition (X = 1, Y = 3) and the initial condition (X = Y = 0) both tend to the same limit cycle proves first that there exists only one such trajectory. Indeed, as a limit cycle necessarily encircles the origin (see Sec. V-4) there would be at least one trajectory which would cut the second limit cycle. This is excluded as it would violate the conditions of the Cauchy–Lipschitz uniqueness theorem (see Sec. IV-B). At the same time these considerations establish the stability of the limit cycle. Let us finally add that these conclusions hold for a wide spectrum of values of B. Keeping A = 1, B has been varied between 3 and 6. The results are identical as before. Moreover, for $2 < B < 4$, the approach to the limit cycle starting from the steady state is oscillatory whereas for $B > 4$ the approach is aperiodic. This is in agreement with Eq. (145) obtained by a linear normal mode analysis.

C. Thermodynamic Analysis of Limit Cycles

We first consider the behavior of the system in the neighborhood of the steady state. The linearized equations of evolution are (see Eqs. (142)):

$$\frac{dx}{dt} = (B - 1)x + A^2 y = v_x$$
$$\frac{dy}{dt} = -Bx - A^2 y = v_y \tag{166}$$

We observe that in this limit the system becomes conservative at the marginal state $B = B_c = A^2 + 1$. Indeed, setting

$$\frac{dx}{dt} = -\frac{\partial V}{\partial y}, \qquad \frac{dy}{dt} = \frac{\partial V}{\partial x} \tag{167}$$

we verify that the existence condition

$$\frac{\partial^2 V}{\partial y\,\partial x} = \frac{\partial^2 V}{\partial x\,\partial y}$$

is satisfied for $B = B_c$.

From (166) and (167), V can be computed (up to an arbitrary constant). The result is

$$-V = (A^2 + 1)\frac{x^2}{2} + A^2\frac{y^2}{2} + A^2xy \geq 0 \qquad (168)$$

The system thus behaves at the marginal state in much the same way as the Volterra–Lotka model, and admits an infinity of closed trajectories corresponding to different values of V (that is, to different initial conditions).

From the point of view of thermodynamics the constant of motion V does not have a special significance in this case. Indeed, expanding entropy to second order as in Sec. IV-E, we find:

$$-\frac{\delta^2(ns)}{k} = \frac{x^2}{2A} + \frac{Ay^2}{2(A^2 + 1)} \neq V \times \text{(constant)} \qquad (169)$$

It follows that $\delta^2(ns)$ is not a constant of motion, in contrast to the result established in Sec. IV-C for the Volterra–Lotka model.

On the other hand, it can be verified that the general requirements of Sec. IV-C for the onset of instability are satisfied at the marginal state:

$$T\,\delta P = \frac{1}{2}\int d\mathscr{V}\sum_\rho (\delta w_\rho^* \,\delta\mathscr{A}_\rho + \delta w_\rho\,\delta\mathscr{A}_\rho^*) = 0 \qquad (170)$$

Moreover $\delta\Pi$ is found to have a definite sign, in agreement with the result that the marginal state is oscillatory. Eq. (170) implies the existence of a constant of motion which is an excess entropy defined in terms of complex normal modes ("mixed" excess entropy according to the terminology of Prigogine and Glansdorff):

$$[\delta^2(ns)]_m = -k\left[\frac{1}{2A}x^*x + \frac{A}{2(A^2 + 1)}y^*y\right] \qquad (171)$$

We do not expand these points here but refer the reader to the monograph by Glansdorff and Prigogine for a thorough discussion.[2]

It should also be emphasized that the property $\delta P = 0$ is a direct consequence of the finite distance from equilibrium and of the nonlinearity introduced by the autocatalytic step (see Ref. 2, Chapter 14 for details).

Let us now study the thermodynamic properties beyond the linear domain of small fluctuations. Adding the two equations (139) we obtain:

$$\frac{d(X + Y)}{dt} = A - X \qquad (172)$$

Averaging this relation over one period, T of the motion beyond the instability we obtain

$$0 = \frac{1}{T} \int_0^T dt \, \frac{d(X + Y)}{dt} = A - \frac{1}{T} \int_0^T dt X(t)$$

or

$$(\overline{X})_T = A \tag{173}$$

In other words, the average production of X over a period remains equal to the steady state value (140). A similar result cannot be established for Y because of the nonlinearity of the equations. A numerical computation of the average $(\overline{Y})_T$ for $B > B_c$ shows that this quantity increases as a function of B more rapidly than the corresponding steady state value. Some results for different values of B and for $A = 1$ are given in Table II

TABLE II

Time Average of Y Beyond Instability

B	$(\overline{Y})_T$	Y_0
2.500	2.552	2.500
3.000	3.176	3.000
4.000	4.553	4.000
5.000	6.149	5.000

From that point of view the behavior of this system is distinct from that observed in the Volterra–Lotka model (see Eq. (122)). We can also express this change by saying that the functional behavior of the system changes beyond instability as the system tends now to produce its chemicals in different (and perhaps more "efficient") proportions.

On the other hand by performing a similar calculation as in Eqs. (128) and (129) one can show easily that the average entropy production over a period is equal to the steady state entropy production:

$$(\overline{\sigma})_T = \sigma_0 \tag{174}$$

Therefore in the new regime beyond instability there is no increase of energy dissipation in the system. The detailed calculation shows that this result remains true in all cases where the intermediate produced by the initial (fixed) chemicals has an average concentration equal to the steady state concentration (as X in the present model).

What is then the principal feature characterizing such periodic behavior beyond instability? We have already stated in the previous subsection, that a limit cycle is a dissipative structure. One would like to go further and consider it as a manifestation of some degree of "temporal organization". However, it seems rather difficult to give a precise significance to this term. Certainly one obvious property of limit cycles, namely their asymptotic orbital stability beyond the bifurcation point, implies that, whatever the initial state, the system will tend to a uniquely determined regime. This "ergodicity" is rather different from the properties of the states on the thermodynamic branch because it implies nonlinearity as a necessary premise. Still one can ask the question: What is the precise thermodynamic criterion for a temporal organization?

It is easily seen that entropy is not always a good measure for temporal organization. For instance, in the model considered in the last two subsections there is no entropy lowering in the time dependent state compared to the steady state entropy, owing to the difference in the average values of Y over a period and at the steady state (see Table II). It is curious to observe that in the Volterra–Lotka model discussed in the previous section there is a decrease of average entropy along each periodic trajectory compared to the steady state entropy value. Indeed, owing to the convexity property of entropy as a function of the concentrations, we have:[35]

$$S((\overline{X})_T, (\overline{Y})_T) > (\overline{S})_T$$

Referring to Eqs. (122), this implies

$$(\overline{S})_T = \frac{1}{T} \int_0^T dt S(X(t), Y(t)) < S(X_0, Y_0) = S_0 \qquad (175)$$

However, it would be erroneous to conclude from this result that in the Volterra–Lotka model there is a degree of temporal organization which is higher than in the system operating beyond instability. Indeed, in the former a physical state cannot be assigned to a well defined trajectory because of the infinitely dense network of such trajectories around a given phase space point. On the contrary, in the scheme exhibiting a limit cycle, the system will attain, with probability one (apart from small fluctuations) a uniquely determined regime. This point will be further discussed in Sec. VIII.

In the next subsection, a partial answer as to the nature of temporal organization beyond instability will be given by a comparative study of the behavior of Volterra–Lotka and unstable systems.

D. Temporal Organization Beyond Instability—Comparison with the Volterra–Lotka Model

In the preceding subsection we obtained the average values of the chemical variables $X(t)$, $Y(t)$ over a period beyond instability, and considered that these averages provide a first order description of the system. On the other hand, between the values of $X(t)$, at different times, there exists some correlation: The value of X at some time t affects the probabilities of different values at time $t + \tau$. A convenient way to estimate this influence is to calculate the *time correlation function* as the average of the product $X(t)\,X(t + \tau)$ both along the motion and over an initial ensemble describing the probabilities for the values which the quantity can take initially.

We thus define the (normalized) concentration correlation function by[11]

$$C(\tau) = \frac{1}{\langle X^2(t)\rangle_T} \, \overline{\langle X(t)X(t + \tau)\rangle}_T, \qquad C(0) = 1 \qquad (176)$$

where we performed a time average and also an additional ensemble average with respect to initial conditions. In fact, because of the uniqueness of the periodic trajectory in our model, and because of the fast approach to this trajectory, starting from a wide variety of initial states, Eq. (87) reduces to

$$C(\tau) = \frac{(\overline{X(t)X(t + \tau)})_T}{(\overline{X^2(t)})_T} \qquad (177)$$

where the average is now over one period of the motion. For the purposes of studying the consequence of the existence of periodic trajectories, it is convenient to calculate instead of (176), the frequency spectral function

$$G(\omega) = \int_0^\infty d\tau\, C(\tau) \cos \omega\tau \qquad (178)$$

This function has been calculated numerically for $A = 1$ and for different values of B. The results are shown in Fig. 4. It should be noticed that for long times $C(\tau)$ becomes periodic, so that strictly speaking the frequency spectrum is discrete and $G(\omega)$ is a superposition of δ-functions weighted by different factors corresponding to the contributions of the different harmonics of $C(\tau)$. For the purpose of our comparison, we have taken the limit of continuous spectrum and represented in Fig. 4. $G(\omega)$ as a continuous function of ω such that for the ωs belonging to the periodic trajectory $G(\omega)$ is equal to the corresponding amplitude occurring in a Fourier series decomposition of $C(\tau)$. It should be noticed that in this representation

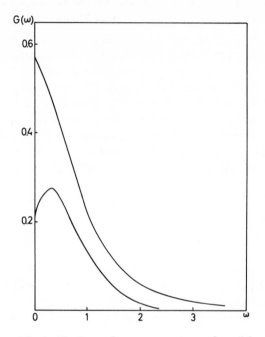

Fig. 4. Nonlinear frequency spectrum of model (139) beyond instability for various values of B.

$G(\omega)$ becomes properly normalized only in the limit of large periods corresponding to large limit cycles with increasing values of B. This is also apparent from Fig. 4, where, for example, the area under the curve corresponding to $B = 3$ is not normalized to one.

It will be observed from Fig. 4 that the shape of the spectrum strongly depends upon B. For B close to the critical value $B_c = 2$, for example, for $B = 3$, the spectrum presents a maximum at zero frequency. The behavior of the system is thus largely determined by the (unstable) steady state. This behavior is an extrapolation on the spectrum for $B < B_c$ which is also peaked around $\omega = 0$ and then falls off quickly as ω increases. For $B > 3$ and somewhat smaller than 4, the spectrum becomes peaked around the fundamental frequency $\omega = 2\pi/T$ corresponding to the period T of the motion. Since T gets larger for larger B, this peak is continuously displaced to the left and at the same time the zero frequency amplitude diminishes. It is interesting to recall that the linear stability analysis of Sec. V-B has shown that for $A = 1$ and $B \geq 4$ the departure from the unstable steady state becomes aperiodic (see Eq. (145)), and therefore the frequency of the periodic motion is due to purely nonlinear effects. It would seem therefore that the appearance of the peak at $\omega = 2\pi/T$ is related to this aperiodic behavior around the steady state. In this case, the system is no longer dominated by the steady state behavior but dissipates mostly in a narrow frequency range around the fundamental frequency.

Therefore it seems that, beyond unstability the system operates with an average frequency close to the fundamental corresponding to the periodic motion and that the frequency dispersion around this average is relatively small. Let us now compare this result with the behavior of the Volterra–Lotka system.

Because of the existence of an infinite number of trajectories in this model, the concentration autocorrelation function has to be defined as

$$C(\tau) = \langle \overline{X(t)X(t+\tau)} \rangle = \sum_m P_m \frac{1}{T_m} \int_0^{T_m} dt\, X_m(t) X_m(t+\tau) \qquad (179)$$

where the index m refers to the mth orbit and P_m is the probability of occurrence of this orbit.

The calculation of the P_ms representative of the Volterra–Lotka model is a matter of stochastic theory and will be discussed in Sec. VIII. As we shall see, the surprising result of this calculation is that for finite perturbations around the steady state the probability distribution is quite different from the "thermodynamic" distribution of small fluctuations and, in addition, it is always time dependent. Therefore strictly speaking (179) has a double time dependence; one through $X_m(t + \tau)$, and a second through P_m. Now in Sec. VIII, we show that for fluctuations around the steady state P_m is a modified Gaussian whose width increases linearly in time. On the other hand X_m is a periodic function; that is, a function which varies more rapidly than P_m is deformed. In calculating (179), therefore, it makes sense to "freeze" P_m at some intermediate form and calculate $C(\tau)$ as if the distribution of orbits were time independent. The form of P_m will first be chosen in such a way that $C(\tau)$ becomes automatically normalized:

$$P_m = \frac{1}{N} \frac{1}{(\overline{X_m^2(t)})_{T_m}} \qquad (180)$$

where N is the number of orbits considered (ultimately $N \to \infty$). Clearly, the results derived below have to be regarded as qualitative since now P_m is not an exact form representative of the model.

Table III shows some of initial conditions used in determining the orbits appearing in Eq. (179), together with their periods which have been com-

TABLE III

Properties of the Orbits Used to Calculate the
Frequency Spectrum of System (107)

Initial X	Initial Y	Period of Orbit
0.968	1	6.288
0.750	1	6.328
0.500	1	6.491
0.250	1	6.971
0.125	1	7.614
0.031	1	9.104

puted numerically. We have set $A = 1$, $B = 1$. In this case, the linear frequency becomes

$$\omega_0 = 1, \qquad T_0 = 2\pi \qquad (181)$$

The results of the calculation are plotted in Fig. 5. It is striking to observe that the frequency spectrum consists of two principal bands (there exist additional high frequency bands but the corresponding amplitudes are practically negligible). A first low frequency band is centered between $\omega = 0$ and $\omega = 1$ corresponding to an orbit with initial ($X \simeq 0.1$, $Y = 1$). It consists of a branch starting from the value 1 at $\omega = 0$ and decreasing slowly until $\omega = 1$. The part between $\omega \simeq 0.5$ and $\omega = 0$ cannot be evaluated accurately, and has been designed in the figure by a dotted line. In general, we can say that this band is dominated by the fundamentals of the orbits relatively close (yet at finite distance) from the steady state. The important point is that this branch is still dominated by the zero frequency behavior and is not peaked around any finite frequency point. It is therefore closer to the behavior given in Fig. 4 below an instability point. The second, high frequency band is centered about $\omega \sim 1.5$ corresponding to an orbit with initial ($X \sim 0.01$, $Y = 1$). In general, this band is dominated by the first harmonics of orbits lying quite far from the steady state. A curious phenomenon is that the junction between the two bands is singular, corresponding to an angular point at $\omega = \omega_0$ where $(dG/d\omega)_0 \to \infty$. This is due to the quasiperiodic character of $C(\tau)$ as a result of which $C(\tau)$ does not tend to zero as $\tau \to \infty$ and therefore the time integral of $\tau C(\tau)$ may be singular.

The exact nonlinear spectrum just calculated is fundamentally different from the linear one which, by virtue of (181) consists of a δ-function at

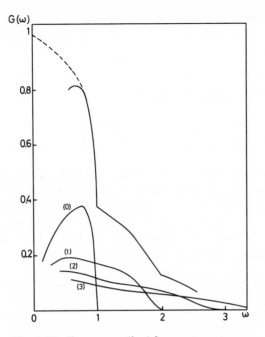

Fig. 5. Nonlinear normalized frequency spectrum
of the Volterra–Lotka model for A = B = 1. (0)
to (3), spectral distribution of the fundamental
and the first three harmonics.

$\omega = \omega_0$. The effect of large orbits (that is, of large fluctuations) appears to
be two fold:

(1) A shift to lower frequency together with a line width giving rise to
a spectrum of roughly the same type as in Fig. 4 below the instability
point.

(2) The appearance of a second high frequency part which is characterized
by a large frequency dispersion and is quite different from the spectrum
of Fig. 4.

In order to study the effect of the initial probability assigned to different
orbits (cf. Eq. (180)), we have performed another calculation of the spectrum of the system using this time a probability distribution of the form:

$$\rho \sim \exp[\alpha(V - V_0)] \tag{182}$$

where α is a suitably chosen constant and V is the constant of motion given
in Eq. (115) (V_0 being its steady state value).

We recall that for orbits close to the steady state V reduces to

$$V = V_0 + \int d\mathcal{V}\, \frac{\delta^2(ns)}{k} \tag{183}$$

In this case, Eq. (182) takes the form

$$\rho \sim \exp\left[\frac{\alpha\delta^2 S}{k}\right] \tag{184}$$

In Sec. VIII we show that Eq. (184) is the probability distribution for small deviations around the steady state. It is therefore interesting to keep in mind that (184) has a simple thermodynamic interpretation in this limit.

The results of the numerical calculation of the spectrum using the probability distribution (182) are given in Figure 6. It will be observed that, in spite of the fact that this probability function is sharply peaked

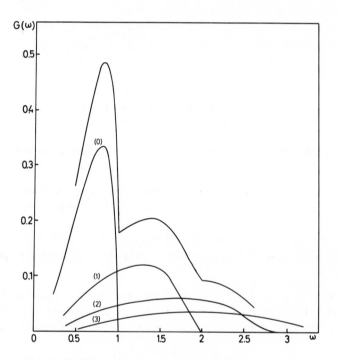

Fig. 6. Nonlinear frequency spectrum of the same system as in Fig. 5 for a probability distribution peaked around the steady state as given by Eq. (182).

around the steady state, the frequency spectrum still consists of two principal bands. The main difference from Fig. 5 is that the low frequency band is now peaked at some frequency between 0 and the linear frequency. It would seem therefore that the results plotted in Figs. 6 and 5 are not accidental and due simply to the choice of a specific form of the probability distribution, but are rather intrinsic to the Volterra–Lotka model.

In conclusion, it appears that a Volterra–Lotka type system operates with a variety of *a priori* probable frequencies and that the average frequency calculated from $G(\omega)$ has no special significance because of the large dispersion around this average. This property is a consequence of the lack of a mechanism of regression of fluctuations, as a result of which there exists no preferred set of orbits (in spite of the fact that, through our choice of initial probability distributions, we have favored the small orbits close to the steady state).

An alternative way to express the same result is that the temporal structure characterizing the Volterra–Lotka model is to a great extent incoherent, and reminiscent rather of effects of the noise type. This is to be contrasted with the coherence manifested by the system operating beyond instability which tends to dissipate mostly in the neighborhood of a well defined frequency. It is conceivable that this *coherence* is a good measure of the temporal organization characterizing a periodic system.

E. Experimental Evidence and Further Examples of Chemical Instabilities

Sustained oscillations in chemical systems have been observed in the domain of metabolic reactions and of catalytic organic reactions. Oscillations have also been reported to occur in the protein synthesis at the cellular level. Among metabolic reactions the best studied example is glycolysis.[36] In a very interesting recent paper, Sel'kov[37] has shown that the self oscillations observed for this reaction can be interpreted in terms of a mechanism involving an unstable transition point and a limit cycle thereafter (for a theoretical study of glycolysis see also Higgins, Ref. 38). Sel'kov's scheme reads

$$
\begin{aligned}
&\xrightarrow{\;v_1\;} X \\
X + E_1 &\rightleftharpoons V_1 \\
V_1 &\longrightarrow Y + E_1 \\
\gamma Y + E_2 &\rightleftharpoons E_1 \\
Y &\xrightarrow{\;k_2\;}
\end{aligned}
\qquad (185)
$$

X represents ATP entering the system at a constant rate while Y stands for ADP which decomposes to a final product. At the same time, ADP is assumed to activate the enzyme phosphofructokinase from the inactive (E_2) to the active form E_1. V_1 is an enzymatic complex. The factor γ implies that ADP may activate the enzyme in a complicated fashion owing to the allosteric character of the latter. Assuming that the enzymes remain at a steady state and that the rate of inflow v_1 of ATP is very low, Sel'kov reduces the chemical kinetic equations describing (185) to

$$\frac{dX}{dt} = v_1 - w_1 E^{(0)} w_2{}^\gamma XY^\gamma$$

(186)

$$\frac{dY}{dt} = w_1 E^{(0)} w_2{}^\gamma XY^\gamma - k_2 Y$$

where E^0 is the total enzyme concentration and w_1, w_2 are combinations of kinetic constants.

He then shows that for v_1, smaller than a critical value v_{1c}, the steady state becomes unstable and a limit cycle appears. Note that the condition $v_1 < v_{1c}$ is only possible for $\gamma > 1$. This again illustrates the importance of nonlinearity for the occurrence of instabilities in the thermodynamic branch.

In the domain of organic reaction, the best known and most thoroughly studied example is the oxidation of malonic acid in the presence of Ce ions (Zhabotinski[39]). Although there is little doubt that these oscillations are of the limit cycle type, no theoretical study has been performed because the detailed mechanism of this reaction is not well established.[39,40] What is certain is that the reaction involves autocatalytic steps. The nonlinearity premise is again found here, as it was repeatedly stressed in the previous sections.

The experimental evidence of oscillations in cellular control processes is not yet established beyond any doubt.[28,41] On the other hand, a number of authors have investigated models mostly inspired from the Jacob–Monod model,[42] which may give rise to a limit cycle type of oscillation. In the previous section, we have already indicated that the particular version of the control equations studied by Goodwin gives rise only to Volterra–Lotka type of oscillations. More recently Griffith[43] and Walter[44] have considered more general schemes. For instance, Griffith considers a repression process involving a protein, E encoded by the mRNA, M and a metabolite, P formed under the catalytic control of the protein and acting as repressor. This gives rise to the equations

$$\frac{d\mathrm{M}}{dt} = \frac{1}{1 + \mathrm{P}^m} - \alpha\mathrm{M}$$

$$\frac{d\mathrm{E}}{dt} = \mathrm{M} - \beta\mathrm{E}$$

$$\frac{d\mathrm{P}}{dt} = \mathrm{E} - \gamma\mathrm{P} \tag{187}$$

Using linear stability analysis and Bendixson's criterion, he finds a strong evidence for one limit cycle as long as $m > 8$. At present, such high values of m seem highly improbable. What seems to be more promising is the possibility of oscillations for low values of m in systems in which two or more genes are coupled.

Walter considers models of chemical control systems of the form

$$\frac{dS_1}{dt} = \frac{b_0}{1 + \alpha(S_{n+1})^\rho} - b_1 S_1$$

$$\cdot \quad \cdot \quad \cdot \quad \cdot \quad \cdot \quad \cdot \quad \cdot \quad \cdot \quad \cdot \quad \cdot$$

$$\frac{dS_i}{dt} = b_{i-1}S_{i-1} - b_i S_i \qquad (i = 2, \ldots, n + 1) \tag{188}$$

He shows that for n and ρ sufficiently large, (in any case $n \geq 2$, $\rho > 2$) the possibility of a limit cycle cannot be excluded. In any case, the whole problem of oscillations is certainly not yet fully understood. Some further comments regarding the biological significance of oscillatory effects will be made in Sec. IX.

VI. SYMMETRY–BREAKING INSTABILITIES

A. General Comments

In this section, we discuss the stability properties of chemical systems with respect to *space dependent* perturbations. In the preceding two sections, the assumption was made that the unperturbed state and the fluctuations were maintained spatially uniform. In some cases this may be realistic, especially when the reactions act on a scale which is much longer than the times involved in diffusion processes. In other types of problems, however, for instance the propagation of chemical waves from localized sources, diffusion plays a prominent role and the effect of space dependent perturbations should be taken into account.

The equation we shall deal with in this section is thus of the form (see Sec. III-A).

$$\frac{\partial n}{\partial t} = D\nabla^2 n + \sum_\rho v_\rho w_\rho \tag{189}$$

For simplicity, we assume that the diffusion coefficient matrix is diagonal, and independent of space and time.

Because of the nonlinear terms due to chemistry we expect Eq. (189) to show, as far as stability is concerned, a behavior which is similar to that discussed in Sec. V. That is, we expect a steady uniform state to become unstable, under certain, far from equilibrium conditions. However, this time because of diffusion, the variety of situations is much greater. For instance:

(1) A steady uniform state may be unstable with respect to nonuniform fluctuations, and evolve to a time independent spatial structure.

(2) A time dependent uniform state, for instance of the limit cycle type, may become unstable with respect to space dependent perturbations and evolve to a time dependent or to a static spatial structure.

(3) Possible interferences between processes (1) and (2) are conceivable. The departure from an unstable steady state due to a space dependent perturbation may be oscillatory. The system then evolves to a final spatial structure through a regime which may approach a behavior of the limit cycle type.

In all these cases, the unstable transition implies a change in the spatial symmetry of the final state with respect to the initial symmetry (*symmetry-breaking* transitions).

We will develop problems mainly of the type (1). The interference between limit cycles and spatial structure is only at the very early stages of investigation, and will only be briefly discussed. Again, the importance of the constraints keeping the system far from equilibrium and the role of the nonlinearity will be stressed.

B. Examples of Symmetry–Breaking Instabilities

We will illustrate the main points raised in the previous subsection on the simple trimolecular scheme (137) discussed in Sec. V. This time we allow for diffusion of the chemicals.[3,45] For simplicity, we consider a one dimensional system and call r the space coordinate. The kinetic equations read:

$$\frac{\partial X}{\partial t} = A - (B + 1)X + X^2Y + D_X \frac{\partial^2 X}{\partial r^2}$$

$$\frac{\partial Y}{\partial t} = BX - X^2Y + D_Y \frac{\partial^2 Y}{\partial r^2}$$

(190)

We want to investigate the stability of the uniform steady state $X_0 = A$, $Y_0 = B/A$. (see Eq. (140)) with respect to space dependent perturbations. As in Sec. V-B, we have taken the limit of infinite affinity by neglecting all back reactions. The reader may consult Ref. 2 for an analysis of the properties of the scheme as the affinity increases from close to equilibrium value to the region where instability becomes possible. The important point is to realize that the steady state solution (140) belongs to the thermodynamic branch of solutions extrapolated in the domain of very large affinities.

We assume that the system obeys periodic boundary conditions. The infinitesimal stability properties of (190) may then be investigated by considering perturbations of the form

$$X(t) = X_0 + x(t), \qquad x(t) = x \exp\left(\omega t + \frac{ir}{\lambda}\right)$$

$$Y(t) = Y_0 + y(t), \qquad y(t) = y \exp\left(\omega t + \frac{ir}{\lambda}\right)$$

(191)

where λ is a wavelength and $(|x/X_0|)$, $(|y/Y_0|) \ll 1$.

From (190) and (191) we easily obtain the secular equation in the form

$$\omega^2 + \left(A^2 + 1 - B + \frac{D_X + D_Y}{\lambda^2}\right)\omega + A^2\left(1 + \frac{D_X}{\lambda}\right)$$

$$+ (1 - B)\frac{D_Y}{\lambda^2} + \frac{D_X D_Y}{\lambda^4} = 0 \quad (192)$$

or, in more compact notation

$$\omega^2 + Tr\omega + \Delta = 0$$

(192a)

In the homogeneous limit ($D_X, D_Y = 0$ or $\lambda \to \infty$), Eq. (192) reduces to (143), and the system presents an oscillatory state of marginal stability. Let us now discuss the different possibilities at finite λ^*. For simplicity, we set $A = 1$, $D_X = 1$, $D_Y = D$. We distinguish between two cases:

(1) real roots, $\mathscr{D} = Tr^2 - 4\Delta > 0$.
(2) complex conjugate roots, $\mathscr{D} < 0$.

* A thorough discussion of the secular equation for a general model involving two intermediates has been carried out by Othmer and Scriven.[46]

After a few algebraic manipulations we find:

$$\mathscr{D} = B^2 + \left(2\,\frac{D-1}{\lambda^2} - 4\right)B + \frac{(D-1)^2}{\lambda^4} \tag{193}$$

This expression is positive definite, provided

$$\lambda^2 < D - 1 \tag{194a}$$

If this inequality is violated, the system may exhibit oscillations. This will actually happen if B lies between the roots of the quadratic:

$$2 - \frac{D-1}{\lambda^2} - 2\sqrt{1 - \frac{D-1}{\lambda^2}} < B < 2 - \frac{D-1}{\lambda^2} + 2\sqrt{1 - \frac{D-1}{\lambda^2}} \tag{194b}$$

Let us now analyse cases (1) and (2) separately.

Case 1

If the roots are real and do not simultaneously vanish, the only possibility for an instability is that $\Delta = 0$. Indeed, in this case ω changes sign and the system is at a point of nonoscillatory marginal stability. Setting $\Delta = 0$ we obtain:

$$B(\lambda) = \frac{1}{\left(\dfrac{D}{\lambda^2}\right)}\left(1 + \frac{1}{\lambda^2}\right)\left(1 + \frac{D}{\lambda^2}\right) \tag{195}$$

Figure 7 represents B as a function of λ:

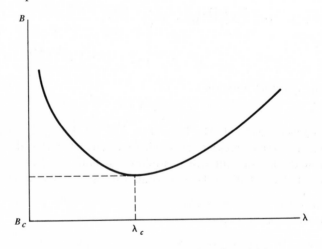

Fig. 7. Graphical representation of Eq. (195).

If we gradually increase B, instability will first arise at a wavelength λ_c such that B is minimum. Eq. (195) then yields:

$$\lambda_c{}^2 = D^{1/2}$$

$$B_c = \left[1 + \frac{1}{D^{1/2}}\right]^2 \tag{196}$$

Referring to (193) and (194), one can verify that for these values $\mathscr{D} > 0$ provided $D \neq 1$. The assumption of a nonoscillatory marginal state is therefore consistent. Also for $D > 1$ one finds $Tr > 0$, and the second root of the secular equations remains negative. The case $D = 1$ ($D_X = D_Y$) is singular, and gives $\mathscr{D} = 0$, that is, $Tr = 4\Delta = 0$. The system then only admits the trivial solution.

For $B > B_c$, the system remains aperiodic as long as (194a) is satisfied.

Case 2

If the roots are complex conjugate and the real and imaginary parts do not vanish simultaneously, instability will arise for $Tr = 0$, that is, when the real parts of ω will change sign. This will be a point of oscillatory marginal stability. Setting $Tr = 0$ we obtain:

$$B(\lambda) = 2 + \frac{1 + D}{\lambda^2} \tag{197}$$

This relation does not give a minimum value for B as a function of λ other than $\lambda = \infty$. This is the limit of homogeneous perturbations. In this case we recover

$$B(\lambda \to \infty) = 2 = B_c \tag{198}$$

where B_c is the same as in Eq. (144) for instability with respect to homogeneous perturbations. As long as $B < 2$ Eq. (197) will never be satisfied; the system will be stable with respect to all kinds of perturbations. For $B > 2$ there will always be a wavelength satisfying (197), but the system will have already undergone an instability corresponding to long wavelength perturbations.

Compiling the results for the two cases (1) and (2), we may say for $\lambda^2 > D - 1$ the system may undergo an oscillatory instability. Whether this will actually happen depends on the values of B and D. For instance for $D = 1$ ($D_X = D_Y$) Eq. (194b) gives:

$$0 < B < 4$$

For $2 < B < 4$, the system will thus always exhibit an oscillatory instability. From (197) we see that the corresponding wavelengths satisfy necessarily

$\lambda \geq 1$. Alternatively, for equal diffusion coefficients, the stability character-istics of the system are the same as in the case of uniform perturbations.

Let us take now $D = 4$. By Eqs. (194a) and (196) the system will exhibit an aperiodic instability for $\lambda \leq \sqrt{3}$, $B \geq \frac{9}{4}$. For $\lambda > \sqrt{3}$ and sufficiently large B given by (197) and within the limits (194b), the instability will be oscillatory.

It would be tempting to think that a nonoscillatory or an oscillatory instability imply, respectively, an evolution to a steady spatial structure or to a space dependent limit cycle. In fact, the results of Sec. V-B and of this subsection show that this is not necessarily so since, for $B > 4$ and $\lambda \to \infty$, the system could still tend, aperiodically, to a limit cycle. Conversely, the oscillatory departure from the steady state does not necessarily indicate an evolution to a final space-time organized structure.

Further examples of chemical mechanisms giving rise to symmetry-breaking instabilities may be found in the literature. Perhaps the first explicit example is the one considered by Turing in 1952.[47] Let A, B be two initial products, D and E two final products whose concentrations are maintained constant, X, Y intermediate products, and C, V, V′, W catalysts or catalytic complexes. Turing's scheme is

$$A \xrightarrow{k_1} X$$
$$X + Y \underset{k_3}{\overset{k_2}{\rightleftharpoons}} C$$
$$C \xrightarrow{k_4} D$$
$$B + C \xrightarrow{k_5} W$$
$$W \xrightarrow{k_6} Y + C$$
$$Y \xrightarrow{k_7} E$$
$$Y + V \xrightarrow{k_8} V'$$
$$V' \xrightarrow{k_9} E + W \tag{199}$$

In a more schematic form one has:[46]

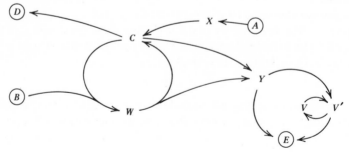

It can be shown that in the limit, as k_8 and k_6 become very large the system may undergo a nonoscillatory instability with respect to diffusion.[48] Similarly, it has been shown that Sel'kov's scheme for glycolysis [Section V, Eqs. (185) and (186)] implies an instability of the same type as Turing's under conditions similar to those derived by Sel'kov for the existence of a limit cycle.[49] Further examples of symmetry breaking instabilities refer to catalytic reactions involving negative feedback.[49] Again, the conditions of finite distance from equilibrium and of nonlinearity are found to be essential for the occurrence of such instabilities.

C. Dissipative Structures Beyond Symmetry-breaking Instabilities

The thermodynamic interpretation of symmetry-breaking instabilities and the behavior of the system beyond the unstable transition has been worked out by Prigogine and Lefever,[45,50,51] and is discussed in much detail in the monograph by Glansdorff and Prigogine.[2] For completeness, we reproduce in this subsection the main results of the calculation of the final dissipative structure.

Consider the trimolecular scheme described by Eq. (190) under conditions of a nonoscillatory instability of the steady state. Lefever has performed a numerical integration of these equations[50] for $A = 2$, $D_X = 0.0016$ (in reduced units), $D_Y = 2D_X$, $B = 5.24$, and fixed concentrations of X and Y at the boundaries. It was shown that the system attains for long times, a *steady non-uniform state.* Figure 8 shows the spatial distribution of the intermediate products X and Y at the final state. Clearly, this is an example of a dissipative structure arising beyond a point of instability of the thermodynamic branch of solutions of the kinetic equations.

A very interesting problem is the stability of the final solution. In Sec. V we have seen that the limit cycle arising for the trimolecular scheme, beyond the instability point with respect to homogeneous perturba-

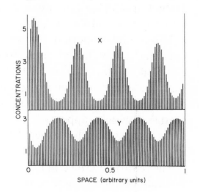

Fig. 8. A spatial dissipative structure for system (190), and for $A = 2$, $B = 5.24$, $D_X = 0.0016$, $D_Y = 0.008$.

tions, was always asymptotically stable. For the space dependent problem discussed in this section it has been shown numerically that the final inhomogeneous state is extremely stable in the limit when diffusion is represented as a flow of matter across the surface of separation between two discrete boxes.[51] For diffusion in a continuous medium, the problem of stability remains open. In this case, it may happen that the system admits more than one inhomogeneous steady state solution for the same boundary conditions. Asymptotic stability would then hold only for fluctuations smaller than the average separation between two states.

Experimentally, dissipative space structures are much less known than are sustained oscillations. It is only very recently that such a structure has been observed in the laboratory. The system considered was the Zhabotinski reaction (see Section V-E). The first experiments were carried out by Busse[52] in a closed system and without a continuous homogenization of the mixture.* It is observed that if the system is perturbed by the addition of small quantities of one of the constituents, it evolves beyond certain limiting values of concentrations, to a state characterized by the appearance of spatial rays corresponding to a spatial distribution of the chemicals. These patterns are maintained for relatively long times, and indicate strongly that the structure observed is indeed a dissipative structure.

More recently Herschkowitz[53] has carried out another type of experiment for the Zhabotinski system. The system is first maintained homogeneous and free of concentration gradients which were implied in Busse's experiment. It is then observed that, at a certain range of values of temperature and concentrations there appear oscillations of the same type as reported by Zhabotinski. After a number of such oscillations, inhomogeneities set up in the system and they finally organize in the form of bands.

This result is important for two reasons. First, the existence of symmetry-breaking instabilities resulting from *fluctuations around a uniform state* is established. This is to be contrasted with Busse's result, where the initial state was subject to a concentration gradient. And second, for Zhabotinski's reaction at least, a highly significant coupling between oscillations and spatial structures appears.[54] It would seem that the system evolves first to a limit cycle which then becomes unstable due to the increasingly important diffusion processes.

The results described in this subsection are all derived in a closed system. It would be interesting to repeat the experiments for an open system. Some preliminary results for oscillations under such conditions have been reported by Zhabotinski.[55]

D. Stability of Time Dependent States and Interference Effects

In this subsection, we investigate the influence of diffusion in the stability of uniform time dependent solutions. We again consider, as example, the trimolecular scheme (137). In Section V we showed that this system admits

---

* Homogenization is a necessary condition for the observation of sustained oscillations in Zhabotinski's experiments.

a single periodic solution $X_0(t)$, $Y_0(t)$ of the limit cycle type. We want to investigate the properties of the evolution equations containing diffusion (see Eq. (190)) around this solution. We set

$$X(r, t) = X_0(t) + x'(r, t)$$
$$Y(r, t) = Y_0(t) + y'(r, t)$$

(201)

and linearize with respect to x', y'. The result is

$$\frac{\partial x'}{\partial t} = -(B + 1 - 2X_0 Y_0)x' + X_0{}^2 y' + D_X \frac{\partial^2 x'}{\partial r^2}$$

$$\frac{\partial y'}{\partial t} = (B - 2X_0 Y_0)x' - X_0{}^2 y' + D_Y \frac{\partial^2 y'}{\partial r^2}$$

(202)

As the coefficients of this system do not depend on r, we may choose periodic boundary conditions and seek for solutions of the form:

$$x'(r, t) = x(t)e^{i(r/\lambda)}$$
$$y'(r, t) = y(t)e^{i(r/\lambda)}$$

(203)

We obtain

$$\frac{dx}{dt} = -(B + 1 - 2X_0 Y_0)x + X_0{}^2 y - \frac{D_X}{\lambda^2} x$$

$$\frac{dy}{dt} = (B - 2X_0 Y_0)x - X_0{}^2 y - \frac{D_Y}{\lambda^2} y$$

(204)

The terms involving diffusion in (204) may be considered as *perturbations* of an autonomous system with periodic coefficients. We distinguish between two extreme cases, D/λ^2 small and D/λ^2 large.

(1) D_X/λ^2, D_Y/λ^2 *small* (D_X, D_Y small and λ large). The following theorem is then valid:[56]

Theorem: If all but one of the characteristic exponents of system (204) for $D_X = D_Y = 0$ have negative real parts, then for small D_X/λ^2, D_Y/λ^2, the system (190) has a unique periodic solution with period $T = T(D_X/\lambda^2, D_Y/\lambda^2)$ which is asymptotically orbitally stable.

In Section V we showed that the limit cycle solution was asymptotically stable. The conditions of the theorem are therefore satisfied. It follows that the system admits, for long wavelengths, a unique space dependent periodic solution.* This implies that the limit cycle can never be asymptotically stable with respect to diffusion as long as D_X/λ^2, D_Y/λ^2 are small.

---

* Note that the conditions of the theorem are not satisfied by a Volterra–Lotka type system where both characteristic exponents of the variational system vanish.

(2) D_X/λ^2, D_Y/λ^2 *large*. We set $D_Y = nD_X$ (n is given small number), $\lambda^2/D_X = \varepsilon$. Multiplying through ε we write system (204) in the form

$$\varepsilon \frac{dx}{dt} = -\varepsilon(B + 1 - 2X_0 Y_0)x + \varepsilon X_0^2 y - x$$

$$\varepsilon \frac{dy}{dt} = \varepsilon(B - 2X_0 Y_0)x - \varepsilon X_0^2 y - ny \tag{205}$$

or, in matrix form

$$\varepsilon \frac{dx}{dt} = A(t, \varepsilon)x$$

$$A_{11} = -\varepsilon(B + 1 - 2X_0 Y_0) - 1$$

$$A_{12} = \varepsilon X_0^2 \tag{206}$$

$$A_{21} = \varepsilon(B - 2X_0 Y_0)$$

$$A_{22} = -\varepsilon X_0^2 - n$$

We now make use of the following theorem:[57]

Theorem: Let $A(t, \varepsilon)$ be holomorphic in t and ε, for $0 < \varepsilon \le \varepsilon_0$. Assume that $A(t, \varepsilon)$ has the real period T, independent of ε, and that it possesses for all t an asymptotic expansion

$$A(t, \varepsilon) = \sum_{r=0}^{\infty} A_r(t)\varepsilon^r \qquad \text{as} \quad \varepsilon \to 0 +$$

If $A_0(t)$ has no purely imaginary eigenvalue, for any real value of t, then the differential system (206) possesses *no* nontrivial solution with period T, if ε is sufficiently small.

Using the definitions of the matrix elements A_{ij}, it is easy to see that the conditions of the theorem are satisfied. For short wavelength perturbations the system therefore fails to admit periodic solutions.

Combining the results of cases (1) and (2), we conclude that beyond some critical D/λ^2, the system does no longer admit periodic solutions. The determination of the relation between this critical value and the parameters describing the system is a difficult problem and is presently under investigation.

An important consequence of this situation may be that a final space dependent dissipative structure may be attained in many possible ways.

The aperiodic departure from the steady state beyond a symmetry-breaking instability (Subsection VI-B) is only one of these ways. In addition, the system seems to admit a highly nontrivial coupling between time periodic and space dependent states. As a result of such an interplay, new spatio-temporal periodic solutions (waves) may appear, which subsist up to a critical wavelength. Beyond this value, the evolution in time is no longer periodic and the system may either go back to or depart from the limit cycle and tend to a steady space dependent dissipative structure.*

The experimental observations on the Zhabotinski reaction described in Sec. VI-C tend to confirm that such an evolution is indeed possible.[53,54] Further remarks on the importance of solutions of this type are made in Section IX.

VII. MULTI-STATIONARY STATES

A. Introduction

In this section we consider again spatially uniform systems, and study situations where the equations of evolution admit more than one steady state solution (that is, singular points). This type of problem may be interesting for at least three reasons; first, we may have more than one simultaneously stable solution *before* the thermodynamic branch becomes unstable; second, in the presence of three solutions, the system may exhibit a hysteresis type of phenomenon, and third, there is possibility of coupling with time dependent phenomena of the limit cycle type.

Hysteresis phenomena have been conjectured a long time ago by Rashevsky in connection with biological systems.[58] More recently, Bierman[60] discussed a model with two or even three stable stationary states. Spangler and Snell[60] have studied a set of coupled catalytic chemical reactions involving inhibition, and demonstrated the possibility of sustained oscillations as well as of multistationary state transitions.

A different type of problem has been studied by chemical engineers.[61] It refers to the stability with respect to temperature fluctuations in chemical reactors. Again, a great variety of new effects is shown to occur; transitions between two steady states, oscillations of the limit cycle type around an unstable steady state, multiple limit cycles, and so on.

The role of nonlinearity in multiple steady state problems, is obviously to give rise to more than one steady state solution.

---

* The existence of chemical waves in the trimoleclar scheme (137) has been established quite recently by Narashima, Herschkowitz and the author (submitted to J. Chem. Phys.).

B. Examples of Multiple Steady States

1. *Bierman's Model*

Bierman has considered systems involving simultaneously autocatalytic and surface reactions which may lead to multiple steady states.[59] In fact, instead of the surface reactions one may also consider enzymatic reactions with product inhibition. The model may be described as follows. A substance X is produced autocatalytically at a velocity $k_1 X/1 + \sigma_1 X$. It either decays at a rate $k_3 X$ or it combines with a second substance Y to form a complex (XY); the latter transforms to Y at a velocity $k_2 XY/1 + \sigma_2 XY$. Finally, Y decays, at a rate $k_4 Y$. This mechanism is described by the nonlinear equations

$$\frac{dX}{dt} = \frac{k_1 X}{1 + \sigma_1 X} - \frac{k_2 XY}{1 + \sigma_2 XY} - k_3 X$$

$$\frac{dY}{dt} = \frac{k_2 XY}{1 + \sigma_2 XY} - k_4 Y$$

(207)

These equations have been studied in detail numerically by Herschkowitz and the author*.[62] We briefly outline here the main results of this investigation. We limit ourselves to steady states with finite concentrations of X and Y. Thus the solutions $(X_0 = Y_0 = 0)$, $(X_0 = [(k_1 A^2 - k_3)/k_3 \sigma_1]$, $Y_0 = 0)$ are excluded. Assuming now $X_0 Y_0 \neq 0$, one can transform (207) into

$$Y_0 = \frac{k_2 X_0 - k_4}{k_4 \sigma_2 X_0}$$

(208)

where X_0 satisfies the cubic equation

$$k_3 k_4 \sigma_1 \sigma_2 X_0^3 - (k_1 k_4 \sigma_2 A^2 - k_3 k_4 \sigma^2 - k_2 k_4 \sigma_1)X_0^2$$
$$- (k_4 \sigma_1 - k_2 k_4)X_0 - k_4^2 = 0 \quad (209)$$

As in the previous sections, the stability of the solutions of these equations may be investigated by a normal mode analysis. One obtains a dispersion equation

$$p\omega^2 + q\omega + r = 0$$

(210)

where p, q, r are complicated expressions of X_0, Y_0 and the parameters σ_i and k_i.

To study Eqs. (208) to (210), we consider three separate domains of values of the parameters.[59]

---

* In his original paper[59] Bierman presented a graphical study of these equations and sorted out qualitative conclusions concerning stability.

a. $k_1 < k_3$. No steady state solutions other than the trivial ones exist.

b. $k_3 < k_1 < k_3 + k_2(\sigma_1/\sigma_2)$. There exists a single nontrivial steady state solution which is always stable. The corresponding singular point is a focus.

c. $k_1 > k_3 + k_2(\sigma_1/\sigma_2)$. A great number of configurations is possible depending on the values of the parameters. Tables IV and V give the

TABLE IV

Stability as a function of k_4 for $k_1 = 27.0$, $k_2 = 16$

k_4	(X_{01}, Y_{01})		(X_{02}, Y_{02})		(X_{03}, Y_{03})	
	$Re\omega_1$	$Re\omega_2$	$Re\omega_1$	$Re\omega_2$	$Re\omega_1$	$Re\omega_2$
2.2	+0.37		+1.1	−0.93		
3.2	+0.43		−0.29			
4.2	+0.42	equal to $Re\omega_1$	−0.63	equal to $Re\omega_1$	stable	stable $Re\omega_1 \neq Re\omega_2$
6.2	+0.10		−1.1			
6.7	−0.06		−1.2			
7.2	−0.29		−1.2			

stability characteristics calculated from (210) as a function of k_4 and k_1 in the region of coexistence of three stationary states. We have set:

$$\sigma_1 = \sigma_2 = 0.5, \qquad k_3 = 1.0 \tag{211}$$

Consider first table IV.

It is significant to observe that the singular point (X_{01}, Y_{01}) is always a focus whereas (X_{03}, Y_{03}) is always a node. Whenever (X_{02}, Y_{02}) is stable, it is also a focus, whereas in the region of instability it is a saddle point. These properties imply that whenever (X_{01}, Y_{01}), becomes unstable, the system does not necessarily jump to one of the other two stable states but can also go to a limit cycle around the unstable focus. Further study is required to show whether this possibility may actually be realized.

Consider now table V.

Again, point (X_{01}, Y_{01}) is always a focus and (X_{03}, Y_{03}) is always a node. Point (X_{02}, Y_{02}) becomes unstable at a bifurcation point, where it behaves like a center. At the instability points of (X_{01}, Y_{01}) and (X_{02}, Y_{02}) the system may therefore evolve to limit cycles as well as to a new steady state. Figure 9 shows the steady state concentration of X as a function of k_1. We observe that the system exhibits a hysteresis loop; in the loop region the system may have two simultaneously stable steady states.

TABLE V

Stability as a function of k_1 for $k_4/k_2 = 0.2$, $k_2 = 46$

k_1	(X_{01}, Y_{01})		(X_{02}, Y_{02})		(X_{03}, Y_{03})	
	$Re\omega_1$	$Re\omega_2$	$Re\omega_1$	$Re\omega_2$	$Re\omega_1$	$Re\omega_2$
42	−0.65	*equal to $Re\omega_1$*	—	—	—	—
52	−0.32		—	—	—	—
62	+0.26		−1.3	−5.9	−0.95	−8.8
72	+1.1		−0.5	−0.5	−0.97	−9.1
77	+1.5		+1.2	+1.2	−0.97	−9.1

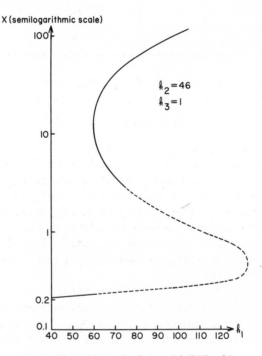

Fig. 9. Plot of X vs k_1 for model (207) with
$k_2 = 46$, $k_3 = 1.0$, $\sigma_1 = \sigma_2 = 0.5$. Dotted parts
of the curve represent unstable states.

In conclusion, Bierman's model is an example of multiple stability, oscillations, and hysteresis effects.

2. Simple Autocatalytic Models

Edelstein[63] has studied a scheme involving an autocatalytic and a simple catalytic step of the Michaelis–Menten type:

$$
\begin{aligned}
A + X &\rightleftharpoons 2X \\
X + E &\rightleftharpoons C \\
C &\rightleftharpoons E + P
\end{aligned}
\tag{212}
$$

E and C are constrained to satisfy the conservation condition

$$E + C = E^{(0)} = \text{Constant} \tag{213}$$

He finds that, for example, for $P = 0.2$, $E^{(0)} = 30$, and $k_i = k_{-i} = 1$, the system exhibits a hysteresis loop with 3 steady states and bistability. In this example, the departure from a steady state at the bifurcation point is always aperiodic. For a detailed presentation of the numerical results see Refs. 2 and 63.

Babloyantz[64] has considered schemes involving more than one autocatalytic step. For example:

$$
\begin{aligned}
A + X &\xrightarrow{\ k_1\ } 2X \\
B &\underset{k_{-2}}{\overset{k_2}{\rightleftharpoons}} X \\
X + 2Y &\underset{k_{-3}}{\overset{k_3}{\rightleftharpoons}} 3Y \\
D + Y &\underset{k_{-4}}{\overset{k_4}{\rightleftharpoons}} P
\end{aligned}
$$

She finds that for $A = 0.5$, $B = 2.3$, $D = 1$, $k_1 = k_3 = k_{-3} = k_4 = k_{-4} = 1$, and $k_2 = k_{-2} = 0.5$, the system exhibits the same behavior as in model (212).

3. Control Processes of the Monod and Jacob type

Monod and Jacob[42] have discussed qualitatively some schemes which may present transitions between multiple steady states (see also Ref. 65). They were interested in regulatory processes at the genetic level; however their models may be used to describe certain types of enzymatic reactions involving product inhibition or activation steps. One of their schemes is as follows:

Substances E_1 and E_2 (which may represent enzymes) are synthesized by SG_1, X_1 and SG_2, X_2 (which may represent two structural genes and

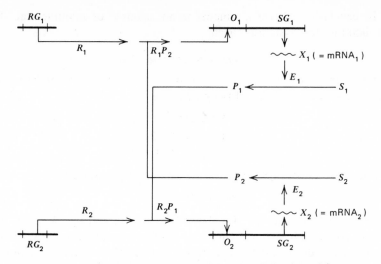

Fig. 10. Schematic representation of the Manod–Jacob model.

two m-RNA species). On the other hand RG_1 and RG_2 (regulatory genes in the case investigated by the authors) synthesize two inactive forms R_1, R_2 (repressors) which combine respectively with the products P_2 and P_1 of two parallel reactions involving E_1, E_2 and external substrates S_1, S_2. The activated complexes R_1P_2, R_2P_1 may stop the synthesis of X_1, X_2, and therefore also of E_1, E_2 by acting on a part O_1, O_2 of the genes (the operators).

The repression reaction may be written as

$$G^+ + Re \; \underset{k-1}{\overset{k_1}{\rightleftharpoons}} \; G^- \tag{215}$$

where G^+ is the probability that the operator is open.

G^- is the probability that the operator is closed.

Re is the concentration of the active repressor.

Cherniavskii et al.[66] have analyzed quantitatively the model by first considering the simple case where a single molecule of the inactive repressor combines with a single molecule of the product of the opposite half of the system. The rate of production of $X = m$RNA is then

$$v_X \sim \frac{A}{B + Re} \tag{216}$$

where A, B are related to the ratio k_{-1}/k_1. It can be shown easily that in this case the system possesses a single stable steady state

One can then go further and consider processes where more than one molecule of the product participates in the synthesis of the active repressor. The rate of synthesis of X becomes

$$v_X \sim \frac{A}{B + (Re)^n}, \qquad n > 1 \qquad (217)$$

The kinetic equations describing the system are, for $n = 2$:

$$\frac{dX_1}{dt} = \frac{A}{B + E_2{}^2} - kX_1$$

$$\frac{dX_2}{dt} = \frac{A}{B + E_1{}^2} - kX_2$$

$$\frac{dE_1}{dt} = \alpha X_1 - \beta E_1 \qquad (218)$$

$$\frac{dE_2}{dt} = \alpha X_2 - \beta E_2$$

These equations bear some resemblance to Goodwin's and Griffith's models (cf. Eqs. (123) and (187)), but are, of course, more complicated.

Cherniavskii et al.[66] have simulated these equations on an electronic device (trigger) and have shown that the system may possess two stable steady states. Schematically

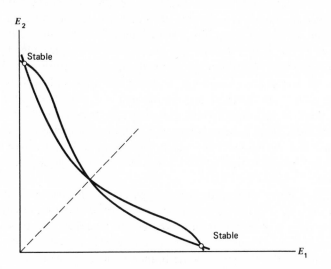

Fig. 11. Phase plane plot of the steady state solution for system (218).

Systems involving more complicated repression processes have also been studied with similar results.[66] Among the most interesting conclusions of the analysis are that the system does undergo transitions between the two stable steady states, and that under certain conditions these transitions are oscillatory.

C. Thermodynamic Properties

Multiple steady states are characterized by a number of interesting and unexpected thermodynamic properties. First, it is easy to show, for all the examples considered in the previous subsection, that at thermodynamic equilibrium (vanishing overall affinities of independent reactions) there is only one steady state and it is always stable. Multistationary state transitions are therefore typically nonequilibrium effects. On the other hand, in contrast to what happens for limit cycle and symmetry-breaking transitions, it is much harder here to determine objectively the thermodynamic branch of solutions. Certainly, if one *fixes* a type of reference states as representative of equilibrium behavior (for example, states at low B in system (212)) the passage to the second steady state at the point of instability of this type of solution may be interpreted as a transition from the thermodynamic to another branch of solutions. Lavenda[67] has studied some of the consequences of such transitions for a number of examples involving autocatalytic steps. He finds, as a general tendency, that the entropy production is lowered as the system jumps from the thermodynamic to the other branch. Now there is no argument against choosing, for example, in system (212) equilibria with high concentrations of B. In this case, one can show that the thermodynamic branch is represented by different types of states, for instance states with high concentrations of X, while previously the thermodynamic branch was that of states of low X concentration. A given state therefore may be reached in at least two ways, among which there is only one involving an unstable transition.[2,63] From this point of view, there is an analogy with equilibrium phase transitions, in the sense that the equilibrium branch is *degenerate*. An attractive but still speculative interpretation of this result has been suggested recently by Kobatake.[68]

The degeneracy referred to here is apparent in the case of a hysteresis loop such that all parts of the curve with positive slope represent stable, and the part with negative slope represents unstable, steady states. In Bierman's model (Figure 9), this is *not* the case. It is possible that in such type of model the thermodynamic branch is defined unambiguously and that instabilities lead us to new, dissipative structures belonging to different nonthermodynamic branches of solutions of the kinetic equations. This point requires further investigation.

VIII. FLUCTUATIONS

A. Introduction

The developments outlined in the foregoing sections should already have pointed up the fundamental importance of fluctuations. Whether the system were to undergo an oscillatory instability, a symmetry-breaking instability, or a transition between two steady states, at some "critical" point the fluctuations began to amplify until the system reached a new regime. The latter was critically dependent upon the nature of the fluctuation; a very long wavelength fluctuation can lead the system to a limit cycle, shorter wavelengths tend to destroy such a state and bring the system to a steady space dependent structure; finally, fluctuations of the order of the average separation between two steady states could be responsible for a macroscopic transition between these two states. In addition, as we saw in Sections IV to VII, at the transition point, the system generally admits an infinity of long living normal modes. The evolution of the fluctuations is therefore characterized by a great variability intrinsic to the system.

These arguments suffice to demonstrate the interest of determining the probability of occurrence of a specific type of fluctuation at a given time. Such a study is a necessary premise to a real understanding of the mechanism of setting up an instability and of the evolution of the system to the new macroscopic state beyond instability.

In physical problems, the problem of fluctuations appears once a *macroscopic* description of a large system containing $N \sim 10^{23}$ degrees of freedom is adopted. In this description, the state of a system is determined in terms of a restricted number of independent variables $\{a_1, \ldots, a_n\}$, $n \ll N$, for instance, temperature, pressure, volume, and so on. Now, a given *macroscopic* state is always associated with rapid transitions between different *atomic* states, due to the slight perturbations necessarily experienced by a macroscopic body. As a result, the macroscopic variables are subject to deviations around some fixed instantaneous values. These deviations, which are mechanical in origin, but which appear to a macroscopic observer as purely random events, are precisely the *fluctuations*.[11,69] In classical thermodynamics, it has always been implied that fluctuations are small except at the points of phase transitions. Irreversible thermodynamics in the nonlinear region also predicts nonequilibrium transitions and instabilities. It is therefore essential to construct a theory of fluctuations around far from equilibrium states which would supplement the predictions of the "average" thermodynamic description, especially in the neighborhood of instabilities.

B. The Markovian Stochastic Approximation

We have mentioned that fluctuations are purely mechanical effects. In principle, therefore, fluctuation theory is a branch of nonequilibrium statistical mechanics. Let $\rho(\mathbf{x}_N, \mathbf{p}_N)$ be the N-particle distribution function of the system ($\mathbf{x}_i$, $\mathbf{p}_i$ are positions and momenta of the N particles). Once ρ is known, the microscopic state of the system is determined.[11] Consider now the ensemble of variables $\{a_1, \ldots, a_n\}$ determining the macroscopic state of the same system. In general, $a_i = a_i(\mathbf{x}_N, \mathbf{p}_N)$. The probability $P(\{a_n\})\{da_n\}$ that the system be at a macroscopic state such that the a_ns take values between $\{a_n\}$ and $\{a_n + da_n\}$ will be given in terms of ρ by the relation

$$P(\{a_n\})\{da_n\} = \int_{[a_n, \, a_n + da_n]} d\mathbf{x}_N \, d\mathbf{p}_N \, \rho(\mathbf{x}_N, \mathbf{p}_N) \tag{219}$$

If $P(\{a_n\})$ is expanded around some average value $\{a_n{}^0\}$ Eq. (219) will provide the probability of a fluctuation around this average value.

Unfortunately, the approach just outlined meets with several difficulties which have not been resolved at the present time. For this reason, we shall follow here a method which is "intermediate" between the macroscopic description and the rigorous statistical mechanical method.

The basic ideas of this theory, which is known as *stochastic theory*, are the following.[70]

(1) In the first place, the variations of a_ns due to a fluctuation are considered as a random, or stochastic process, that is, as a phenomenon where the a_ns do not depend on the independent variable (time) in a well defined manner. An observation of the different members of a representative ensemble of systems will therefore yield different functions $a_n(t)$. Then, the best one can do is to study certain probability distributions. For instance, it will be possible to determine the functions:

$P_1(a, t) \, da$ = Probability to find a within $(a, a + da)$ at time t

$P_2(a_1 t_1; a_2 t_2) \, da_1 \, da_2$

$\qquad$ = Probability to find a_1 within $(a_1, a_1 + da_1)$ at time t_1 $\quad$ (220)

$\quad$ then

$$a_2 \text{ within } (a_2, a_2 + da_2) \text{ at time } t_2$$

Next, the *conditional* probabilities $W_2(a_1 t_1 \mid a_2 t_2)$, and others, are introduced. These functions are related to the P_ns by

$$P_1 = W_1$$

$$P_2(a_1, t_1; a_2, t_2) = P_1(a_1, t_1)W_2(a_1t_1|a_2t_2) \tag{221}$$

and so on.

The problem therefore reduces to the determination of an infinite hierarchy of functions P_n or W_n.

(2) In a number of interesting cases, it happens that W_2 contains all necessary information. We say then that we have a *Markov process*:

$$W_3(a_1t_1; a_2t_2|a_3t_3) = W_2(a_2t_2|a_3t_3) \tag{222}$$

and so on.

In this case, it is easy to show that the probability distribution satisfies an integral equation which is usually referred to as the *Smoluchowski equation*:

$$P_2(a_1; a_2, t) = \int da W_2(a|a_2, s)P_2(a_1; a, t-s) \qquad 0 \le s < t \tag{223}$$

In writing this equation, we have considered the limit of *stationary* Markov processes, wherein the doublet probabilities depend only on the time differences.

We see that the problem of fluctuations amounts to constructing and solving an equation of the form (223), adapted to the type of problem considered. The important point now is that the structure of this equation depends critically on the macroscopic state around which fluctuations are studied. This can be easily seen when one considers fluctuations around equilibrium and around a steady state far from equilibrium. In both cases, one must solve an equation of more or less the same structure. Indeed, the transition probabilities W_2 depend on the nature of the system, which remains unchanged. On the other hand, it is necessary to realize that a system may attain a nonequilibrium permanent state if, and only if, it is subject to well defined constraints, otherwise the only solution of Eq. (223) will be the equilibrium one. We see that *far from equilibrium the solution of (223) is subject to conditions*, whereas, close to equilibrium one can solve this equation (at least for an isolated system) without additional conditions.

Because of this simplification, the problem of fluctuations around an equilibrium state can be solved in its most general form, and one can say that it is currently a classical subject. There is also another reason for this. It is well known that the equilibrium state is characterized by a number of thermodynamic potentials which attain at that point their extremal values (entropy for an isolated system, free energy at constant T and $\mathscr{V}$, etc.).

One can show that the equilibrium solutions of (223) may be expressed in terms of these thermodynamic potentials. For instance in an isolated system, it can be shown that small fluctuations are described by

$$P \propto \exp \left[\frac{1}{2k} (\delta^2 S)_e \right] \tag{224}$$

where $(\delta^2 S)_e$ is the second order variation of entropy around equilibrium due to a fluctuation (the first variation $(\delta S)_e = 0$ as S is maximum at equilibrium). Relation (224) is the celebrated Einstein formula. Eq. (224) and its relation with (223) is now firmly established, thanks to the important work by Onsager,[9,71] Callen,[69,72] van Kampen et al. in the case of purely dissipative systems, and the works of Landau and Lifshitz,[73] Rytov, Kadomtsev et al. in the case of fluid dynamical systems. Among the consequences of these analyses, we may quote that fluctuations are shown to be small except the neighborhood of phase transition points.

C. Fluctuations Around Nonequilibrium States—General Ideas and a Simple Application

We shall now discuss the basis of the theory of fluctuations around states far from equilibrium. As these states are no longer described by the extremum of a thermodynamic potential, one should not expect to derive formulae of the same type as (224). We shall therefore try to solve Eq. (223), and then see to what extent the result may be expressed in terms of thermodynamic quantities. We recall that the main difficulty is to introduce into Eq. (223) the conditions allowing for the occurrence of a steady nonequilibrium state.[74]

We shall work out the theory in the simplest limit of an open uniform system involving chemical reactions[75] and refer to the literature for problems involving transport processes.[76] The choice of this type of system is motivated by several reasons:

(1) The problem may be formulated in terms of a finite number of *discrete* variables, for instance, the number of particles of different species. In fluid dynamical and transport theoretical problems, the variables are necessarily continuous and this gives Eq. (223) the character of a functional equation.

(2) Situations arbitrarily far from equilibrium may be realized by varying the affinity of the overall reactions through the ratio of the initial to final product concentrations.

(3) This type of system very often involves nonlinear processes and thus possesses a great intrinsic variability. Therefore, the problem of instabilities

and abnormal fluctuations is not compromised in this model. In fact, the theory of dissipative structures and the examples of Sections V and VII show that such systems exhibit several types of change of regime beyond unstable transition points.

(4) A number of physically interesting systems, for instance, biological systems belong to the category of open systems, the chemical reactions being in this case the metabolic or the biosynthetic reactions (see also comments in Section IX).

We shall limit this discussion to the case of ideal solutions, and discuss first the simplest example of two monomolecular reactions:

$$A \underset{k_{21}}{\overset{k_{12}}{\rightleftarrows}} X \underset{k_{32}}{\overset{k_{23}}{\rightleftarrows}} E \qquad (225)$$

The overall reaction

$$A \rightleftarrows E \qquad (226a)$$

has an affinity

$$\mathscr{A} = RT \ln \frac{kA}{E} \qquad (226b)$$

with

$$k = \frac{k_{12} k_{23}}{k_{21} k_{32}} = \text{equilibrium constant}$$

The system $\{X\}$ is supposed to be in contact with two large reservoirs of matter containing A and E. Our problem is to study, using the theory of Markov processes, the fluctuations around the macroscopic state which will be established in the system. Classical chemical kinetics would describe this state by

$$\overline{A}, \quad \overline{E} : \text{constant}$$

$$\frac{d\overline{X}}{dt} = (k_{12} \overline{A} + k_{32} \overline{E}) - (k_{21} + k_{23})\overline{X} \qquad (227)$$

We have introduced the (statistical) average values of the concentrations of chemicals to express that in the macroscopic chemical kinetic description one necessarily deals with averages. Eq. (227) predicts that the system will tend, asymptotically, to the steady state

$$X_0 = \frac{k_{12} \overline{A} + k_{32} \overline{E}}{k_{21} + k_{23}} \qquad (228)$$

In this classical description, fluctuations are neglected. In order to study this problem, it is necessary to undertake a more refined description in terms of the probability distributions introduced in the previous subsections. For model (225), the probability distribution will be,[77]

$$\rho = \rho(A, X, E, t)$$

where now A, X, E stand for the numbers of particles of the different chemicals. Once ρ is known, average values may be computed. For instance,

$$1 = \sum_{A, X, E = 0}^{\infty} \rho(A, X, E, t) \qquad \text{(normalization}|$$

$$\overline{X}(t) = \sum_{A, X, E = 0}^{\infty} X\rho(A, X, E, t)$$

$$\overline{\Delta X^2} = \overline{(X - \overline{X})^2} = \sum_{A, X, E = 0}^{\infty} (X - \overline{X})^2 \rho(A, X, E, t) \qquad (229)$$

The quantity $\overline{\Delta X^2}$ could be used as a measure of the importance of fluctuations in the system.

In order to construct an equation of evolution for ρ, we observe that, in agreement with the Markovian stochastic assumption, the contribution of the four reactions in (225) will be additive.*

Consider the first reaction

$$A \xrightarrow{ k_{12} } X$$

We have:

$$\rho(A, X, E, t + \Delta t) = \text{Probability of conversion at } t$$
$$+ \text{Probability of no conversion at } t$$

or   (230)

$$\rho(A, X, E, t + \Delta t) = W_{12}(A + 1)\rho(A + 1, X - 1, E, t)$$
$$+ (1 - W_{12}(A))\rho(A, X, E, t)$$

Now in an ideal solution it is reasonable to assume

$$W_{12}(A) = l_{12}A\Delta t + 0(\Delta t^2) \qquad (231)$$

---

* In chemical kinetics, the Markovian stochastic assumption should fail for extremely fast reactions.

where l_{12} is some proportionality constant independent of A, X, E. Introducing (231) into (230), and taking the limit $\Delta t \to 0$ we obtain:

$$\frac{d\rho(t)}{dt} = l_{12}(A + 1)\rho(A + 1, X - 1, E, t) - l_{12}A\rho(A, X, E, t) \quad (232)$$

or, including the effect of all four reactions:

$$\frac{d\rho(t)}{dt} = l_{12}(A + 1)\rho(A + 1, X - 1, E, t) - l_{12}A\rho$$

$$+ l_{21}(X + 1)\rho(A - 1, X + 1, E, t) - l_{21}X\rho \quad (233)$$

$$+ l_{23}(X + 1)\rho(A, X + 1, E - 1, t) - l_{23}X\rho$$

$$+ l_{32}(E + 1)\rho(A, X - 1, E + 1, t) - l_{32}E\rho$$

Eq. (233) is a finite difference equation with linear coefficients. The finite difference property is a result of the discreteness of the stochastic variables, while the linearity of the coefficients is a consequence of the monomolecular character of reaction scheme (225).

The study of equations of the type (233) is performed most conveniently in the generating function representation.[77] We define the generating function of ρ by

$$F(s_A, s_X, s_E, t) = \sum_{A, X, E = 0}^{\infty} s_A{}^A s_X{}^X s_E{}^E \rho(A, X, E, t) \quad (234)$$

The convergence of this series requires that

$$|s_A|, \quad |s_X|, \quad |s_E| \le 1 \quad (235)$$

We also note the following properties of F (see Eqs. (229)):

$$(F)_{s=1} = 1$$

$$\left(\frac{\partial F}{\partial s_X}\right)_{s=1} = \sum_{A, X, E = 0}^{\infty} X\rho = \overline{X}$$

$$\left(\frac{\partial^2 F}{\partial s_X{}^2}\right) = \sum_{A, X, E = 0}^{\infty} X(X - 1)\rho = \overline{X^2} - \overline{X} \quad \text{etc.} \quad (236)$$

Introducing (234) into (233), we obtain the following equation for F:

$$\frac{\partial F}{\partial t} = l_{12}(s_X - s_A)\frac{\partial F}{\partial s_A} + l_{21}(s_A - s_X)\frac{\partial F}{\partial s_X} + l_{23}(s_E - s_X)\frac{\partial F}{\partial s_X}$$

$$+ l_{32}(s_X - s_E)\frac{\partial F}{\partial s_E} \quad (237)$$

We observe that F obeys a first order partial differential equation with linear coefficients (owing to the monomolecular character of the reactions).

If one tries to solve Eq. (237) using a physical initial condition, one will find that as $t \to \infty$ the only steady state for the system (A, X, E) is the state of thermodynamic equilibrium. However, our problem is to study the evolution of a system (X) in contact with large external reservoirs (A) and (E). To this end we introduce the reduced probability distribution

$$\rho(X) = \sum_{A, E=0}^{\infty} \rho(A, X, E)$$

and

$$f(s, t) = \sum_{X, A, E=0}^{\infty} s^X \rho(A, X, E)$$

$$= F(s_A = 1, s_E = 1, s_X = s, t) \tag{238}$$

Eq. (237) reduces to (see also Eqs. (236)):

$$\frac{\partial f}{\partial t} = l_{12}(s - 1) \sum_X s^X \left(\sum_{A, E} A\rho \right) + l_{32}(s - 1)$$

$$\times \sum_X s^X \left(\sum_{A, E} E\rho \right) \tag{239}$$

$$+ (l_{21} + l_{23})(1 - s) \frac{\partial f}{\partial s}$$

This equation for f will be closed only when $\sum_{A,E} A\rho$, $\sum_{A,E} E\rho$ will be expressed in terms of f. In general, such a reduction is not possible. We shall now appeal to physical assumptions which are expected to be valid for the type of system considered.[75] What we wish to describe here is a steady nonequilibrium state; under these conditions we have to make sure that the state of the reservoirs (A), (E) varies in a much slower scale that the state of the system (X). This scale separation now permits the assumption that $\sum_{A,E} A\rho$ etc., which are conditional averages, do not depend on the state of the subsystem (X). In other terms the state of (A) and (E) is not influenced by the internal state of the system:

$$\sum_{A, E=0}^{\infty} A\rho = \bar{A} \sum_{A, E} \rho = \bar{A}\rho(X, t) \tag{240}$$

Eq. (239) reduces to

$$\frac{\partial f}{\partial t} = (1 - s) \left[(l_{23} + l_{21}) \frac{\partial f}{\partial s} - (l_{12} \bar{A} + l_{32} \bar{E})f \right] \tag{241}$$

where we suppressed the average value symbol for the initial and final product concentrations, which appear now as simple parameters. At the steady state $\partial f/\partial t = 0$, Eq. (241) yields a uniquely determined, properly normalized solution:

$$f(s) = \exp\left[(s - 1)\frac{l_{12}A + l_{32}E}{l_{21} + l_{23}}\right] \tag{242}$$

This solution predicts a steady state average value

$$\overline{X}_0 = \left(\frac{\partial f}{\partial s}\right)_{s=1} = \frac{l_{12}A + l_{32}E}{l_{21} + l_{23}} \tag{243}$$

If we want the stochastic description to reduce, on the average, to the macroscopic description, we should identify Eqs. (243) and (228). It follows

$$l_{ij} = k_{ij} \quad (i, j = 1, 2, 3) \tag{244}$$

Eq. (242) reduces to

$$f(s) = \exp\left[(s - 1)X_0\right] \tag{245}$$

Going back to physical variables, and using (238), we obtain a Poisson distribution for ρ^{11}:

$$\rho(X) = e^{-X_0}\frac{X_0^X}{X!} \tag{246}$$

It is interesting to study the limit of this equation for small fluctuations:

$$\delta X = \frac{X - X_0}{X_0} \ll 1 \tag{247}$$

It is easily seen that Eq. (246) gives:

$$\rho(X) = (2\pi X_0)^{-1/2} \exp\left[-\frac{\delta X^2}{2X_0}\right] \tag{248}$$

We obtain a Gaussian distribution for X. The important point now is that, by Eq. (67), $-\delta X^2/2X_0$ is equal to the second order excess entropy around the steady nonequilibrium state X_0. Eq. (248) therefore reduces to

$$\rho \propto \exp\left[\frac{1}{2k}(\delta^2 S)_0\right] \tag{249}$$

We obtain a formula which has the same structure as the Einstein formula (224). The difference is that here *the entropy excess is computed around a nonequilibrium reference state.* It is only in the limit of zero affinity that (228) gives

$$X_0 = X_{eq} = \frac{k_{12}}{k_{21}}$$

and the nonequilibrium probability function (249) becomes identical to the Einstein equilibrium formula.

In the regime described by Eq. (224) or (246), the average effect of fluctuations is small. For instance a straightforward calculation gives $\overline{\Delta X^2} = X_0$, that is,

$$\left(\frac{\overline{\Delta X^2}}{X_0^2}\right)^{1/2} = \frac{1}{X_0^{1/2}} \ll 1 \tag{250}$$

This result is completely analogous to the equilibrium result.[11]

Therefore, we reach the following, highly nontrivial conclusion; the theory of fluctuations around nonequilibrium states can be formulated in terms of thermodynamic functions, in fact the same ones used in equilibrium fluctuation theory. This conclusion had been conjectured by Prigogine fifteen years ago on the basis of the local equilibrium assumption, which has been the starting point for the extension of thermodynamics to nonequilibrium situations. The result can also be extended to more general cases; fluctuations around time dependent nonequilibrium states, nonlinear model systems involving bimolecular reactions, and so on. Eq. (249) is always recovered at least for a large class of non trivial models, once the assumption of separation of the system and reservoir time scales is made *and the models chosen are free of unstable transition points.*

These conclusions permit establishing a very interesting relation between fluctuations and stability, provided we limit ourselves in the domain of a local equilibrium theory. In Sec. III, it was shown that in this limit the stability properties of the system are expressed in terms of the excess entropy $(\delta^2 S)_0$ (see Eqs. (67), (68)), which also appears in the probability function (249). Suppose now one starts, at some $t = t_0$, at a state where fluctuations are limited to be small. Eq. (249) is valid, and predicts (as $(\delta^2 S)_0 < 0$) that this state is more probable than all neighboring states which may be reached by fluctuations. Let us now modify the constraints gradually up to the value where the state is becoming unstable. First, before the point where formula (249) breaks down, fluctuation theory still predicts that the macroscopic state is the most probable of all neighboring states. According to Eqs. (67), (68), however, this property is no longer sufficient to guarantee stability of the state; as soon as (68) is compromised, fluctuations will increase and will drive the system to a new regime where ρ will again attain a maximum value, around which fluctuations will be small. Schematically, that is:

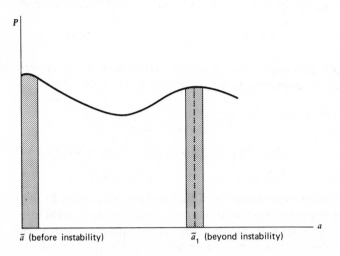

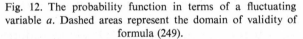

Fig. 12. The probability function in terms of a fluctuating variable a. Dashed areas represent the domain of validity of formula (249).

The evolution of ρ beyond the regime of "normal" thermal fluctuations corresponding to (250) will be discussed in the following subsection.

D. Evolution of Large Fluctuations—the Volterra–Lotka Model

We have already emphasized that the system departs from an unstable state and tends to a new regime by a mechanism of large fluctuations. The latter will determine to a large extent the nature of the new structure, depending on the most probable type of fluctuation which will occur first. The evolution of the system around an instability point therefore takes an essentially statistical character.

The direct computation of the probability function for systems undergoing a chemical instability of the type discussed in Secs. V to VII is a difficult problem, and is still under investigation. As we have mentioned, however, a system at a marginal state behaves in a way very similar to the Volterra–Lotka model. The latter may be considered as a system which is permanently at a marginal state, where asymptotic Lyapounov stability does not apply and orbital stability is the only form of stability which still characterizes the system. Therefore, we expect that the study of fluctuations in this model will provide valuable indications for the behavior of systems undergoing unstable transitions.

Using the Markovian stochastic theory outlined in Sec. VIII-B one can easily write an equation for the reduced probability distribution*

$$\rho(X, Y, t) = \sum_{A, B, E = 0}^{\infty} \rho(A, B, X, Y, E, t) \qquad (251)$$

We apply the same method as in Sec. VIII-C to separate the variables of the external reservoirs (A, B, E). The result is (see Eq. (233))

$$\mathscr{V} \frac{d\rho}{dt} = A(X - 1)\rho(X - 1, Y, t) - AX\rho(X, Y, t)$$

$$+ (X + 1)(Y - 1)\rho(X + 1, Y - 1, t) - XY\rho(X, Y, t) \qquad (252)$$

$$+ B(Y + 1)\rho(X, Y + 1, t) - BY\rho(X, Y, t)$$

The volume factor appearing in the left hand side is due to the fact that the rate of the bimolecular reactions in (102) is proportional to $1/\mathscr{V}$.

The nonlinearity of the coefficients of the equation is also a result of the bimolecular character of reaction scheme (102). It is easy to show that (252) generates the two chemical kinetic equations (107) as average equations of conservation for the mean values $\overline{X}$ and $\overline{Y}$.

As in Section VIII-C, the study of Eq. (252) will be performed in the generating function representation. We define

$$f(s_X, s_Y, t) = \sum_{X, Y; |s_i| \leq 1} s_X^X s_Y^Y \rho(X, Y, t) \qquad (253)$$

From (252) and (253), we derive the following partial differential equation for f:

$$\frac{\partial f}{\partial \tau} = (\eta + 1)(\eta - \xi) \frac{\partial^2 f}{\partial \xi \, \partial \eta} + \xi(\xi + 1)A \frac{\partial f}{\partial \xi} - \eta B \frac{\partial f}{\partial \eta} \qquad (254)$$

where we have set

$$\xi = s_X - 1 \leq 0$$

$$\eta = s_Y - 1 \leq 0 \qquad (255)$$

and

$$\tau = t/\mathscr{V}$$

* In a nonlinear system, a straight-forward application of the stochastic theory outlined in the previous subsections may give rise to inconsistencies. A more complete theory based on a Boltzmann equation description is presently under investigation. The study of fluctuations reported in this subsection has therefore only an indicative character and is meant to illustrate the mechanism of creation of large fluctuations.

Let us study the properties of Eqs. (252) or (254) in the asymptotic limit of a large system:

$$A, B \to \infty, \qquad \mathscr{V} \to \infty$$

$$B/\mathscr{V} \qquad A/\mathscr{V} = \text{finite} \tag{256}$$

We take the time average of both sides of Eq. (107) over one period T of the motion:

$$\overline{(XY)}_T = (A\overline{X})_T - \overline{\left(\frac{dX}{dt}\right)}_T \tag{257}$$

We have seen (see Eqs. (122)) that

$$A = (\overline{Y})_T, \qquad \overline{\left(\frac{dX}{dt}\right)}_T = 0$$

It follows that

$$\overline{(XY)}_T = (\overline{X})_T(\overline{Y})_T \tag{258}$$

Eq. (258) is valid for the infinity of all possible trajectories that is, for a continuous interval of values of T. On the other hand, in order that the stochastic formulation of the problem be meaningful, the system has to exhibit some kind of ergodic property. This suggests that Eq. (258) also could be assumed to hold for the statistical averages:

$$\overline{(XY)} = \overline{X}(t)\overline{Y}(t) \tag{259}$$

This equation will be the fundamental assumption of this subsection.

It is now convenient to average Eq. (254) over one of the variables, Y and construct, using (259) a closed equation for the reduced distribution $\phi(X)$. In the space of generating functions this equation reads

$$\frac{\partial g}{\partial \tau} = -\xi \overline{Y}(t) \frac{\partial g}{\partial \xi} + \xi(\xi + 1)A \frac{\partial g}{\partial \xi} \tag{260}$$

where we defined the reduced generating function

$$g(\xi) = \lim_{\eta \to 0} f(\xi, \eta) \tag{261}$$

and $\overline{Y}(t)$ is a solution of the macroscopic equations (107) Eq. (260) is a first order partial differential equation which can be solved by the method of characteristics. We obtain

$$\frac{d\tau}{-1} = \frac{d\xi}{A\xi^2 + [A - \overline{Y}(\tau)]\xi} \tag{262}$$

This is a Bernoulli equation whose general solution is[78]

$$\frac{1}{\xi \overline{X}(\tau)} - \int_0^\tau dt \, \frac{A}{\overline{X}(t)} = C \tag{263}$$

where C is a constant. In going from (262) to (263), we have also employed the equations of evolution (107) for $\overline{X}(t)$ and $\overline{Y}(t)$. The solution of Eq. (260) subject to a quasi-equilibrium initial condition is

$$g(\xi, \tau) = \exp\left[\frac{\xi\overline{X}(\tau)}{1 + \xi\overline{X}(\tau)\Delta(\tau)}\right] \tag{264}$$

with

$$\Delta(\tau) = -\int_0^\tau d\tau\, \frac{A}{\overline{X}(\tau)} \leq 0 \tag{265}$$

At this point it is instructive to study the behavior of small fluctuations around the steady state X_0. Let us first observe that $\tau = t/\mathscr{V}$ remains a small quantity for times t not too long. As a result Δ is approximately equal to (cf. also Eqs. (108) and (265)):

$$\Delta(\tau) \sim -\frac{t}{\mathscr{V}}\frac{A}{X_0} \tag{266}$$

Eq. (264) becomes:

$$g(\xi, t) = \exp\left[\frac{\xi X_0}{1 - \xi t\, \dfrac{A}{\mathscr{V}}}\right] \tag{267}$$

Consider now the interval of times

$$t < \left(\frac{A}{\mathscr{V}}\right)^{-1} \tag{268}$$

such that t is smaller than the inverse of the rate of the first of the reactions (102). Eq. (267) may then be expanded as follows:

$$g(\xi, t) = e^{\xi X_0}\left[1 + \xi^2 t\frac{A}{\mathscr{V}} + \cdots\right] \tag{269}$$

Transforming back to physical variables we obtain:

$$\rho(X, t) = \rho_P\left[1 + \frac{A}{\mathscr{V}}t + \frac{(X - X_0)^2}{X_0}\frac{A}{\mathscr{V}}t + 0\left(\frac{(X - X_0)}{X_0}\right)\right] \tag{270}$$

where ρ_P is the Poisson distribution (246). Eq. (270) remains a meaningful expansion as long as $X - X_0 \sim 0(X_0^{1/2})$, i.e., as long as fluctuations remain small. At the same time we see that such fluctuations are approximately described by a Poisson distribution provided inequality (268) remains valid. Furthermore, as in Sec. VIII-C, in the limit of small fluctuations the Poisson distribution reduces to:

$$\rho \propto \exp\left[\frac{(\delta^2 S)_0}{2k}\right] \tag{271}$$

Again $(\delta^2 S)_0$ remains a negative definite quadratic form, in much the same way as in fluctuations around equilibrium. Small fluctuations behave therefore "thermodynamically" in the Volterra-Lotha model provided t satisfies inequality (268).

As t approaches the inverse of the rate $A/\mathscr{V}$ the generalized Einstein distribution (271) breaks down and one has to go back to the exact formula (269). The distribution function given by the latter differs appreciably from the Poisson form since $\Delta \overline{X}$ is no longer a small quantity. In order to represent better the difference, let us calculate from (264) the correlations of the one time fluctuations of X. Using the properties of the generating functions outlined in Sec. VIII-C, we obtain

$$\overline{\Delta X^2} = \overline{X^2} - \overline{X}(\tau)^2$$

$$= \left(\frac{\partial^2 g}{\partial \xi^2}\right)_{\xi=0} - \overline{X}(\tau)^2 + \overline{X}(\tau) \tag{272}$$

Calculating the second derivative of g and substituting into Eq. (272) we obtain

$$\overline{\Delta X^2} = \overline{X}(\tau) - 2\overline{X}(\tau)^2\Delta$$

$$= \overline{X}(\tau) + 2\overline{X}(\tau)^2 \int_0^\tau dt \, \frac{A}{\overline{X}(t)} \tag{273}$$

The important point about this result is that large fluctuations around states arbitrarily far from the steady state behave "nonthermodynamically". Indeed, the generalized Einstein fluctuation formula (249) or (271) predicts that in an ideal system

$$\overline{\Delta X^2} = \overline{X}(\tau) \tag{274}$$

whereas (273) predicts fluctuations larger than (274) by a factor $\overline{X}(\tau)\Delta$. As a consequence of Eq. (273) the time dependent distribution g is considerably more "flat" in phase space (around $\overline{X}(t)$) compared to the Poisson distribution. This in turn expresses the fact that in the Volterra–Lotka system there is no mechanism for the regression of fluctuations; as a result, a set of orbits arbitrarily far from a given orbit has a considerable probability of occurrence, and no particular orbit is in any sense preferential.

The time dependence of the correlation of fluctuations also has some peculiar properties. Since $\overline{X}(\tau)$ is periodic, the time integral over the path in Eq. (273) diverges almost linearly in t. In other words, if at $t = 0$ we require the fluctuations to be "thermodynamic" (see Eq. (274)), for $t > 0$ the system will sweep, by thermal fluctuations, an ever broader set of states to finally cover, as $t \to \infty$, the whole space of physical states available. Again this is an alternative expression of the fact that in the Volterra–Lotka system there is no mechanism for the decay of fluctuations.

The result about the time dependence of fluctuations also has another interesting thermodynamic interpretation. According to the thermo-dynamic stability theory of nonequilibrium states discussed in Section III, a sufficient condition for the macroscopic state of a chemical system to be stable with respect to small fluctuations is

$$(\delta^2 S)_0 < 0$$

$$\frac{\partial}{\partial t}(\delta^2 S)_0 > 0 \tag{275}$$

The first inequality determines the probability of occurrence of a fluctuation around the macroscopic state. We have shown previously (cf. Eq. (271)), that in the limit of small fluctuations the probability function was indeed given by a generalization of the classical Einstein formula.

The second inequality of (275) determines the *regression* of fluctuations. According to our result (273) fluctuations may increase in time. This implies that for the Volterra–Lotka problem

$$\frac{\partial}{\partial t}\delta^2 S \leq 0 \tag{276}$$

This is not surprising, since the motion in the Volterra–Lotka system is *not* stable in the sense of Lyapounov unless there is isochronism, that is, the period is the same for different orbits.[21] The only orbits satisfying the isochronism condition are those corresponding to small perturbations around the state (X_0, Y_0) (see Eq. (111)). We have shown that indeed whenever fluctuations are restricted to these orbits, the system behaves "thermodynamically". It is only for large deviations from the steady state that the system jumps to nonisochronic orbits and thereby violates the stability condition (275).

It is instructive to look on the results of this section in a somewhat different way. Let us recall that our main conclusion has been that, for finite fluctuations, the only physical solutions of the master equation are *time dependent solutions* of the form (264). In other terms, even as $t \to \infty$, the system cannot attain a physically reasonable steady state. Clearly, this is a singular case corresponding to the absence of a regression mechanism for the fluctuations. Imagine now that some inverse reactions, for instance, the third reaction in (102), are switched on and let k' be the rate of this back reaction. For k' arbitrarily small, a linear stability analysis of Eqs. (102) shows that the system, if perturbed, exhibits an oscillatory approach to the steady state. In other terms for $t \to \infty$ the system now approaches a uniquely determined macroscopic steady state which is easily found to be

$$X_0 = B - \frac{k'E}{A}, \qquad Y_0 = A \qquad (277)$$

Let us now investigate the consequences of this back reaction in the master equation. It is easily shown that in the generating function space, the equation which now describes the system takes the form

$$\frac{\partial f}{\partial \tau} = (\eta + 1)(\eta - \xi)\frac{\partial^2 f}{\partial \xi \, \partial \eta} + \xi(\xi + 1)A\frac{\partial f}{\partial \xi} - \eta B\frac{\partial f}{\partial \eta} + k'\eta E f \qquad (278)$$

where we let k' be very small. Eq. (278) should now admit a physically reasonable steady state; at the same time for $k' \to 0$, or previous solution (270) should be recovered. Unfortunately, the steady state solution of (278) cannot be found in a closed form. For this reason, we adopt an approximation scheme in order to show how our previous result is inserted into this more general scheme.

The main point is to realize that the back reaction introduces a new time scale

$$\lambda = \frac{1}{k'} \qquad (279)$$

which, for $k' \ll 1$ is long with respect to the previous time scale (see Eq. (111))

$$\lambda_0 = \frac{1}{\omega} \qquad (280)$$

This suggests a multiscale perturbation solution of (278) which follows the ideas of the Poincaré-Bogoliubov theory of nonlinear oscillations.[20,21] We assume that f depends on time in different scales:

$$f = \sum_{n=0}^{\infty} \varepsilon^n f_n(t_0, \varepsilon t_1, \ldots, \varepsilon^s t_s, \ldots) \qquad (281)$$

where we defined

$$\varepsilon = \frac{\lambda_0}{\lambda} = \frac{k'}{\omega} \ll 1 \qquad (282)$$

and

$$\frac{dt_s}{dt} = 1 \qquad (s = 0, 1, \ldots,) \tag{283}$$

together with the "initial" conditions

$$t_s(0) = 1$$

Similarly, the time derivatives can be expressed in the form

$$\frac{\partial f}{\partial t} = \sum_{n=0}^{\infty} \varepsilon^n \frac{\partial}{\partial(\varepsilon^n t_n)} f = \sum_{n,\,m=0}^{\infty} \varepsilon^{n+m} \frac{\partial}{\partial(\varepsilon^n t_n)} f_m \tag{284}$$

Substituting this expansion into Eq. (278), we obtain the set of equations

$$\frac{\partial}{\partial \tau_0} f_s - (\eta + 1)(\eta - \xi) \frac{\partial^2 f_s}{\partial \xi\, \partial n} - \xi(\xi + 1)A \frac{\partial f_s}{\partial \xi} + \eta B \frac{\partial f_s}{\partial \eta}$$

$$= -\frac{\partial}{\partial(\varepsilon \tau_1)} f_{s-1} + k'\eta E f_{s-1} - \sum_{r=2}^{\infty} \frac{\partial}{\partial(\varepsilon^r \tau_r)} f_{s-r} \tag{285}$$

Because of the partial differential character of Eq. (285), we are not allowed to annul the left hand side of this equation. We can only look on Eq. (285) as consisting of two parts. One (the l.h.s.) developing in the short time scale, the second (r.h.s.) in the long time scale. It is therefore only a *first approximation* to set the l.h.s. of (285) equal to zero. The resulting equation is identical to (254) *provided t is now limited to the short time scale.* As time lengthens, even for k' being very small, the right hand side of (285) becomes more and more important and the system tends asymptotically, to a steady state solution.

Alternatively, for k' sufficiently small, the solution established previously in the form (264) is valid for arbitrarily long times, and only becomes unsatisfactory for times $t \gtrsim (k')^{-1}$. The system then tends slowly to a steady state, but the approach to that state is still characterized by "nonthermodynamic" fluctuations.

The statistical properties of the Volterra–Lotka system have also been investigated by Kerner from a quite different standpoint. His theory is based on the observation that a Volterra–Lotka system with an arbitrary number of species is conservative in the same sense as in Section IV.

Starting from this, Kerner sets up a statistical mechanical theory of the system by considering the kinetic equations (107) (suitably generalized to N species, N even number $\rightarrow \infty$) as *microscopic* equations of motion. He defines his phase space variables $\{u\}$ as (see Eq. (113)):

$$\text{"coordinate"} \; u_x = \ln \frac{X}{B}, \qquad \text{"momentum"} \; u_Y = \ln \frac{Y}{A} \qquad (286)$$

and so on, over all species pairs and derives a Liouville equation in the space of $\{u\}$s. The equilibrium solution of this equation is a function of the constants of motion. Among all possible such functions, Kerner studies almost exclusively the "canonical" distribution

$$\rho = Z_N^{-1} \exp\left[-\beta \sum_i V(X, Y, \ldots) \right] \qquad (287)$$

where Z_N is a suitable normalization factor, and β is to be interpreted as a "generalized reciprocal temperature" which provides a measure of the importance of statistical fluctuations. Because of the additivity of V over species, ρ is factorizable.

It should be pointed out that the arguments used in deriving Eq. (287) are based on the analogy between the Volterra–Lotka problem and mechanics. The original Volterra–Lotka equations cannot by themselves yield information about fluctuations, even when they are considered as microscopic equations. It is therefore necessary to introduce an additional element, for example, an equal *a priori* probability assumption, in order to justify (287).

On the other hand, in the stochastic theory outlined in this subsection the kinetic equations (107) are considered as *macroscopic* average equations of evolution. The master equation itself is considered as a reduced description of the behavior of a subsystem (X, Y) of a large system (A, B, X, Y, E) with suitable nonequilibrium boundary conditions. Clearly, this description is motivated by the concepts and methods of continuum mechanics and thermodynamics of irreversible processes. Irreversibility is introduced here by the stochastic assumption, whose validity is beyond any doubt, at least for the chemical kinetic problem considered herein. We want now to compare the master equation and the Kerner approaches and to investigate the implications of the latter as far as irreversible thermodynamics are concerned.

Let us first construct the distribution function which would describe, in Kerner's theory, the behavior of the simple two component system considered in this paper. Upon integrating (287) over N-2 species we obtain (see also Eq. (115b))

$$\rho = Z_2^{-1} \exp\left[-\beta V(X, Y) \right]$$

$$= Z_2^{-1} \exp\left[-\beta(X + Y) \right] X^{\beta B} Y^{\beta A} \qquad (288)$$

It is this reduced function which has to be compared with the results obtained earlier in this section. Let us first consider the behavior in the neighborhood of the steady state. Eq. (288) can then be expanded in powers of $X - B = \delta X$, and $Y - A = \delta Y$. According to Eq. (116), when the first nontrivial terms are retained, one obtains $V - V_0 = (\delta^2 S)_0/2k$. Clearly, if one chooses $\beta = 1$, one obtains a distribution of the form

$$\rho = Z_2^{-1} \exp \left[\frac{(\delta^2 S)_0}{2k} \right] \qquad (289)$$

in agreement with our previous result (271).

Let us now consider the case of states arbitrarily far from the steady state. In principle, Kerner's function (288) should still provide a description of these states. In order to compare (288) with the predictions of the master equation, we check whether it is an exact steady state solution of Eq. (252). After a few algebraic manipulations, we find that the condition for this to be the case is

$$Ae(X - 1)(1 - X^{-1})^B - AX + (X + 1)(Y - 1)(1 + X^{-1})^B(1 - Y^{-1})^A$$
$$- XY + Be^{-1}(Y + 1)(1 + Y^{-1})^A - BY = 0 \qquad (290)$$

For X, Y being not close to A, B (that is, for arbitrarily large fluctuations) and A, B $\rightarrow \infty$ Eq. (290) cannot be an identity in X and Y. It is only in the asymptotic limit of small fluctuations that Eq. (290) is satisfied. The conclusion therefore is that Kerner's distribution function is not an exact steady state solution of the master equation. The reason for this is that as we have shown previously, situations where statistical fluctuations are finite cannot be described by a steady state solution and are therefore beyond the domain of validity of Kerner's theory.

In conclusion, stochastic theory enables us to obtain results for the behavior of "nonthermodynamic" fluctuations for systems very far from equilibrium. We expect that the really important problem of unstable transitions may be treated by the same type of method, and that the results will be not be very different, for the reasons explained in the beginning of this subsection.

IX. APPLICATIONS IN BIOLOGY

A. Introduction

In the preceding sections, we have shown that the equations of chemical kinetics, eventually coupled with transport processes such as diffusion, admit certain types of solution showing some degree of temporal, spatial, and functional organization. Many fundamental biological processes consist primarily of chemical reactions involving macromolecules, and of transport of small molecules through membranes. Essential for the correct evolution of these processes is their subtle coupling with the biological underlying structures such as mitochondria, membranes, etc. The result of this coupling is a *functional order* which may be manifested on an intracellular, cellular or intercellular scale. It is therefore meaningful to inquire about the relation between this functional order and the phenomena analyzed in Sections IV to VII.

To avoid misunderstanding, it should be stressed that the form of dissipative structures discussed in this review is probably not yet the exact prototype of biological structures in the usual sense of the term (macromolecules, membranes, cells as a whole, and so on). Indeed, the drastic lowering of entropy and the increase of entropy production which characterize these structures[79,80] does not occur as a general rule in the dissipative structures analyzed previously (see discussion in Sec. V).[3] What the theory of dissipative structures can explain, however, are some aspects of functional order, once the existence of macromolecules and their interactions is taken for granted. It is conceivable that future work may prove that the concept of dissipative structures is even more fundamental, and is one of the necessary prerequisites for life.

We shall now briefly present separately illustrations of the different types of dissipative structures in concrete biological examples.

B. Oscillatory Processes in Biology

Once it is realized that different control mechanisms (activation, inhibition, or cross-catalysis) are essential for the occurrence of sustained oscillations, it is natural to expect that this type of phenomena will be important in biology. Indeed, it is well known that living systems possess regulatory mechanisms which permit an optimal functioning compatible with the maintenance of life.

It is convenient to distinguish between two types of periodic phenomena in biological systems:

(1) Oscillations in processes involving the synthesis of macromolecules or metabolites.

(2) Rythmic activity of the nervous system.

Type (1) may further be divided into three categories:

(1a) *Oscillations in enzymatic reactions*, usually in the concentrations of metabolites participating in this reaction. This phenomenon is due to a regulatory process in the enzymatic level. A great number of theoretical models have been proposed to describe this regulation. Usually, one distinguishes between negative feedback (inhibition) and positive feedback.

Models involving inhibition have been worked out by Spangler and Snell,[60] Morales and Mac Kay,[81] Walter,[44] Sel'kov,[37] Griffith et al.[43] A general scheme for these models has been given in Sec. V-E. The result is that, under certain conditions, oscillation of the limit cycle type become possible. On the other hand, if the number of inhibiting molecules of the final product is small, the model predicts oscillations of the Volterra–Lotka type. Models involving activation have been worked out by Higgins[38] and Sel'kov.[37]

Experimentally, sustained oscillations in enzymatic reactions have been observed and established beyond doubt for glycolysis. Theoretical studies attribute these oscillations to the enzyme phosphofructokinase, which is both activated by the products ADP and FDP and inhibited by the substrate ATP. Experiments carried out by Hess,[82,83] Betz,[84] Chance et al.[36] have produced first damped, then sustained, oscillations which seem to be perfectly reproducible.

(1b) *Oscillations in the synthesis of enzymes in the cellular level.* On a theoretical basis, such oscillations are attributed to the induction and repression mechanisms of the Jacob–Monod type. The first systematic study of such oscillations is due to Goodwin.[28] However, the models he studied principally can only lead to Volterra–Lotka type of oscillations. The possibility of sustained oscillations of the limit cycle type has been discussed by Griffith,[43] Koch,[85] Knorre,[86] etc.

Anong the recent experimental works, Knorre[41] reports oscillations on the rate of synthesis of β-galactosidase in *Escherichia Coli* in asynchronously growing cultures, which subsist for a few periods. Masters and Donachie[87] observe periodic enzyme synthesis in synchronous cultures of *Bacillus subtilis* which subsist in the absence of DNA synthesis, and seem therefore not to be directly dependent on gene replication. These authors attribute the effect to end product repression.

(1c) *Oscillations in the supercellular level.* These so called circadian rythmic phenomena are very slow ones; the complexity of the processes involved does not permit a precise physico-chemical quantitative analysis[88].

Rythmic phenomena have long been known to exist in the nervous system. Electroencephalograms and electrocorticograms provide a spectacular proof of the importance of time dependent effects. On the other hand, the extraordinary complexity of the nervous system is such that all theories which attempt to describe neural networks have a speculative character. For a survey of the different approaches the reader may consult

Ref. 89. We here mention explicitly two types of approaches which are more closely related to the ideas developed in this review:

(2a) *Statistical mechanical approach to nervous nets.* This study, due to Cowan,[29] is based on equations of evolution for suitably defined variables* which are conservative and therefore predict oscillations of the same type as the Volterra–Lotka model.

(2b) *Analysis in terms of limit cycles.* In this approach, it is considered that the stability and the sharp character of nervous activity necessarily implies an autonomous oscillatory process of the limit cycle type.[90] Once this is admitted, it is possible to construct simple models involving only a few groups of neurons and imagine neural connections resulting in limit cycles.

From a more general viewpoint, it is important to realize that periodic phenomena such as (1a) and (1b), with sharply defined reproducible frequencies or amplitudes, should be necessarily of the limit cycle type (cf. also analysis of Sec. V-D and VIII-D). Now a limit cycle necessarily arises beyond a bifurcation point of a steady state solution which is the continuation of the close to equilibrium regime. In other words, a system presenting oscillations as in (1a) and (1b) is necessarily a dissipative structure, which is only maintained by suitable *nonequilibrium* conditions.

It is much more difficult to formulate a similar conclusion for the rythmic activities of the nervous system. However, it seems legitimate to think that the analyses of Sections V and VIII cast some doubts about the validity of an explanation based on coupled Volterra–Lotka oscillators. The whole problem requires further investigation.

C. Symmetry-Breaking Instabilities and Interference Effects

The emergence of spatial order in a previously homogeneous system is a central problem in embryogenesis, and more generally in all problems involving cell differentiation. The difficulty of this problem is that it involves "morphogenetic fields" with a great number of cells. On the other hand, the problem of space order on the cellular level is much more profound and has not yet received a satisfactory explanation.

In a discussion of pattern formation in embryogenesis Wolpert[91] considers a model where a gradient of a chemical substance develops between two prescribed cells of an array. As the chemical propagates by diffusion

---

* These variables are related to the fraction of time during which a neuron is not refractory, and can be fired.

in the other cells of the array, it may reach "critical" values beyond which differentiation begins in the cells concerned, by a suitable mechanism other than the propagation of the chemical.*

Wolpert considers primarily static structures. Waddington[92] and more recently Goodwin and Cohen[93] have considered the problem of the mechanism of establishment of a spatial organization. Their approach is based on the premise that cells are to be regarded as fundamentally oscillatory systems. It is then possible to expect that in an aggregate composed of cells, oscillations of one entity will affect those of the neighbors, for instance by diffusion, and some discrete overall patterns of oscillations will be built up. If one further assumes that this space dependent rythym will influence the type of synthetic processes in the cell, one will have a spatial pattern of regions resulting from an underlying temporal organization.

In all these types of problems, the concept of dissipative structures appears quite naturally, and helps to clarify further the different kinds of processes. The problem of differentiation, beyond a critical value of some substance, may be considered as a phenomenon arising beyond a symmetry-breaking instability of the type discussed in Section VI-B.† In addition, the theory of dissipative structures gives criteria about the type of metabolic processes necessary for differentiation, and about the precise role of diffusion. The problem of "chemical waves" arising from oscillating cells, and propagating to yield spatial patterns appears to be related to the interference effects discussed in Section VI-D. The latter may provide a model for spatial and temporal organization which is related to the phase shift model of Goodwin and Cohen.

On the cellular level, it is difficult to avoid the feeling that spatial dissipative structures have not contributed in an essential way to the first biogenetic steps, and therefore also implicitly to the formation of the cellular structure itself. This point requires further study and will not be developed here.

D. Multiple Steady States

The problem of multiple steady states is of primary importance in all phenomena involving "an all or nothing" type of transition. For instance, for suitable critical values of parameters, a system may depart abruptly

---

* It is assumed that differentiation occurs on a slow time scale compared to the time necessary for the complete establishment of the gradient.

† An analysis of morphogenetic processes in terms of the stability properties of the evolution equations is also made by Thom.[94]

from a steady state and go to a new one, which has roughly speaking the same symmetry as the former but differs in the level of concentration of certain characteristic substances.

There exist at least two types of all or nothing effects in biology: The process of differentation in higher organisms and the functioning of excitable membranes.

(1) *The problem of differentiation.* Monod and Jacob have proposed a number of models describing differentiation.[42] One of these models which has also been referred to in Section VII-B has been analyzed by Cherniavskii et al.[66] They find indeed that there exist critical regions where the system can switch to a regime such that a given type of substance is produced preferentially. A different type of model, also suggested by Monod and Jacob, is completely independent of the metabolic activity of the enzymes, and describes a system which is switched on from one "inactive" state to a second "active" one by contact with a specific inducer. All or nothing effects which are believed to be of this type have been observed in different types of phages.

(2) *Excitable membranes.* Roughly speaking, a biological excitable membrane such as the membrane of a nervous cell may exist in two permanent states; one polarized (associated with the maintenance of different ionic charges in the two sides), and one depolarized state resulting from the former upon passage of a pulse or upon a change in permeability. Blumenthal, Changeux and Lefever have shown recently[95] that this depolarization may be quantitatively interpreted as a transition arising beyond the point of instability of the polarized state, which lies on the thermodynamic branch. This instability is due to the difference in the ionic concentrations, which here plays the role of the constraint keeping the system in a far from equilibrium state. On the other hand, experimental observations establish that the transition is indeed in the form of an all or nothing effect both for biological[96] and artificial membranes.[68]

X. GENERAL CONCLUSIONS

In this review, we have shown that thermodynamics of irreversible processes provides a basis for the interpretation of a great number of complex physico-chemical phenomena arising in open systems undergoing chemical reactions and transport processes. One particularly important class of such phenomena is instability and the evolution of the system thereafter to states showing a certain degree of "ordering". We have pointed out in many instances the analogy between this "order" and some types of biological organized structures. The important point to be drawn from the results of this comparison is that biological order seems to be perfectly compatible with the principles of thermodynamics; there exists no law in physics prohibiting the formation of organized systems. Even more interesting, beyond a certain regime (bifurcation point) the occurrence of order is imposed, with a probability of one, by the nature of the system itself, provided the exchanges with the surroundings attain a certain minimum level. This was the very essence of *dissipative structures*, a concept which has been repeatedly discussed in the preceding paragraphs; it is also the principal result of the theory of Prigogine and coworkers presented in this review.

This successful development, also supported by the experimental results described in Secs. V, VI and IX, gives us confidence about the fundamental importance of dissipative structures in biology. At the same time, it makes us realize that the whole approach is still at its very early stages of development as far as the study of applications is concerned. First, the examples of spatially, temporally, or functionally organized systems which have been worked out are probably not yet good prototypes of what is commonly considered to be a biological structure (cf. also comments in Sec. IX-A) but rather correspond to examples of "metabolic order". Second, the mechanism of establishment and the detailed properties of spatially organized structures are not known to a degree of generality and rigor comparable to the theory of limit cycles. Third, a number of questions of fundamental importance in biology have not even been approached in the light of the theory of dissipative structures; differentiation, evolution, the functioning of the central nervous system are only a few examples of such problems. Yet one has the feeling that the theory should be able to say something new concerning these questions. Fourth, the precise mechanism involved in the amplification of fluctuations in the neighborhood of the unstable transition is practically unknown. Some progress in this direction has been made by the analysis of fluctuations in the Volterra–Lotka model given in Sec. VIII. However, the problem of fluctuations in systems undergoing one of the instabilities discussed in Secs. V to VII has not yet been solved. Of particular interest is the study of the behavior of the system with respect to *finite* fluctuations. From this point of view, a thermodynamic stability theory with respect to such fluctuations is also to be worked out.

Finally, it should be pointed out that the concept of dissipative structures is not only of primary importance in biology. Hydrodynamics, especially in the domain of flow instabilities, appears to be another branch where order and dissipation are closely linked. Also nonbiological chemical reactions, for instance like the Zhabotinski reaction, may give rise to dissipative structures. We believe that further progress in the theory should be directed in part to this domain, where experimental verifications are more easily carried out.

Acknowledgments

We wish to express our gratitude to Professors I. Prigogine and P. Glansdorff and Drs. A. Babloyantz, R. Lefever, J-P. Changeux, M. Herschkowitz-Kaufman, and B. Lavenda for interesting discussions and critical comments.

This research has been supported, in part by the Air Force Office of Scientific Research (S.R.P.P.) through the European Office of Aerospace Research, OAR, United States Air Force under grant number EOOAR 69-0058.

References

1. I. Prigogine, *Introduction to Thermodynamics of Irreversible Processes*, 3rd. Edition, Interscience, Wiley, New York (1967).
2. P. Glansdorff and I. Prigogine, *Thermodynamic Theory of Structure, Stability and Fluctuations*, Interscience, Wiley, New York (in press).
3. I. Prigogine, *Structure, Dissipation and Life*, in *Theoretical Physics and Biology*, Ed. M. Marois, North Holland Publ. Co, Amsterdam (1969).
4. P. Duhem, *Energétique*, Gauthier-Villars, Paris (1911).
5. I. Prigogine, *Etude Thermodynamique des Phénomènes Irréversibles*, Desoer, Liége (1947).
6. I. Prigogine, *Physica*, **14**, 272 (1949).
7. G. Nicolis, J. Wallenborn and M. G. Velarde, *Physica*, **43**, 263 (1969).
8. S. de Groot and P. Mazur, *Non-Equilibrium Thermodynamics*, North Holland Publ. Co, Amsterdam (1961). This provides a very detailed and clear presentation of irreversible thermodynamics in the linear region.
9. L. Onsager, *Phys. Rev.*, **37**, 405 (1931); ibid., **38**, 2265 (1931).
10. H. B. G. Casimir, *Rev. Mod. Phys.*, **17**, 343 (1945).
11. L. D. Landau and E. M. Lifshitz, *Statistical Physics*, Pergamon Press, Oxford (1959).
12. P. Glansdorff and I. Prigogine, *Physica*, **20**, 773 (1954); I. Prigogine and R. Balescu, *Bull. Cl. Sci. Acad. Roy. Belg.*, **41**, 917 (1955).
13. P. Glansdorff and I. Prigogine, *Physica*, **30**, 351 (1964).
14. *Non-Equilibrium Thermodynamics, Variational Techniques and Stability*, Edts. R. Donnelly, R. Herman and I. Prigogine, Chicago Univ. Press, Chicago (1965).
15. I. Prigogine and P. Glansdorff, *Physica*, **31**, 1242 (1965).
16. P. Glansdorff, *Physica*, **32**, 1745 (1966).
17. R. S. Schechter, *The Variational Method in Engineering*, McGraw-Hill Book Co, New York (1967).
18. G. Nicolis, *Adv. Chem. Phys.*, **13**, 299 (1967).
19. P. Glansdorff and I. Prigogine, *Physica*, **46**, 344 (1970).
20. L. Cesari, *Asymptotic Behavior and Stability Problems in Ordinary Differential Equations*, Erg. Mathem. New Series, **16**, Springer Verlag, Berlin (1962).
21. N. Minorsky, *Non-linear Oscillations*, Van Nostrand Co., Princeton, N.J. (1962).

322   G. NICOLIS

22. V. I. Zubov, *Methods of A.M. Lyapounov and their Applications*, U.S. Atomic Energy Comm., AEC-Tr-4439 (1961).
23. T. R. Fowler, *J. Math. Phys.*, **4**, 559 (1963).
24. This property is demonstrated in textbooks of Electromagnetic Theory, for example W. Panofsky and M. Philips, *Classical Electricity and Magnetism*, Addison-Wesley Publ. Co., Reading, Mass. (1955).
25. R. Lefever, G. Nicolis and I. Prigogine, *J. Chem. Phys.*, **47**, 1045 (1967).
26. A. Lotka, *Elements of Mathematical Biophysics*, Dover Publ. Inc., New York (1956).
27. V. Volterra, *Leçons sur la Theorie Mathématique de la Lutte pour la Vie*, Gauthier-Villars, Paris (1931).
28. B. Goodwin, *Temporal Organization in Cells*, Academic Press, London (1963).
29. J. Cowan, *Statistical Mechanics of Nervous Nets*, in *Neural Networks*, Springer Verlag, Berlin (1968).
30. E. H. Kerner, *Bull. Math. Biophys.*, **19**, 121 (1957).
31. E. H. Kerner, *Bull. Math. Biophys.*, **21**, 217 (1959).
32. E. H. Kerner, *Bull. Math. Biophys.*, **26**, 333 (1964).
33. H. T. Davis, *Introduction to Nonlinear Differential and Integral Equations*, Dover Publ., New York (1962).
34. A. Liénard, *Rev. gén Electr.*, **23**, 901 (1928).
35. A. M. Yaglom and I. M. Yaglom, *Probabilité et Information*, Dunod Ed., Paris (1959).
36. B. Chance, R. W. Eastbrook and A. Ghosh, *Proc. Natl. Acad. Sci.* (U.S.A.), **51**, 1244 (1964).
37. E. E. Sel'kov, *Europ. J. Biochem.*, **4**, 79 (1968); *Molec. Biol.*, **2**, 252 (1968).
38. J. Higgins, *Proc. Natl. Acad. Sci.* (U.S.A.), **51**, 989 (1964).
39. A. M. Zhabotinski, *Biofizika*, **9**, 306 (1964).
40. H. Degn, *Nature*, **213**, 589 (1967).
41. W. A. Knorre, *Biochem. Biophys. Res. Commun.*, **31**, 5 (1968).
42. F. Jacob and J. Monod, *J. Mol. Biol.*, **3**, 318 (1961); J. Monod and F. Jacob, in *Cold Spring Harbor Symposia on Quantitative Biology* (1961).
43. J. S. Griffith, *J. Theoret. Biol.*, **20**, 202 (1968); ibid, **20**, 209 (1968).
44. C. Walter, *J. Theoret. Biol.* **25**, 39 (1969); *Biophys. J.* **9**, 863 (1969).
45. I. Prigogine and R. Lefever, *J. Chem. Phys.*, **48**, 1695 (1968).
46. H. G. Othmer and L. E. Scriven, *Ind. and Ec. Fundam.*, **8**, 302 (1969).
47. A. M. Turing, *Phil. Trans. Roy. Soc. Lond.*, **B237**, 37 (1952).
48. I. Prigogine and G. Nicolis, *J. Chem. Phys.*, **46**, 3542 (1967).
49. I. Prigogine, R. Lefever, A. Goldbeter and M. Herschkowitz-Kaufman, *Nature*, **223**, 913 (1969).
50. R. Lefever, *J. Chem. Phys.*, **49**, 4977 (1968).
51. R. Lefever, *Bull. Cl. Sci. Acad. Roy. Belg.*, **54**, 712 (1968).
52. H. Busse, *J. Phys. Chem.*, **73**, 750 (1969).
53. M. Herschkowitz-Kaufman, *Comptes Rendus Acad. Sci.* (Paris), **270C**, 1049 (1970).
54. A. M. Zhabotinski, *Nature*, **225**, 535 (1970).
55. A. N. Zaikin and A. M. Zhabotinski, *Russ. J. Phys. Chem.*, **42**, 1649 (1968).
56. E. A. Coddington and N. Levinson, *Theory of Ordinary Differential Equations*, McGraw-Hill, New York (1955).
57. W. Wasow, *Asymptotic Expansions for Ordinary Differential Equations*, Interscience, Wiley, New York (1965).

58. N. Rashevsky, *Mathematical Biophysics*, Dover Publications, Inc., New York (1959).
59. A. Bierman, *Bull. Math. Biophys.*, **16**, 203 (1954).
60. R. A. Spangler and F. M. Snell, *Nature*, **191**, 457 (1961); *J. Theoret. Biol.*, **16**, 381 (1967).
61. R. Aris, *Chem. Engin. Science*, **24**, 149 (1969), and references quoted therein.
62. M. Herschkowitz-Kaufman, *Mémoire de Licence*, Université Libre, Brussels (1968).
63. B. Edelstein, *J. Theoret. Biol.*, **29**, 57 (1970).
64. A. Babloyantz, to be published.
65. L. Szilard, *Proc. Natl. Acad. Sci.* (U.S.A.), **46**, 277 (1960).
66. D. C. Cherniavskii, L. N. Grigorov and M. C. Poliakova, in *Oscillatory Processes in Biological and Chemical Systems* (in Russian), Nauka, Moscow (1967).
67. B. Lavenda, *Ph. D. Thesis*, University of Brussels (1970).
68. Y. Kobatake, *Physica*, **48**, 301 (1970).
69. H. B. Callen, *Thermodynamics*, Wiley, New York (1960).
70. M. Kac, *Probability and Related Topics in Physical Sciences*, Interscience, New York (1959).
71. L. Onsager and S. Machlup, *Phys. Rev.*, **91**, 1505 (1953).
72. H. B. Callen, in *Non-equilibrium Thermodynamics, Variational Techniques and Stability* (Chicago Univ. Press (1965)).
73. L. D. Landau and E. M. Lifshitz, *Fluid Mechanics*, Pergamon Press, Oxford (1958).
74. A. Fokker-Planck equation approach to the problem of fluctuations around non-equilibrium states has been discussed in detail by M. Lax, *Rev. Mod. Phys.*, **32**, 25 (1960).
75. G. Nicolis and A. Babloyantz. J. Chem. Phys., **51**, 2632 (1969).
76. A. Babloyantz and G. Nicolis, *J. Stat. Phys.*, **1**, 563 (1969).
77. The Markovian Stochastic Approach to chemical kinetic problems is described, for example, in D. Mc Quarrie, *Suppl. Review Series in Appl. Probability*, Methuen and Co, London (1967).
78. See, for example, E. L. Ince, *Integration of Ordinary Differential Equations*, Oliver and Boyd, Edinburgh (1952).
79. I. Prigogine and J. M. Wiame, *Experientia*, **2**, 451 (1946).
80. A. Lwoff, *Biological Order*, M.I.T. Press, Cambridge, Mass. (1960).
81. M. Morales and D. Mc Kay, *Biophys. Journ.*, **7**, 621 (1967).
82. B. Hess, in *Funktionelle und Morphologische Organisation der Zelle*, Springer Verlag, Berlin (1963).
83. B. Hess and A. Boiteux, in *Regulatory Functions of Biological Membranes*, Elsevier Publ. Co. (1968).
84. A. Betz and B. Chance, *Arch. Biochem. Biophys.*, **109**, 579 (1965).
85. A. L. Koch, *J. Theoret. Biol.*, **16**, 166 (1967).
86. W. A. Knorre, *Studia Biophysica*, **6**, 1 (1968).
87. M. Masters and W. D. Donachie, *Nature*, **209**, 476 (1966).
88. See, for example, E. Bünning, *The Physiological Clock*, Academic Press, New York (1964).
89. *Neural Networks*, E. R. Caianiello, Editor, Springer Verlag, Berlin (1968).
90. See, for example, L. L. Boyarsky, *Currents in Mod. Biol.*, **1**, 39 (1967).
91. L. Wolpert, in *Towards a Theoretical Biology*, Vol. **1**, C. H. Waddington Ed., Edinburgh Univ. Press, Edinburgh (1968).
92. C. H. Waddington, *J. Theoret. Biol.*, **8**, 367 (1965).

93. B. C. Goodwin and M. H. Cohen, *J. Theoret. Biol.*, **25**, 49 (1969).
94. R. Thom, *Stabilité Structurelle et la Morphogenèse*, Benjamin and Co, New York (1969).
95. R. Blumenthal, J. P. Changeux and R. Lefever, Comptes Rendus Acad. Sci. (Paris), **270**, 389 (1970); *J. Membrane Biol.*, **2**, 351 (1970).
96. See, for example, Tasaki, *Nerve Excitation*, C. C. Thomas, Springfield, Illinois (1698).

STATISTICAL–MECHANICAL THEORIES IN BIOLOGY

EDWARD H. KERNER

Physics Department, University of Delaware, Newark, Delaware

CONTENTS

I. INTRODUCTION

However crude or refined a mathematical model one makes in biology, there seems always to bestride it a higher order of complexity due to sheer numbers of coupled elements. The multitude finally has to be reckoned with if the model is to bear upon significant observation. So for instance in ecology, remarkable historically for its theoretical wealth at an intrinsically biological level, one finds even simple models for interacting species running into trouble as soon as the number of species associated together is more than a few.

The type of trouble so quickly encountered is in many cases fundamentally like that in particle dynamics when one steps from the completely analyzable two-particle problem to the unmanageable many-particle (say 10^{23}-fold) one; it is the problem of differential laws so grossly numerous and intricate that our mathematical powers are swamped.

The marvelous instrument of statistical mechanics (classical Gibbs ensemble theory) not only walks around the dynamical problem; it parlays the deep ignorance into global laws of the whole by means of quite novel theoretical constructions like temperature and entropy that are not originally in the picture. These have a significance that reaches even beyond an ultrafine knowledge of dynamical orbits, for such knowledge (if we had it) would also swamp us, likely leading us (after baffling us) to average over its enormity in one way or another so as to make some sense of it. Further, and stunningly, it is of the statistical essence of ensemble theory that its thermalistic insights are the sharper, not the duller, as the complexity increases.

What makes statistical mechanics work has nothing to do at base with mechanics. The cornerstones are (a) Liouville's theorem and (b) some one or a few conservation laws of suitable type and (c) ignorance otherwise. The Hamiltonian format of mechanics merely places (a) and (b) into transparent view, being sufficient to this purpose but not at all necessary. Many another format for differential laws will do just as well in regard to (a), (b), (c). It is on this ground that systems of differential equations modeling biological phenomena, where typically great numbers of degrees of freedom are interlocked, may be brought under Gibbsian surveillance. In short, ensemble theory is a statistical theory of differential equations, and by so abstracting its core, it may serve biology and doubtless other sciences equally with physics. Just from this standing as theoretical method rather than specific theory do Gibbs ensembles gain their unusual versatility.

The present essay reviews some examples of statistical-mechanical attempts into biological questions, specifically the examples of highly multicomponent ecological networks; neural networks; and, much more faintly, biochemical-kinetic networks, the most important and challenging and un-understood of all, for which questions are raised rather than answers given. In a final section is broached the overall mathematical issue, When is any differential theory statistically mechanizable? The answer appears to be, Virtually always, theoretically. From the same mathematical context it is indicated that quantization, too, is a wide theoretical instrument and not just a specifically microphysical scheme. The macrostochastic (stamped by introduction of a temperature parameter) and the micro-stochastic (characterized by a Planck-type parameter), which exfoliate historically from Hamiltonian particle mechanics, both have a deeper tap-root, at least formally capable of reaching out to greater classes of differential statements. One might be glad enough to statistically mechanize a differential model from general principles; also to be able to quantize it would seem to be gratuitous in many cases, perhaps, however, not all.

Several general features of ensemble biology may be observed.

First, the statistically scanned biotheories as noted are phenomenological and are deterministic; the coordinates entrained into differential equations are counted as continuous variables,—biological species population sizes, or neuron firing frequencies, or *in vivo* biochemical concentrations. This order of construction is a frank concession to the difficulties of theorizing up from below into biology. But the range of observation is so large that phenomenology, even when highly simplified into provisional models, clearly has to be given much free play. Granting a theoretical corpus of this type that eventually gets refined into a good representation of selected facts (a native theoretical biology, that is, speaking to biological issues) it becomes an interesting question whether the biological coordinates finally are, or need be, completely coherent with the ultimate physical ones. The separation between the two is entrancingly close in biochemical kinetics, where the phenomenology in the form of mass action laws is relatively sharp and is also substantially universal.

Throughout, the determinism of differential laws may appear to be obnoxious, for basically stochastic effects are covered over by them, as in fluctuations of low chemical concentrations or of low ecological population counts that are beyond the reach of deterministic description. Yet, in the spirit of theorizing from the top down, such models would seem worth exploring to their limit, if only to see what may be their limits, including especially those collective features arising from the always numerous biological degrees of freedom. If the determinism have some useful outer limit, it may then serve as a touchstone to the more complete stochastic theory, and conceptual elements (like a generalized temperature) intro- duced on the deterministic base may well survive a stochastic elaboration of this base. Physical statistical mechanics, after all, did not have to await quantum theory, but proceeded nicely on Newtonian ground to invent lasting conceptions; and it was from thermal vantage points that a critical survey of this ground was afforded. Although differential equations are no doubt a severely limited means of theoretical expression, they have persisted remarkably through centuries of mathematical sketchworks of nature.

Next, the frequently asserted multiplicity of biocoordinates is not all so great, by physical standards, at any rate. In ecology, the entire biosphere may have perhaps 10^6–10^8 species (heavily weighted with beetles); the neural network in higher animals, some 10^9–10^{11} neurons; the intracellular biochemical apparatus, possibly 10^3–10^6 chemical types.

In these ranges, the statistical force of the laws of large numbers is bound to be noticeably less than for 10^{23}-fold complexity. Supposing the differential laws in all cases to be given, the distinction between smaller and larger numbers of degrees of freedom shows up clearly in the motion of a single coordinate, and in our knowledgeability of the motion; for the single coordinate coupled to a few others, we do not admit statistics at all, the motion is predictable or "orbital"; when coupled to a trillion trillion others it is as orbital as ever in principle, but we admit that principle fails us and call the motion "chaotic", implanting orbital ignorance into ensembles.

Where does the orbital stop and the chaotic begin? The biocoordinates broadly seem to lie sprawled between the two. With complexity on the order of 10^9, 10^6, even 10^3, there would seem to be room for the ensemble approach, but room too for questioning and bettering it. It is of course the single "microscopic" biocoordinate that stands for what is observable in the phenomenological models, its more or less noisy behavior in the course of long times being the object of statistical assay through ensembles. Unlike the case in physics where the thermostatics of bulk matter is the main statistical mechanical issue, the biomodels have no "bulk" but what the micronoise reveals; but that may be attunement enough into the whole of the model, or, with luck, into some small part of nature.

II. GENERALITIES ON ENSEMBLES

Let us rehearse briefly the core of ensemble theory without the usual Hamiltonian distractions.

Suppose an autonomous differential system

$$\dot{x}_i = X_i(x) \qquad i = 1, 2, \ldots n \tag{1}$$

is prescribed. In the Cartesian (phase-) space $x_1, x_2, \ldots x_n$, this defines a velocity field

$$\mathbf{V}(x) \equiv (X_1, X_2, \ldots X_n) \tag{2}$$

for a fluid of system-points set into the space and cut loose to move according to the equations of motion. The flow is incompressible whenever (Liouville)

$$\operatorname{div} \mathbf{V} \equiv \sum \frac{\partial X_i}{\partial x_i} = 0, \tag{3}$$

and then the fluid density $\rho(x, t)$ is a constant of motion under (1) due to the conservation of fluid:

$$0 = \frac{\partial \rho}{\partial t} + \operatorname{div}(p\mathbf{V}) = \frac{\partial \rho}{\partial t} + \mathbf{V} \cdot \nabla \rho + \rho \nabla \cdot \mathbf{V} = \left(\frac{\partial}{\partial t} + \mathbf{V} \cdot \nabla\right)\rho = \frac{d\rho}{dt} \tag{4}$$

Conversely, if ρ is a constant of motion, the flow has to be incompressible. For instance, for a conserved quantity $G(x, t)$,

$$\frac{\partial G}{\partial t} + \mathbf{V} \cdot \nabla G = 0 \tag{5}$$

the density could be a function of it, $\rho = \rho(G(x, t))$, provided $\nabla \cdot \mathbf{V} = 0$, including the case of stationary flow

$$\frac{\partial G}{\partial t} = 0, \qquad \rho = \rho(G(x)) \tag{6}$$

when the currents $\rho(x)\mathbf{V}(x)$ may run powerfully but are everywhere steady. The fluid mass looks as if it were standing still, but is really flowing around upon itself.

In a small time ε, x_i becomes $x_i' = x_i + \varepsilon X_i$, and a volume element $d\tau = dx_1 \ldots dx_n$ becomes

$$d\tau' = dx_1' \cdots dx_n' = \frac{\partial(x_1', \ldots x_n')}{\partial(x_1, \ldots x_n)} dx_1 \cdots dx_n$$

$$= (1 + \varepsilon \operatorname{div} \mathbf{V}) d\tau, \tag{7}$$

so that the rate of change of volume is

$$\frac{d}{dt}(d\tau) = \frac{d\tau' - d\tau}{\varepsilon} = (\operatorname{div} \mathbf{V}) d\tau. \tag{8}$$

Or, under incompressible flow, a sac $\delta\tau$ of fluid travelling along with it will suffer no change in volume although likely changing appreciably in shape.

If the sac is confined to swim in some limited total volume τ, then presently in a limited time a part of it $(\delta\tau)'$ has to intersect the starting volume $(\delta\tau)^\circ$, merely because of the confinement. For if $\delta\tau$ is photographed at many discrete instants, more than $\tau/\delta\tau$ of them, at least two photographs must show volume overlap; running backward in time until the earlier photo assumes the configuration $(\delta\tau)^\circ$, the later photo (also time displaced) represents some later configuration of $(\delta\tau)^\circ$ that overlaps $(\delta\tau)^\circ$. Then a secondary fragment $(\delta\tau)''$ must later hit the $(\delta\tau)'$ part of $(\delta\tau)^\circ$, and so along in infinite sequence down to at least a point that then courses through $(\delta\tau)^\circ$ infinitely often (Poincaré recurrence).

It is visible, too, that the time of passage of a point through a volume $\delta\tau$ is proportional to $\delta\tau$; a streamline entering $\delta\tau$ at α and leaving at β, enters the time-displaced (by a time t^*) and (say) nonoverlapping volume $(\delta\tau)^* = \delta\tau$ at α^* and leaves at β^*; the entry-to-entry or α to α^* time is that for [α to β plus β to α^*], while the exit-to-exit or β to β^* time is [β to α^* plus α^* to β^*]; both these portal-to-portal times, the [ ]s, are one and the same flow time t^*; so, owing to the common leg β to α^* in each, the α to β

passage time through $\delta\tau$ comes out equal to the α^* to β^* passage time through $(\delta\tau)^*$. Now thinking of $\delta\tau$ and $(\delta\tau)^*$ as fixed geometrical skeletons that get traversed in the course of time by the streamline, we have the result that a doubling of phase volume means a doubling of passage time.

Poincaré recurrence and the time-volume proportionality tell now, that the fraction $\delta t/T$ of an ultralong time T that a phase point spends in recurrently traversing a volume $\delta\tau$ is proportional to $\delta\tau$. The time fraction must in fact be the volume fraction $\delta\tau/\tau$ in order that when $\delta\tau$ is all of τ, then δt is all of T,

$$\frac{\delta t}{T} = \frac{\delta\tau}{\tau} \tag{9}$$

Here τ is the whole of the finite volume that is just available to flow, not more and not less. This dynamical-geometrical connection is the germ of ergodicity.

The simple general hydrodynamical principles yield quickly to statistical ones; the fluid, however disposed, is already an ensemble, that is, $\rho(x, t)$ scans across the evolution of all copies of (1) originally distributed over initial data according to $\rho(x, 0)$.

Let the sole knowledge of the nature of the motion under (1) be the conservation law

$$\mathbf{V} \cdot \nabla G(x) = 0 \qquad \left(\frac{\partial G}{\partial t} = 0\right) \tag{10}$$

and let the surfaces, $G = $ constant, describe in phase space, single closed sheets everywhere within a finite distance of the origin. The evolution of one phase point from a starting position x_0 is on $G(x) = G(x_0) = G_0$. Not only is the course of motion $x(t; x_0)$ unknown, but the starting point is unknown; so, to be even-handed about the ignorance, naught will do but to pepper the surface (or strictly the shell $G_0 \leqslant G \leqslant G_0 + \delta G_0$) with points quite uniformly by way of constructing the appropriate probability distribution. In short, the only admissible probability hypothesis is (micro-canonical ensemble)

$$\rho = \rho_0 \delta(G - G_0). \tag{11}$$

This is tenable however only if the flow is incompressible, since this $\rho(G)$ is a constant of motion and the equation of continuity (here expressing probability conservation) then requires $\nabla \cdot \mathbf{V} = 0$.

Expectation values alone are now available for quantities $f(x)$ of interest,

$$\bar{f} = \frac{\int f \, \delta(G - G_0) \, d\tau}{\int \delta(G - G_0) \, d\tau}$$

$$= \frac{\int f \, \dfrac{dS}{|\nabla G|}}{\int \dfrac{dS}{|\nabla G|}} \tag{12}$$

$$\left(d\tau = \frac{dG}{|\nabla G|} \cdot dS; \quad dS = \text{element of surface on } G = G_0 \right).$$

But these averages are out of hand computationally because $dS/|\nabla G|$ is usually so intractable. It is better to smooth and simplify $\delta(G\text{-}G_0) \, d\tau$ and just hold to the main feature of sharp peaking at $G = G_0$. There is no unique way to do this, only more or less convenient ways. When the surfaces $G = G_0$ run continually outward for increasing G_0, the volume between neighboring G-shells is apt to be like

$$d\tau \sim G^p \, dG \tag{13}$$

where p is a power comparable to phase-space dimensionality, varying slowly as one goes outward. So that if this dimensionality is high, the microcanonical density will be well approximated by

$$\rho' \sim e^{-\beta G} \tag{14}$$

since the shell-by-shell fluid contents are

$$\sim G^p e^{-\beta G} \, dG \tag{15}$$

with an acute peak at $G = p/\beta$ that selects β as some p/G_0.

The choice $\exp(-\beta G)$ has no special standing here. Many others of the lot of monotone decreasing functions will do as well, just so long as $\rho(d\tau(G)/dG)$ has the simple strong maximum that can imitate the delta function. The essential parametrization is phase-space dimensionality in $d\tau/dG$, and one parameter in ρ to respect the one point of knowledge that $G = G_0$ is the main G. A more primitive flexibility is that, from the outset conserved G has no distinction above any function of it, $\phi(G)$. Selection of ϕ, but not uniquely, is what may be used here to gain the convenience of ascending phase volume with increasing ϕ (should it fail for increasing G), so as to place the role of phase-space dimensionality into clearer view.

When G (or some now distinguished function of it) is separable into $G_y(y) + G_z(z)$, and when G_y, G_z are closely (not exactly) conserved separately, the composition law

$$\rho(G_y + G_z) \, dy \, dz = \rho_y(G_y) \, dy \cdot \rho_z(G_z) \, dz \tag{16}$$

comes to hand as the only one available for making up stationary probability flows totally and also singly. Then uniquely (canonical ensemble)

$$\rho_i \sim e^{-G_i/\theta} \qquad (17)$$

and the "temperature" θ communally characterizes the components, which are "heat baths" to each other. It is essential that the G_i not be strictly conserved separately in order that there can be a trickle between y- and z-spaces that smears out what otherwise would be separate microcanonicities. Different $G = $ constant shells in the x-space can be cast in the preceding y, z roles, admitting different copies of $\dot{x} = X$ to be faintly coupled. Thus a significant expansion of the probability rule of the microcanonical ensemble, speaking not only to ignorance of location of a system-point on one shell but to an uncertainty as to which shell it should be placed on under heat bath conditions.

Finally, the crucial issue of ergodicity. Only one system is ordinarily open to inspection, and on the present view only some one or a few of its Brownian–like microcoordinates x_i are objects of observation in the course of time. The path of the single phase point is itself a species of ensemble, since any point of it can be reckoned an initial point because of the time-translational invariance of the equations of motion. By (9) is found the basic link between time-averages and these "orbital-ensemble" averages,

$$\frac{1}{T} \int_0^T f(x(t))\, dt = \frac{1}{\tau} \int_\tau f(x)\, d\tau, \qquad (18)$$

where τ is the just-available domain of flow of the single phase point. This orbital-ensemble average *could* come out the same as the microcanonical one, given that the motion is constrained to the surface $G = G_0$. It is a question whether the one phase point generally goes sufficiently comprehensively over this surface, or whether, to the contrary, the surface has appreciable inaccessible regions which then are microcanonically mis-weighted. There can be no answer short of greater (and mainly unobtainable) knowledge of the motion. It stands as an assumption, sanctioned by some weight of experience and argument, but principally by ignorance to do anything else, that orbital-ensemble and microcanonical averages agree.

On this basis, the overall ensemble probe of complex differential systems is the scheme: time averages of measurable noisy coordinates = orbital-ensemble averages = microcanonical averages = canonical (or other approximational) ensemble averages set out to foster explicit calculation. Physically one goes directly to the last step appropriately to interpret gross thermostatic observation, for which the canonical ensemble has primary, not just computational status. Biologically, it is the first step that is central but can be surveyed only from the far end of a possibly tenuous chain.

III. ECODYNAMICS

The ecological model due to Volterra[1] and Lotka,[2] which alone will be discussed here, has a generality which distinguishes it above most others in that few or many species equally may be surveyed. Only two, but quite central, points are grasped: a species, population N_i (a) grows or decays on the physical environment alone and (b) eats or is eaten by other species:

$$\dot{N}_i = \varepsilon_i N_i + \frac{1}{\beta_i} \sum_j \alpha_{ji} N_j N_i. \tag{19}$$

The autoincrease parameter ε_i is either positive or negative and tells how species i fares if left to itself without other species around. The coupling coefficients α_{ji} tell the strength of interspecies interaction. Assuming just binary collisions, the number of such per unit time between i and j is some $\lambda_{ij}N_iN_j$. If i is predatory on j, N_j diminishes in time dt by some $[p_{ij}\lambda_{ij}N_iN_j \, dt]$ while N_i gains by (β_j/β_i) [ ], where β_j/β_i measures the gain of i per j consumed. Very crudely, one can think of j-biomass β_j[ ] (with β_j the biomass of a j-individual) being "transformed" into β_j[ ]$/\beta_i$ i-individuals each of biomass β_i. Then α_{ji} stands for $\beta_j\lambda_{ij}p_{ij}$ and in the interaction the N_i, N_j increments are

$$dN_i = \frac{\alpha_{ji}}{\beta_i} N_j N_i \, dt$$
$$dN_j = -\frac{\alpha_{ji}}{\beta_j} N_i N_j \, dt. \tag{20}$$

With the understanding now that

$$\alpha_{ij} = -\alpha_{ji},$$

(including $\alpha_{ii} = 0$, that is, no self-interaction), the inherently reciprocal character of predator-prey coupling is expressed summarily by the $\sum_j$ in (19) for all i.

This is ecologically simplistic to be sure, but it would appear to touch some major nerve of "ecology in the large". It is, however, already nonsimple mathematically, as the elementary case of two coupled species quickly shows,—there is an oscillation between an $\varepsilon > 0$ prey and an $\varepsilon < 0$ predator, but the motion cannot be obtained in terms of known functions. Such oscillation has been observed microecologically in Gause's[3,4] classic experiments,—predatory paramecia eating yeast growing in sugar solution (Figure 1). Other experiments show a voracious predator sweeping up all prey and then perishing; but if the predation is cut down by giving hiding places to the prey, an oscillation becomes possible. This tells that the notion "species" is inseparable from that of "niche". It is the en-niched species that enters the Volterra dynamics, Eq. (19). In still other experiments (competing species of flour beetles) one species wins out in the long term, but not always the same one. Such a stochastic effect is clearly out of reach of Eq. (19). The latter in fact presumes low population densities (so that self-interaction can be neglected) but appreciable total population levels N_i that can be sensibly reckoned as continuous without fear from those chance effects that may rock microecology.

The transition to ensemble theory,[5] in order to get at what Volterra dynamics has to say about the very-many-species problem, follows from a reorganization of Eq. (19). First locate the stationary levels $N_i = q_i$ that secure all $\dot{N}_i = 0$,

$$\varepsilon_i \beta_i + \Sigma \alpha_{ji} q_j = 0. \tag{21}$$

It is supposed that unique positive qs are allowed. This requires non-singular α, requiring in turn that the number of associated species be even, since odd-order antisymmetric α is always singular; also not all ε_i can be of the same sign, as $\Sigma \varepsilon_i \beta_i q_i = 0$. Then Eq. (19) can be brought to

$$\beta_i \frac{\dot{N}_i}{N_i} = \sum_j \alpha_{ji}(N_j - q_j) \tag{22}$$

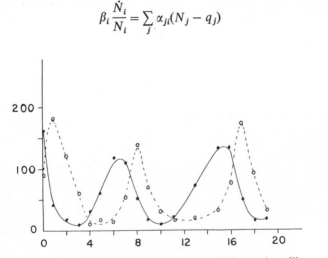

Fig. 1. Paramecium (broken curve), yeast (full curve) oscillations in Volterra–Lotka pattern (Gause[3,4]).

or, by $v_i \equiv \log N_i/q_i$

$$\beta_i \dot{v}_i = \sum_j \alpha_{ji} q_j (e^{v_j} - 1). \tag{23}$$

Multiply by $q_i(e^{v_i} - 1)$ and sum on i to produce zero on the right because of $\alpha_{ij} = -\alpha_{ji}$, and thence the conservation law

$$\sum q_i \beta_i (e^{v_i} - 1)\dot{v}_i = 0$$

$$\sum q_i \beta_i (e^{v_i} - v_i) \equiv G = \text{conserved.} \tag{24}$$

Finally, v-space flow is incompressible

$$\text{div } \mathbf{V} \equiv \sum \frac{\partial \dot{v}_i}{\partial v_i} = 0 \tag{25}$$

as each $\dot{v}_i$ is independent of v_i.

From time-averaging Eq. (22), there comes an important meaning for the qs,

$$\frac{\beta_i}{T} \log \frac{N_i(T)}{N_i(0)} = \sum_j \alpha_{ji} \left(\frac{1}{T} \int_0^T N_j \, dt - q_j \right). \tag{26}$$

The G-conservation tells that every $v(t)$ is always bounded between finite limits, or every $N(t)$ between positive limits, so letting $T \to \infty$ nullifies the left and prescribes for the right that

$$\frac{1}{T} \int_0^T N_j \, dt = q_j \tag{27}$$

that is, the stationary q-levels, which are independent of initial data, coincide with the long-time averages of the populations.

The simplicity of the conservation law (24) throws a good qualitative light on the complex motion. A surface $G = G_0$ for large values of G_0 protrudes only a relatively short way into the positive sector $v > 0$ $(N > q)$ of phase space because e^v in $e^v - v$ rises so quickly to get to $\sim G_0/q\beta$; in the negative sector $v < 0$ $(N < q)$, the surface balloons out into a great lobe owing to $-v$ having to run out far to achieve $\sim G_0/q\beta$. Then the incompressibility of flow means relatively fast passage through positive regions and much slower passage through negative ones. A wandering (and Poincaré recurrent) orbit over $G = G_0$ signifies in all that the Ns go through successive long shallow troughs and abrupt high peaks, holding in this way to Eq. (27). For small values of G_0, the lopsided surface quiets down to a symmetric ellipsoid $(e^v - v \sim 1 + \frac{1}{2}v^2)$, and apart from pecularities of axis lengths ($q\beta$ values), the Ns swing modestly upwards and comparably downwards, in comparable times, around their mean levels q, satisfying (27) on a quite different basis.

The stipulation above of an even number of species in the ecosystem looks strange. First, it raises the issue, what is a species? Second, it can be circumvented through the hindsight of the developed form (22) of the primitive (19); that is, one could take (22), unrestrictedly, to be the starting model with the stationary point $N = q > 0$ admitted as primary model hypothesis, then recover (19) as a derived form with the autoincrease parameters ε_i being the composites $-\sum_j \alpha_{ji} q_j / \beta_i$. In short, the parametrization α, β, ε can well be replaced by α, β, q. Using q as primary in this way would be legitimately to inject the model with prescribed desired features.

But, thirdly, what after all does happen on the basis of (19) when the speciation is odd? An eminent possibility is a decay to evenness through exhaustion of at least one species. This can be made specially evident when a pair of species are close ecological neighbors (nearly equal α, β, ε); one of the pair, even if present initially in low numbers, slowly but surely supplants the other. Such supplantation, recognized by Darwin a century ago as a principal factor in evolution, is biology's "competitive exclusion principle", that is, two species cannot indefinitely occupy the same ecological niche. The principle may be fully clothed in Volterra dynamics where a secularly drifting conservation law like (24), and a drifting Gibbs ensemble built on it, allows the struggle to be seen on the wide ecological stage.[6] Insofar as genetical creation of new species and ecological removal of old ones can be counted main forces of evolution, Volterra theory would appear capable of a form of evolutionary address, the ecogenetic system gliding through perpetual transitions to parity under genetically infused imparities (or parities). Openness of speciation would call for grand canonical ensembles, as yet little explored.

Now returning to (24), the separation of G into private components, one for each species, makes the ensemble theory extraordinarily easy. At once, the fundamental probability law emerges (dropping subscripts and normalization)

$$P(v) \, dv \sim \exp\left\{-\frac{q\beta}{\theta}(e^v - v)\right\} dv \qquad -\infty \le v \le \infty$$

$$P(n) \, dn \sim n^{\gamma-1} e^{-\gamma n} \, dn \qquad 0 \le n \le \infty \tag{28}$$

$$(n \equiv N/q, \qquad \gamma \equiv q\beta/\theta)$$

as a gamma distribution in n for each species. The arbitrarily tight coupling of species does not interfere in any way with this ensemble view of the system as a kind of eco-gas.

On the one parameter γ that combines the species-specific $q\beta$ and the global θ (like specific molecular mass m and system temperature kT physically), is based the reckoning of most ensemble- (or, ergodically, time-) averages bearing upon the single noisy population-time curve. A host of such averages can be worked out[5] to reveal the statistics of horizontal and vertical spreads of the curve. Only a few need be mentioned here.

The canonical average of $e^v - 1$ is zero, recovering $\overline{N} = q$, as is necessary according to (27). Temperature θ is measured in two simple ways,

$$\frac{\theta}{q\beta} = \frac{\overline{(N - q)^2}}{q^2} = \overline{\left(\frac{N}{q} - 1\right) \log \frac{N}{q}}, \tag{29}$$

or, roughly, by the amplitude of oscillation about the mean, with the totally cold system $\theta = 0$ meaning every N_i static at q_i. The entropy here naturally sinks without limit. Canonical mean G (single species) is

$$\overline{G} \simeq \theta \qquad (\theta \gg q\beta)$$

$$\simeq q\beta + \tfrac{1}{2}\theta \qquad (\theta \ll q\beta) \tag{30}$$

and increases monotonically from the former to the latter, implying total G-equipartition at the two temperature extremes, and a heat capacity per species $\partial \overline{G}/\partial\theta$ rising from $\tfrac{1}{2}$ initially to 1 asymptotically. From the heat capacity, it can be told what is the final equilibrium temperature of two ecosystems mixed together. All the formal thermodynamics is explicitly computable from the Gibbs phase integral

$$\prod \gamma^{-\gamma} \Gamma(\gamma)$$

It is not out of the question to introduce the concept of work and to proceed beyond simple calorimetry.

From (28), comes a finer indication of the quality of motion than was obtained before from the gross geometry of $G = G_0$. Those species with $q\beta < \theta$ can be found often at $N < q$, since $P(n)$ rises without bound at small n in this case. They are a sort of excited phase marked by long or many subaverage swings in N and short superaverage bursts. For the others ($q\beta > \theta$), making up a "condensed phase", the maximum in $P(n)$ tells of a much more even swinging of N, both in period and amplitude. At temperatures less than the smallest $q\beta$ the whole ecosystem is "condensed", all populations rippling more or less slightly and regularly around $N = q$; at sufficiently high temperatures, exceeding the largest $q\beta$, all are "excited" into some showing of sharp infrequent bursts.

Let us turn briefly to observation. The outstanding fact evident to the most casual glance is simply that natural populations in the field do fluctuate appreciably and continually. Some of this, but apparently far from all of it, has to do with fluctuation in season and climate and the like. The conservative character of Volterra dynamics answers at least qualitatively to inherently ecological fluctuation. If self-interaction had been allowed in (19) (Verhulst-Pearl or $\alpha_{ii}N_i^2$ terms), a kind of frictional damping would ensue, finally giving static populations in disaccord with observation on the gross scale.

Moths and butterflies caught in a trap set into a field might be expected to show Poisson statistics in the catch, the Poisson distribution being characterized by a mean-population parameter. In fact, this was not so according to Corbet, Fisher and Williams[7] in a classic study in Malaysia. Instead, a distribution of mean-population parameters had to be introduced to explain observation. The distribution needed was (28) with $\gamma \ll 1$.

Long population-time curves are difficult to obtain. One is shown in Figure 2 (the first 40 years of a 90-year period), based upon Elton's[8] compilation of Labrador fox catch data. The catches (N) are one-year totals and represent intra-annual averages that Elton suggests reasonably measure fox population. In Table I are given assorted time averages on the unsmoothed polygonal curve, together with the value of the parameter γ in (28) that will account for them theoretically, and in the last column the γ-spread corresponding to $\pm 10 \%$ errors in the time averages. Still further statistics on the frequency with which the curve crosses a family of horizontally drawn lines may be found in the original source.[5] These depend not only on (28) but on the structure of Volterra dynamics.

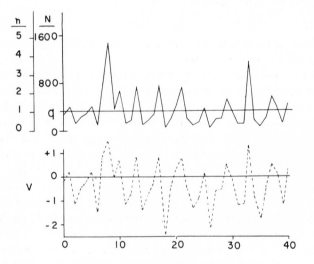

Fig. 2. Fox-catch data from Elton[8], taken as representing econoise (Kerner[5]).

TABLE I

Time–Averages and γ for Elton's[5] Fox Catch Data

Quantity Averaged	Time–Average	γ	γ-spread
Fraction of time that $n < 1$	0.60	1.9	0.7–15.0
Amplitude of n during $n < 1$	0.45	2.1	1.0– 2.8
Amplitude of n during $n > 1$	0.73	1.7	1.4– 2.1
$(n - 1)^2$	0.54	1.9	1.7– 2.1
$(n - 1) \log n$	0.49	2.0	1.9– 2.3
$\log n$	-0.26	2.1	2.3– 1.9

The single γ in the vicinity of 2.0 seems to suffice for quite a few statistics. The population curve (or apparent population curve) is something evidently much like Volterra noise (28).

IV. NEURAL NETWORKS

Some aspects of the behavior of a great network of interconnected neurons have been brought out in a model due to J. D. Cowan,[9] now to be sketched very briefly. Main electro-physiological features of single neurons are drawn so as to tell in differential equations (then in ensembles) how their firing frequencies control each other in the net (then how the net performs globally). The firing frequencies themselves, treated as smoothed variables, are the primary network coordinates.

Looking at one neuron in the system, multiple branches (dendrites) stemming out of its central cell body (soma) are receivers of spiked voltage impulses originating from other neurons. A cable (axon) from the soma later itself branches, as transmitting apparatus, into filaments terminating in endbulbs that link electrochemically across fine gaps (synapses) to sites on other neural dendrites. One neuron's dendrite tree may have thousands of such sites. A pulse from neuron k incident on a synapse with neuron l generates, by a complex process, what can be modeled as a discharging or charging voltage, $-V_{kl}$ or $+V_{lk}^*$ (according as the synapse is "excitatory" or "inhibitory") delivering in a characteristic time T_{kl} or T_{kl}^*, through effective conductances g_{kl} or g_{kl}^*, a current that tends to discharge or charge l's dendrite-soma membrane. If k is firing pulses into the k, l synapse at the rate v_k, the current into l may be taken as

$$I_{kl} = \Lambda_{kl} v_k$$

$$\Lambda_{kl} = \begin{array}{ll} -V_{kl} g_{kl} T_{kl} & \text{(excitatory)} \\ +V_{kl}^* g_{kl}^* T_{kl}^* & \text{(inhibitory)} \end{array} \qquad (31)$$

The receiver l sums the signals coming into its membrane, and when the latter's potential drops sufficiently, the l soma, in another complex process, fires its own strong pulse out its axon to excite or inhibit other neurons synapsed to l's spray of endbulbs. The spikes are not greatly different from one another, their spacing, the frequency v, is what is of controlling interest. Following emission of a spike there is a refractory period δ_0 during which the cell recovers and no spike can be generated. Taking all inputs to l to be in parallel, and allowing a delay time δ_{kl} for the pulse from k to get down k's axon and to actuate the k, l synapse, the net current stimulating l is

$$I_l(t) = \sum_k \Lambda_{kl} v_k(t - \delta_{kl}) \tag{32}$$

Now, following Cowan, there must be brought in the final link, an empirical connection between the smoothed variables I and v, stating how input current I produces firing frequency v. A reasonable representation of an assortment of facts appears to be that

$$v = v_0 \frac{1}{1 + \exp - \gamma \dfrac{(I - I_0)}{I_0}} \qquad \left(v_0 \equiv \frac{1}{\delta_0}\right) \tag{33}$$

where γ, I_0, v_0 are neuron-specific parameters. This says that the refractory time δ_0 sets the limit $1/\delta_0$ to any neural firing rate, and that the neuron-as-frequency-generator is "hard" or "soft" according as γ is large (being then essentially an all-off generator for $I < I_0$ and an all-on one, $v = v_0$, for $I > I_0$) or else is small. This much flexibility is in keeping with the main facts of a smooth rise of v up to a saturation limit, the precise functional form $v = v(I)$ being without fundamental import so far. The hard neurons are much like the formal neurons of McCulloch and Pitts that model the network as a logical or computing apparatus.

Inverting for $I(v)$, and taking some fixed $\delta_{kl} = \delta$ and some fixed v_0, instead of families of them, gives the finite-difference network equations (after advancing t to $t + \delta$ and simplifying)

$$\log \frac{x_l(t + \delta)}{1 - x_l(t + \delta)} = \varepsilon_l + \frac{1}{\beta_l} \sum_k \Lambda_{kl} x_k(t) \tag{34}$$

$$\varepsilon_l \equiv \gamma_l - \frac{\gamma_l v_0}{I_{0l}} \sum_k \Lambda_{kl}$$

$$\beta_l \equiv \frac{I_{0l}}{\gamma_l v_0}$$

$$x \equiv 1 - \frac{v}{v_0} \qquad 0 \le x \le 1.$$

In the last line, the variable x, replacing v, measures the fraction of time that a neuron is sensitive, that is, nonrefractory. Now in differential approximation, for small δ,

$$\log \frac{x_l(t + \delta)}{1 - x_l(t - \delta)} \simeq \log \frac{x_l(t)}{1 - x_l(t)} + \delta \frac{d}{dt} \log \frac{x_l}{1 - x_l}$$

$$= \log \frac{x_l}{1 - x_l} + \delta \frac{\dot{x}_l}{x_l(1 - x_l)}. \quad (35)$$

Furthermore, when (multiplying through by $x_l(1 - x_l)$ in (34) and (35)) $|x_l(1 - x_l) \log x_l/(1 - x_l)|$ is at its largest (about 0.225) exceeded appreciably by $|\varepsilon_l x_l(1 - x_l)|$ owing to appreciable ε_l values, the opening term in (35) can in effect be dropped in an additional approximation, resulting in

$$\frac{d}{dt} \log \frac{x_l}{1 - x_l} = \varepsilon_l + \frac{1}{\beta_l} \sum_k \Lambda_{kl} x_k, \quad (36)$$

where t is measured in units of δ. This is somewhat more in the nature of an heuristic, novel suggestion from (34) rather than any close reduction of it.

A clear parallel to Volterra dynamics is now evident, as the notation indicates, and the ecological pattern may be invoked *mutatis mutandis*. Admit by hypothesis only mutually excitatory ("predator") and inhibitory ("prey") neurons, expressible roundly as $\Lambda_{kl} = -\Lambda_{lk}$. The reciprocity touches a significant facet of central-nervous-system activity in that midbrain (thalamic) neurons are broadly so coupled to outer-brain (cortical) neurons in a major nervous-system interplay. With stationary x-levels x_0 between 0 and 1, and with revised network coordinates

$$v_k \equiv \log[x_k/x_{k0}(1 - x_k)]$$

there follows, as before, a reduction to

$$\beta_l \dot{v}_l = \sum_k \Lambda_{kl} x_{k0} \left(\frac{e^{v_k}}{1 + x_{k0} e^{v_k}} - 1 \right) \quad (37)$$

with divergenceless flow in v-space ($\Lambda_{kk} = 0$) and the conservation law

$$\sum \beta_l \dot{v}_l x_{l_0} \left(\frac{e^{v_l}}{1 + x_{l_0} e^{v_l}} - 1 \right) = 0$$

$$\sum \beta_l \{\log (1 + x_{l_0} e^{v_l}) - x_{l_0} v_l\} \equiv G = \text{conserved}. \quad (38)$$

Thence individual neural oscillators are controlled statistically by $\exp(-G_i/\theta) \, dv_i$ or, going back to the "sensitivities" x,

$$P(x) \, dx \sim x^{a-1}(1 - x)^{b-1} dx$$

$$a \equiv \frac{x_0 \beta}{\theta} \qquad b \equiv \frac{(1 - x_0)\beta}{\theta}. \quad (39)$$

Temperature θ now turns out to be measured on single coordinate behavior by

$$\frac{\theta}{\beta} = (x - x_0) \log \frac{x/x_0}{1 - x} = \frac{\overline{(x - x_0)^2}}{x_0(1 - x_0) - \overline{(x - x_0)^2}} \tag{40}$$

from taking averages of $v_i \, \partial G/\partial v_i$ and $(\partial G/\partial v_i)^2$; and again the stationary x_0-levels are both time-averages and ensemble-averages of the x.

In short, the model neural oscillator jostled in its network is beta-distribution noisy in the sensitivity x or as well in the firing frequency $v/v_0 = 1 - x$. The "condensed" members of the "neuron gas" (so termed again from total Gs separation into components), namely those members with β/θ large enough to make one maximum in $P(x)$, fire roughly comparably above and below the mean frequency $\bar{v} = v_0(1 - x_0)$; the "excited" members (β/θ small enough to make $P(x)$ diverge at $x = 0$ and $x = 1$) are mainly firing either slowly or else in a burst of tightly spaced spikes near the saturation limit $\bar{v}_0$. When the net is sufficiently cool, all members are firing steadily around their $\bar{v}$s, while in the high temperature limit all are spluttering or "motor-boating" in alternating fits of quiet and outburst. It seems clear that the input of the sensorium must be standing as a principal developer of network temperature. Strongly nonsteady inputs would more or less rock the network into transient states that cannot in general be treated as thermal-equilibrium states, or characterized through one temperature parameter.

The transform of the beta-distribution in x is a Pareto–Levy distribution in $y \equiv x^{-1} - 1$, that is, in

$$y = \exp \gamma \left(\frac{I - I_0}{I_0} \right) \tag{41}$$

with a stability such that linear superpositions of y variables also are Pareto–Levy distributed. Here, there is some hint to the observables of encephalography, concerned in effect with superpositions of neural currents, insofar that at least for currents $I = I_0 + i$ in the immediate vicinity of I_0, when y is linear in i, the distribution in superimposed i has simple character.

For a critique of the model and its further experimental connections the work of Cowan[9] must be consulted.

V. BIOCHEMICAL KINETICS

In first approximation, looking at the cellular interior with a purely chemical eye that steadily records only overall concentrations of selected species in the single cell as they fluctuate in time on through cellular generations, one can model the cell as a homogeneous chemical vat which is materially and energetically open through its boundary membrane, forgetting temporarily all the intricate cell architecture. With due reservation also about the possible excessive smallness of concentrations, the rules of chemical kinetics make the model mathematically tight. This is to raise the question: As a prime theory of chemical nonequilibrium in which the time-evolution of concentrations is held up to some direct view, can classical kinetics afford at least a starting point to the problem of describing the outstanding observational fact of *in vivo* biochemical fluctuation?

Let us state the model dynamics. The fundamental reaction types are

isomerization	$A \rightarrow A'$,	$\dot{A} = -\dot{A}' = -k(a, a')A$
composition	$B + C \rightarrow D$,	$\dot{B} = \dot{C} = -\dot{D} = -K(bc, d)BC$
decomposition	$E \rightarrow F + G$,	$\dot{E} = -\dot{F} = -\dot{G} = -k(e, fg)E$
entry	$M(\text{ext}) + W_M(\text{wall})$	
	$\rightarrow M(\text{int})$	$\dot{M}(\text{int}) = K'M(\text{ext}) \cdot W_M$
exit	$N(\text{int}) + W_N(\text{wall})$	
	$\rightarrow N(\text{ext})$	$\dot{N}(\text{int}) = -K''N(\text{int}) \cdot W_N$

None others but these basic ones, relying on the simplest concepts of molecular transition and binary collision, need be considered.

In the first three, for interior reactions, are comprised the base for more complicated overall reaction types, when suitable if short-lived intermediate complexes are admitted. With just these three (closed system) any number of reactants have to settle down from any initial concentrations to final static or equilibrium values.

In the last two, the external M concentrations $M(\text{ext})$ and M-specific wall-site concentration W_M by which M can be spirited to the interior by unspecified membrane action, may be counted constant. Similarly, in the passage of N in the interior, $N(\text{int})$, through N-specific wall sites W_N to the outside, W_N may be taken fixed. Then a species in the interior is fed or depleted by membrane coupling to the outside according to

$$\text{entry} \quad \dot{M} = \sigma_m = \text{constant}$$

$$\text{exit} \quad \dot{N} = -\bar{k}_n N.$$

This kinetically represented transport alone is to (possibly) save the system from decay to a static fate.

Altogether, letting z denote concentrations, the network dynamics, summarizing all kinetic laws, is of structure

$$\dot{z}_i = \sigma_i + \sum_j k_{ij} z_j + \sum_{jl} K^i_{jl} z_j z_l \equiv Z_i(z) \tag{42}$$

with particular constraints on the input vector σ and rate-constant tensors k, K as follows

$$\sigma_i \geq 0$$

$$k_{ii} = -\sum_{a, b} k(i, ab) - \sum_{i'} k(i, i') - \bar{k}_i \leq 0$$

$$k_{ij} = \sum_{a'} k(j, ia') + \sum_{i'} \delta_{m_j m_{i'}} k(i', i) \geq 0 \qquad (m_j \geq m_i) \tag{43}$$

$$= 0 \qquad (m_j < m_i)$$

$$K_{jl}^i = -\sum_{a'} \delta_{ij} K(il, a') + K(jl, i)$$

where the species are ordered $1, 2, \ldots i, i', \ldots$, according to increasing molecular weight $m_1 < m_2 < \ldots$ (including isomers $m_i = m_{i'} = \ldots$) and the rate constants $k(i, i')$, $K(il, a)$, $k(i, ab)$ for isomerization, composition, decomposition, are understood to be defined only for $m_i = m_{i'}$, $m_i + m_l = m_a$, $m_i = m_a + m_b$. The stream velocity $(Z_1, Z_2 \ldots)$ in z-space has at each boundary plane $z_i = 0$, a normal component directed into the positive sector $z > 0$ for a point approaching the plane from this sector, so all motion, as required, is confined to this sector when started there.

While ensemble theory would seem to be a desirable sort of tool to survey the network, this does not seem to be in sight. It is true that special versions of the general format (42) can admit a conservative-Liouville regime, for instance Volterra–Lotka kinetics and others also. Indeed Lotka's[2] early considerations were both chemical-kinetic as well as ecologic, under allowance of autocatalytic chemical species, and this theme was later enlarged by Moore.[10] But while ecological notions of molecular predator and prey species can to a degree be developed under a widening of the idea of chemical species, the issue is thereby not joined as to what the close rules of chemical kinetics, bearing parameters σ, k, K within some sort of reach from first principles, may have to say.

The strict chemical kinetics in fact has contents that are strikingly at odds with the conservative-Liouville outlook. This is the outcome of the studies of Chance,[11] Higgins,[12] and others. In a biochemically apposite formulation of the glycolysis of glucose, for instance, going according to (42), (43) with nine coupled species, Higgins has shown by computer that there is periodic motion in a limit-cycle for ranges of initial concentrations and rate constants (Fig. 3); and biochemical ringing of this type has been experimentally observed.

It is clearly a mathematically well set and biologically most significant question to find the necessary and sufficient conditions on σ, k, K, and on initial data, that can yield such limit-cycles. This is to seek of the biochemical

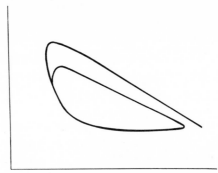

Fig. 3. Limit-cycle oscillation of fructose-6-phosphate (abscissa) and fructose-2, 6-diphosphate (ordinate) in a nine-fold biochemical net (Higgins[12]).

network "orbital" features in-the-large, ahead of "statistical" ones. But since cells are at best noisy clocks, it is in point to inquire further as to how far some order of "noisy limit-cycle" may lie in the very large network and by what analytical methods be describable.

Let us mention briefly also a conservative-Liouville attempt at interior enzyme-synthesis kinetics due to Goodwin,[13] seeking only "extremely coarse functional relationships" between the elements of synthesis. The scheme is, (a) nuclear template T + repressor $R \rightleftarrows$ composite T'; (b) T + precursor $P \rightleftarrows$ composite T''; (c) messenger M supposed generated by T'' according to $\dot{M} = aT'' - b$; (d) enzyme E formation from M supposedly by $\dot{E} = cM - d$; (e) repressor formation from E by presumed kinetics $\dot{R} = eE - fR$. Under (a) and (b) at equilibrium, $T_0'/T_0 R_0 =$ equilibrium constant α_0, and $T_0''/T_0 P_0 = \beta_0$, zero subscripts designating equilibrium. Then total template "concentration" T^0 is $T_0 + T_0' + T_0''$ or

$$T_0 = T_0'' \left(1 + \frac{1}{\beta_0 P^0} + \frac{\alpha_0}{\beta_0 P_0} R_0 \right) = T_0''(\gamma_0 + \delta_0 R_0), \qquad (44)$$

in which R_0 may be replaced by $(e/f)E_0$ from (e) at equilibrium. Goodwin supposes now that in (c) the non-equilibrium T'' can be replaced by $T^0/[\gamma_0 + \delta_0(e/f)E_0]$, wherein, still further, the equilibrium E_0 is replaced by the time-varying E, forcing then some species of drastically reduced E, M kinetics

$$\dot{E} = cM - d$$

$$\dot{M} = \frac{a_0}{\gamma_0 + \varepsilon_0 E} - b \qquad (45)$$

Accordingly

$$\dot{E} = c\dot{M} = \frac{ca_0}{\gamma_0 + \varepsilon_0 E} - cb$$

$$= -\frac{\partial}{\partial E}\left\{cbE - \frac{ca_0}{\varepsilon_0}\log(\gamma_0 + \varepsilon_0 E)\right\} \qquad (46)$$

so E and $\dot{E}$ (or M) motions are akin to those of coordinate and velocity in a one-dimensional mechanics problem with potential well { }. Of course, one may go to a Hamiltonian statement now, including a "gas" of such oscillators in weak coupling, and then into ensemble theory. But already the high distortion of kinetics here is showing itself, among other places, in the domains

$$-\frac{\gamma_0}{\varepsilon_0} \leq E \leq \infty, \qquad -\infty \leq M \leq \infty$$

that permit "concentrations" to run negative.

It must be doubted altogether that ensemble theory can speak cogently to biochemical kinetics so far as presently understood.

VI. MATHEMATICAL OVERVIEW

Let us return finally to (1) and ask[14] whether and when a central conservation law and Liouville's theorem can be evoked out of the general system $\dot{x}_i = X_i(x)$.

Basing on Hamiltonian experience, a way to proceed is first to place the question into the framework of the inverse problem of the calculus of variations: what is the variational principle having prescribed differential laws that define the extremals? If there be any such, it must be of the form

$$\delta \int \left\{\sum_1^n U_i(x)\dot{x}_i - U_0(x)\right\} dt = 0 \qquad (47)$$

giving first-order Euler–Lagrange equations (under fixed endpoint conditions)

$$\frac{d}{dt}\frac{\partial}{\partial \dot{x}_k}\{\} = \frac{\partial\{\}}{\partial x_k} \qquad \text{or}$$

$$\frac{d}{dt}U_k = \sum_i \frac{\partial U_k}{\partial x_i}\dot{x}_i = \sum_i \frac{\partial U_i}{\partial x_k}\dot{x}_i - \frac{\partial U_0}{\partial x_k} \qquad \text{or}$$

$$\sum_i \left(\frac{\partial U_k}{\partial x_i} - \frac{\partial U_i}{\partial x_k}\right)\dot{x}_i = -\frac{\partial U_0}{\partial x_k} \qquad \text{or}$$

$$\sum_i \Gamma_{ki} \dot{x}_i = -\frac{\partial U_0}{\partial x_k} \tag{48}$$

$$\left(\Gamma_{ki} \equiv \frac{\partial U_k}{\partial x_i} - \frac{\partial U_i}{\partial x_k} = -\Gamma_{ik} \right).$$

If these are to be $\dot{x}_i = X_i$, then, first, Γ must be nonsingular, necessitating an even number n of degrees of freedom (for an odd system, an extra coordinate may be supposed added, $\dot{x}_{n+1} = X_{n+1}$, to even it). Second, U's must be obtained satisfying

$$\sum_i \Gamma_{ki} X_i = -\frac{\partial U_0}{\partial x_k} \quad \text{or}$$

$$\sum_i X_i \frac{\partial U_k}{\partial x_i} - \frac{\partial}{\partial x_k}(X_i U_i) + U_i \frac{\partial X_i}{\partial x_k} = -\frac{\partial U_0}{\partial x_k} \quad \text{or}$$

$$\sum_i X_i \frac{\partial U_k}{\partial x_i} = \sum \frac{\partial}{\partial x_k}(X_i U_i - U_0) - U_i \frac{\partial X_i}{\partial x_k}. \tag{49}$$

Let U_0 be chosen as $\Sigma X_i U_i$ and select any convenient coordinate x_a for isolating the principal derivatives

$$\frac{\partial U_k}{\partial x_a} = -\frac{1}{X_a} \sum_{i \neq a} X_i \frac{\partial U_k}{\partial x_i} - \frac{1}{X_a} \sum_i U_i \frac{\partial X_i}{\partial x_k} \tag{50}$$

in terms of all the other (parametric) ones. As a partial differential system for the U_i, this is of classical Cauchy–Kowalewski type for which there is a local existence theorem in the vicinity of a regular point in x-space.

At least in this local sense, then, $\dot{x}_i = X_i$ can always be represented in the fundamental form

$$\dot{x}_i = \sum_l \gamma_{li}(x) \frac{\partial U_0}{\partial x_l} \quad (\gamma \equiv \Gamma^{-1}) \tag{51}$$

from which, by γ's antisymmetry, follows the conservation law

$$\sum_i \frac{\partial U_0}{\partial x_i} \dot{x}_i = \sum_{il} \frac{\partial U_0}{\partial x_i} \gamma_{il} \frac{\partial U_0}{\partial x_l} = 0 \tag{52}$$

$$U_0 = \text{constant of motion}$$

The models for ecological and neural networks were brought to just this form (U_0 having been called G and γ_{ij} having been $\alpha_{ij}/\beta_i \beta_j$ or $\Lambda_{ij}/\beta_i \beta_j$)

through introduction of the v coordinates in the respective contexts. Hamiltonian dynamics is of the same stripe,

$$
\begin{pmatrix} \dot{q}_i \\ \dot{p}_i \end{pmatrix} = - \begin{bmatrix} 0 & -1 \\ 1 & 0 \end{bmatrix} \begin{pmatrix} \dfrac{\partial H}{\partial q_i} \\ \dfrac{\partial H}{\partial p_i} \end{pmatrix} \tag{53}
$$

when coordinates and momenta are all lined up as x_i in the order $q_1, p_1,$ $q_2, p_2, \ldots,$ and γ in (51) is the "canonical" form for antisymmetric matrices that has [ ]s running down the diagonal (zeros elsewhere). It will be observed that the notion of reciprocity, seen in biological illustration in predator-prey and cortex-thalamus couplings, has a very great breadth in the statement of general antisymmetry of γ according to (51). Virtually every differential system has in principle this sort of reciprocity, being an expansion of the idea of canonical-conjugacy of Hamiltonian theory.

Exact connections of reciprocity to Hamiltonian conjugacy come from the variational principle. The differential form occurring there,

$$
\sum_1^{2m} U_i \, dx_i
$$

having $n = 2m$ elements, is reducible always to

$$
\sum_1^m P_i(x) \, dQ_i(x)
$$

with half as many; this is the well known Pfaff's Problem; the reduction is unique up to an exact differential $d\phi(x)$. Now the variational principle reads

$$
\delta \int \sum P_i \, dQ_i - H \, dt = 0, \qquad H \equiv U_0(x(P, Q)), \tag{54}
$$

so that $\dot{x}_i = X_i$ is further brought to complete Hamiltonian form under the transformation $x = x(Q, P)$. This is the Lie–Koenigs theorem:[15] any differential system can in principle be Hamiltonized. An interesting application[16] of these ideas comes up in relativistic many-particle theory, where Lorentz-covariant equations of motion, but still of Newtonian type $\ddot{r}_i = F_i(r_i, \dot{r}_i)$, do not have suitable formulations through a Lagrangian $L(r_i, \dot{r}_i)$ but do have a Lie–Koenigs Hamiltonian representation, starting from $\dot{r}_i = v_i,$ $\dot{v}_i = F_i(r_i, v_i)$.

It is worth noting that the basic machinery of Hamiltonian theory can be brought out directly in the primitive coordinates x without going through the difficult Pfaff reduction to Q, P. For, supposing such a reduction, the Poisson bracket (A, B) is

$$(A, B) = \sum_i \frac{\partial A}{\partial Q_i} \frac{\partial B}{\partial P_i} - \frac{\partial A}{\partial P_i} \frac{\partial B}{\partial Q_i} \tag{55}$$

or, going back to Q, P, A, B as functions of x,

$$(A, B) = \sum_{ilk} \frac{\partial A}{\partial x_l} \frac{\partial x_l}{\partial Q_i} \frac{\partial B}{\partial x_k} \frac{\partial x_k}{\partial P_i} - \frac{\partial A}{\partial x_l} \frac{\partial x_l}{\partial P_i} \frac{\partial B}{\partial x_k} \frac{\partial x_k}{\partial Q_i}$$

$$= \sum_{kl} \frac{\partial A}{\partial x_l} (x_l, x_k) \frac{\partial B}{\partial x_k}. \tag{56}$$

Now

$$\dot{x}_l = (x_l, H) = \sum_k (x_l, x_k) \frac{\partial U_0}{\partial x_k} = \sum_k \gamma_{kl} \frac{\partial U_0}{\partial x_k} \tag{57}$$

so $(x_l, x_k) = \gamma_{kl}$ and

$$(A, B) = \sum \frac{\partial A(x)}{\partial x_l} \gamma_{kl}(x) \frac{\partial B(x)}{\partial x_k} \tag{58}$$

where only the original coordinates enter.

As to Liouville's theorem, we do not have any incompressible flow (or volume conservation) in general, since

$$\text{div } \mathbf{V} = \sum_i \frac{\partial X_i}{\partial x_i} = \sum_{i,l} \frac{\partial \gamma_{li}}{\partial x_i} \frac{\partial U_0}{\partial x_l} + \gamma_{li} \frac{\partial^2 U_0}{\partial x_i \, \partial x_l}$$

$$= \sum_{i,l} \frac{\partial \gamma_{li}(x)}{\partial x_i} \frac{\partial U_0}{\partial x_l} \tag{59}$$

need not vanish. Instead, it can be found how to weight the volume element $dx_1 \cdots dx_n = d\tau$ so that weighted volume $J(x) \, d\tau$ is conserved, as follows.

The line integral

$$\oint \sum U_i \, dx_i$$

taken around a closed curve moving under $\dot{x}_i = X_i$ in x-space is a (relative) integral invariant; for letting the points on the curve run from x_i to $x_i + \varepsilon X_i$ in the infinitesimal time ε, the $\oint$ becomes

$$\oint \sum U_i(x + \varepsilon X) \, d(x_i + \varepsilon X_i) = \oint \sum U_i \, dx_i + \varepsilon \sum_{il} \left(X_l \frac{\partial U_i}{\partial x_l} \, dx_i + U_l \, dX_l \right) \tag{60}$$

or

$$\Delta \oint U_i \, dx_i = \varepsilon \oint \sum_{il} X_l \left(\frac{\partial U_i}{\partial x_l} - \frac{\partial U_l}{\partial x_i} \right) dx_i + d(X_l U_l)$$

$$= \varepsilon \oint \sum_i - \frac{\partial U_0}{\partial x_i} \, dx_i + d(X_i U_i)$$

$$= \varepsilon \oint \sum_i d(X_i U_i - U_0)$$

$$= 0. \tag{61}$$

That is, the line integral remains unaltered as the phase fluid moves. From this, it follows by Stokes' theorem that

$$\int_s \sum \left(\frac{\partial U_i}{\partial x_j} - \frac{\partial U_j}{\partial x_i} \right) dx_i \, dx_j \equiv \int_s \sum \Gamma_{ij} \, dx_i \, dx_j \tag{62}$$

is an (absolute) integral invariant also, where the integral extends over a region S bounded by the curve. The quadratic differential form here is the exterior derivative[17] of the preceding linear form,

$$d^*(\sum U_i \, dx_i) \equiv d^*\Omega \equiv \sum \frac{\partial U_i}{\partial x_j} \, dx_j \times dx_i = \sum \Gamma_{ij} \, dx_i \, dx_j, \tag{63}$$

where $dx_j \times dx_i$ is the exterior product such that $dx_j \times dx_i = -dx_i \times dx_j$. The Stokes integral invariant can be seen directly as a consequence of $d^*(d^*\Omega) = 0$ or effectively

$$\frac{\partial \Gamma_{ij}}{\partial x_k} + \frac{\partial \Gamma_{ki}}{\partial x_j} + \frac{\partial \Gamma_{jk}}{\partial x_i} \equiv 0. \tag{64}$$

Now a chain of invariants can be found from exterior products of $d^*\Omega$ with itself,

$$\int d^*\Omega \times d^*\Omega, \qquad \int d^*\Omega \times d^*\Omega \times d^*\Omega, \dots$$

the invariance coming from $d^*[d^*\Omega]^p = 0$, that states simply differential identities of the type (64). The last member of this sequence is

$$\int [d^*\Omega]^m = m! \int |\det \Gamma|^{1/2} \, dx_1 \, dx_2 \cdots dx_n \tag{65}$$

as a calculation shows, the square root being the square root of the perfect square det Γ. This may also be proved directly from asking when div $(J\mathbf{V})$

vanishes, assuming that J is some function of det Γ and using (64) to obtain $J = \text{const} \cdot |\det \Gamma|^{1/2}$. A corollary here is that if two different integral invariants $\oint \Sigma U_i \, dx_i$ and $\oint \Sigma \overline{U}_i \, dx_i$ are known (the two integrands not differing by merely an exact differential) then the two Js (Jacobi last multipliers) provide an integral of motion

$$\frac{\det \Gamma}{\det \overline{\overline{\Gamma}}} = \text{conserved} \tag{66}$$

according to the well known theorem that the ratio of two last multipliers is a constant of motion.

We have altogether now the Liouville theorem

$$\frac{d}{dt}\left(|\det \Gamma|^{1/2} \, d\tau\right) = 0 \tag{67}$$

telling how to weight phase volume to get an invariant measure of it. Given that $U_0 = \text{constant}$ is a suitable surface in x-space, canonical ensemble theory is expressed by the density

$$\rho(x) = \rho_0 \, |\det \Gamma|^{1/2} e^{-U_0(x)/\theta} \tag{68}$$

In summary, $\dot{x} = X$ very generally can be brought to $\Gamma \cdot \dot{x} = -\nabla U_0$ wherefrom the fundamental elements det Γ and U_0 stand forth in principle for attempting the Gibbsian probe.

Conversely, there is the possibly practical hint to look to the format $\dot{x} = -\gamma \cdot \nabla U_0$ quite directly in making models, forgetting the stringencies of building the U_i, but playing upon only the antisymmetry of γ for holding to conserved U_0, and then leaving enough flexibility in γ and U_0 to satisfy the *one* condition (59) div $V = 0$. In the structure of U_0 and then $\exp(-U_0/\theta)$, one can perhaps think of embracing into dynamics what clews experiment affords from microstatistics, letting the latter be a guide to the former through the mediation of the ensemble.

References

1. Volterra, V., (a) *Leçons sur la Théorie Mathematique de la Lutte pour la Vie*, Gauthier-Villars, Paris, 1931; (b) *Acta Biotheoretica*, **3**, 1 (1937).
2. Lotka, A. J., (a) *Elements of Physical Biology*, Williams & Wilkins, Baltimore, 1925; (b) *J. Am. Chem. Soc.*, **42**, 1595 (1920).
3. Gause, G. F., *The Struggle for Existence*, Williams & Wilkins, Baltimore, 1934.
4. D'Ancona, U., *The Struggle for Existence*, E. J. Brill, Leiden, 1954.
5. Kerner, E. H., (a) *Gibbs Ensemble: Biological Ensemble*, Gordon and Breach, New York, to be published 1971; (b) *Bull. Math. Biophy.*, **19**, 121 (1957); (c) *Bull. Math. Biophy.*, **21**, 217 (1959).

6. Kerner, E. H., *Bull. Math. Biophy.*, **23**, 141 (1961).
7. Corbet, A. S., Fisher, R. A., and Williams, C. B., *J. Animal Ecology*, **12**, 42 (1943).
8. Elton, C., *Voles, Mice, and Lemmings*, Oxford: Clarendon Press, 1942.
9. Cowan, J. D., (a) in *Lectures on Mathematics in the Life Sciences*, v. **2**, *Some Mathematical Problems in Biology*, Am. Math. Soc., Providence, 1970; (b) in *Neural Networks*, Ed. E. R. Caianiello, Springer-Verlag, New York, 1968; (c) *Thesis*, Dep't of Electrical Engineering, University of London, 1967.
10. Moore, M. J., *Trans. Far. Soc.*, **45**, 1098 (1949).
11. Chance, B., Ghosh, A., Higgins, J., and Maitra, P., *Ann. N.Y. Acad. Sci.*, **115**, 1010 (1964).
12. Higgins, J., (a) *Proc. Nat. Acad. Sci.*, **51**, 989 (1964); (b) *Ind. and Eng. Chem.*, **59**, 19 (1967).
13. Goodwin, B. C., *Temporal Organization in Cells*, Academic Press, New York, 1963.
14. Kerner, E. H., *Bull. Math. Biophy.*, **26**, 333 (1964).
15. Whittaker, E. T., *Analytical Dynamics*, Cambridge University Press, 1937.
16. Kerner, E. H., *J. Math. Phy.*, **6**, 1218 (1965); **9**, 222 (1968).
    Hill, R. N., *J. Math. Phy.*, **8**, 201 (1967); **8**, 1756 (1967).
17. Cartan, E., *Leçons sur les Invariants Integraux*, Gauthier-Villars, Paris, 1922.
    Flanders, H., *Differential Forms*, Academic Press, New York, 1963.

PHOTOCHEMICAL REACTION CENTERS AND PHOTOSYNTHETIC MEMBRANES

RODERICK K. CLAYTON

Division of Biological Sciences and Department of Applied Physics, Cornell University, Ithaca, New York

CONTENTS

I. INTRODUCTION

The purpose of this review is to describe two areas of research in photosynthesis that have become particularly illuminating or promising in recent years. These are; first, the development of photosynthetic reaction centers as objects of photochemical study, and the investigation of related "dark" chemical systems of bacterial photosynthesis; second, study of the electrochemical behavior of photosynthetic membranes by optical techniques. The specific ionophorous actions of certain antibiotics have been decisively helpful in this area.

Photosynthetic tissues emit delayed fluorescence, thermoluminescence, and chemiluminescences having the spectrum of the prompt fluorescence of chlorophyll (Chl) or bacteriochlorophyll (BChl). These emissions from the singlet excited state are involved, in intriguing ways, in the foregoing areas of study, and will be discussed in a separate section.

This is not a general review[1-3] of photosynthesis. If it were, the mechanisms of coupling between electron flow and phosphorylation would demand far more attention, and so would the remarkable recent advances[4-8] in our understanding of oxygen evolution and the fate of oxidizing equivalents made by Photosystem II of green plants. The references to the literature are not comprehensive; some have been selected to illustrate the main points of discussion. Almost all of the discussion pertains to photosynthetic bacteria.

II. REACTION CENTERS AND RELATED CHEMICAL SYSTEMS IN THE PHOTOSYNTHETIC BACTERIA

A. Reaction Center Preparations

Emerson and Arnold[9] discovered that the molecules of Chl in plants act cooperatively to absorb light. This suggested the existence of photochemical reaction centers served by light-harvesting "antennas" of many Chl molecules. Van Niel[10] had developed convincing arguments that the primary photochemical act is a light-driven separation of oxidizing and reducing entities, suggesting that if reaction centers exist, they might be the sites of photochemical oxido-reduction. The discovery by Duysens[11] of a small reversible light-induced change in the absorption spectrum of BChl *in vivo*, together with Goedheer's demonstration[12] that this change could be mimicked by chemical oxidation, paved the way for the idea[13,14] that the purple photosynthetic bacteria might have reaction centers in which a molecule of BChl becomes oxidized by light, donating an electron to an unspecified acceptor molecule.

The unmasking and ultimate purification of such a photochemical reaction center began with methods[15] for the selective alteration or destruction of the major part of the BChl, which serves only a light-harvesting function. The most convenient of these methods, oxidation by chloroiridate,[16] causes apparently complete destruction of the light-harvesting BChl in some organisms with little loss of the "reaction center" BChl.[17] More recently it has proved possible to separate "reaction center particles" from the photosynthetic membranes and their associated light-harvesting BChl by attacking the subcellular membrane fragments (*chromatophores*) with detergents, and then fractionating the product by some combination of density gradient centrifugation, gel filtration, ammonium sulfate fractionation, and adsorption-elution chromatography.

Various species and strains of photosynthetic bacteria are highly individualistic with respect to the methods required to produce purified reaction centers. At present, there are three well characterized reaction center preparations; those from blue-green (carotenoidless) mutant strains of *Rhodopseudomonas spheroides*[18-20] and *Rhodospirillum rubrum*,[21] and that from wild type *Rhodopseudomonas viridis*.[22]

The "cleanest" reaction center preparations to date, showing the smallest ratio of protein to BChl, are made[19] with chromatophores from blue-green mutant *Rps. spheroides*, treated with the detergent Ammonyx-LO,* centrifuged over $0.5M$ sucrose, and then fractionated with ammonium sulfate. Absorption spectra of this material, measured in weak and in strong light, are shown in Figure 1. The BChl probably exists as a trimer; two molecules designated P800 and one called P870, responsible for the absorption maxima at 803 and 867 nm. These components also give the absorption band at 598 nm and some of the Soret absorption near 370 nm. The band at 760 nm and the one at 535 nm (resolved into two components at liquid nitrogen temperature; unpublished observation by H. F. Yau in our laboratory) can be attributed to bacteriopheophytin.

B. Photochemical Properties of Reaction Centers

Illumination or chemical oxidation of reaction centers causes the 867 nm band to disappear ("bleaching of P870") and the 803 nm band to move 5 nm toward the blue (blue shift of P800). The redox couple

$$P870 \rightleftharpoons P870^+ + e^- \tag{1}$$

behaves like a one-electron system (as written) with midpoint potential about $+470$ mV, independent of pH, when titrated with ferri-ferrocyanide mixtures (the scale is used in which a stronger oxidant is more positive).

The bleaching of P870 and blue shift of P800 is attended by loss of a double Cotton effect near 800 nm in the optical circular dichroism spectrum.[23] This suggests that P870 mediates an electric dipole interaction between the molecules of P800, and that this interaction is weakened when the P870 is bleached:

$$\text{Trimer, P800}\cdot\text{P870}\cdot\text{P800}\cdot \longrightarrow \text{2 monomers of P800} + \text{P870}^+ \tag{2}$$

The oxidation of P870 is also manifested by an EPR signal resembling that of an oxidized BChl radical *in vitro*[24] and attributed to P870[+].[24,25] One quantum absorbed by either P800 or P870 causes bleaching of one

---

* Dimethyl lauryl amine oxide, made by Onyx Chemical Company, Jersey City, New Jersey.

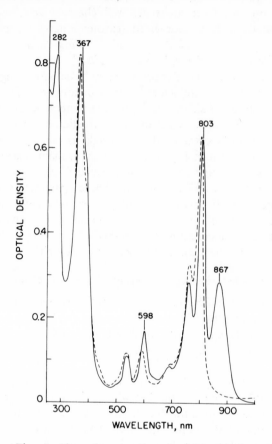

Fig. 1. Absorption spectra of photosynthetic
reaction centers prepared from *Rhodopseudomonas
spheroides* by the method of Clayton and Wang.[19]
The spectra in weak and in strong light (solid and
dashed curves, respectively) were obtained with the
IR-1 and IR-2 modes of operation of a Cary 14R
spectrophotometer. Note the light-induced (reversible) bleaching of the 867 nm band, P870, and
blue-shift of the 803 nm band, P800. These bacteriochlorophyll components also give the 598 nm
band and most of the band near 370 nm. The bands
at 760 and 535 nm are due to bacteriopheophytin.

molecule of P870 and an EPR signal corresponding to one unpaired electron.[25] Until recently there was no direct evidence for the reduction of the primary electron acceptor, the reaction partner in the photochemical oxidation of P870:

$$P870 \cdot A \underset{dark}{\overset{light}{\rightleftarrows}} P870^+ \cdot A^- \tag{3}$$

Now a very broad light-induced EPR signal has been detected, in reaction centers from *Rps. spheroides*, that might correspond to A^-.[26] Aside from that, the nature of the acceptor has been inferred only from indirect observations.

The fluorescence of P870 can be measured in reaction centers; it seems to indicate the states of the reaction centers.[27-29] The state $P870 \cdot A$, which is able to do photochemistry, should be fluorescent, but with small yield as befits an efficient photochemistry. The states $P870^+ \cdot A$ and $P870^+ \cdot A^-$ should be nonfluorescent, since the optical transition (867 nm absorption band) is missing. The state $P870 \cdot A^-$ should be more strongly fluorescent than $P870 \cdot A$ since the photochemistry cannot occur. These expectations are evidently fulfilled. Under mildly oxidizing conditions (untreated reaction centers in air), illumination causes loss of the 867 nm absorption band of P870 and loss of the corresponding fluorescence band with identical kinetics.[28] This shows that during the reaction, the unbleached fraction maintains a constant yield of fluorescence, which is the yield characteristic of $P870 \cdot A$. No appreciable $P870 \cdot A^-$ accumulates under these conditions. More reducing conditions (anaerobic, with oxidation-reduction buffers added) change the fluorescence in two ways; the yield measured at the start of illumination is greater, and the fluorescence grows stronger during illumination, as shown in Figure 2.[28,29] Meanwhile, the bleaching of P870 is suppressed; as quickly as $P870^+$ is formed it becomes restored to P870 by electrons from the surroundings:

$$P870 \cdot A \xrightarrow{h\nu} P870^+ \cdot A^- \xrightarrow{e^-} P870 \cdot A^- \tag{4}$$

As $P870 \cdot A^-$ accumulates, the fluorescence grows correspondingly stronger. The greater initial fluorescence, compared with aerobic preparations, indicates that the chemical reduction system has converted some A to A^- in the dark. Thus the initial fluorescence indicates the "dark" redox properties of A, while the light-induced increase in fluorescence tells about the electron turnover, and the filling of secondary electron pools, during illumination. Analysis of the secondary pools from the rise curve of the fluorescence, which has been pursued extensively with green plant System II,[30,31] has not yet been carried out adequately with these reaction centers.

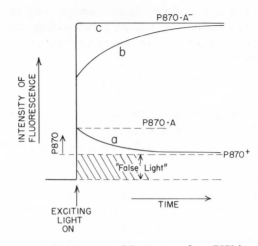

Fig. 2. The intensity of fluorescence from P870 in photosynthetic reaction centers changes during constant illumination. Curve *a* shows how the fluorescence declines as P870 becomes bleached (oxidized; P870 → P870⁺) in an aerobic environment. The basal level labeled "false light" (scattered exciting light; fluorescence from inert components) is reached when all of the P870 has become bleached. The initial level in curve *a* is characteristic of the photochemically competent combination of P870 with electron acceptor, P870 · A. Under more reducing conditions (anerobic; reductant added) the P870⁺ is restored quickly to P870 and the state P870 · A⁻ accrues in the light. This state shows higher fluorescence (curves *b* and *c*) because it cannot perform the photochemical act. The higher initial fluorescence under these conditions shows that some A has been reduced to A⁻ chemically, in the dark. Curve *b* is representative of a potential of −50 mV, and curve *c* of −400 mV.

Titration of the initial fluorescence[29] shows that A behaves like a one-electron agent of midpoint potential −50 mV, independent of pH. This rules out a quinone as primary acceptor. Cramer[32] has performed similar titrations of the yield of fluorescence from light-harvesting BChl in whole cells of photosynthetic bacteria, where the radiative process competes with transfer to and quenching by the reaction centers. The results, −60 to −90 mV in *Rps. spheroides* cells, are reasonably compatible with those for P870 fluorescence in reaction centers. In other photosynthetic bacteria,

Cramer found more negative values (stronger reductant), -160 mV in *Chromatium* and -145 mV in *R. rubrum*. The titration curves all fit Nernst equations for single electron exchanges. Cramer concedes that his measurements could conceivably have shown the behavior of a secondary electron acceptor rather than the primary one, because the cells were exposed continuously to weak light during measurement. At a potential below the midpoint of the secondary pool, most of that pool is in a reduced state. Then the primary acceptor can be driven to its reduced form even in weak light, regardless of its own midpoint potential, and a high fluorescence will be observed. This reservation does not apply to the redox titration of P870 fluorescence in reaction centers, where the fluorescence intensity was observed at the onset of illumination.

The absolute quantum yield of P870 fluorescence in the photochemically active state (P870· A) is 0.4×10^{-3}.[28] We can estimate the intrinsic lifetime of the fluorescence, from the integrated area of the 867 nm absorption band, to be 18 nsec. Multiplied by the quantum yield, this gives[33] a lifetime of 7×10^{-12} sec for the singlet excited state P870*· A. Since the quantum efficiency for P870 $\rightarrow$ P870$^+$ is nearly 100 per cent, we conclude[28] that nearly every quantum leads to an act that quenches fluorescence of P870 in 7×10^{-12} sec, and results in the appearance of P870$^+$. This rapid act could be the movement of an electron (with tunneling?) from P870* to a virtual state in A, combined with a relaxation process (nuclear and electronic rearrangement) that consolidates the charge separation so as to give stable P870$^+$ and A$^-$.

Before settling into their final and most stable states, both P$^+$ and A$^-$ might pass through transient states of higher energy. There might exist a transitory "excited P$^+$", a stronger oxidant than the stable form showing a midpoint potential of $+470$ mV in chemical titrations. Similarly, there might be a short-lived form of A$^-$ with stronger reducing potential than -50 mV. Such stronger oxidizing and reducing states of P$^+$ and A$^-$ could conceivably interact with closely coupled molecules in the living cell, and yet escape detection in chromatophores and reaction center preparations. A redox titration of P870 fluorescence could then show quite legitimately that the primary acceptor *can* be reduced chemically (and lose its photochemical function) at midpoint potential -50 mV, and at the same time this titration could give a misleadingly poor impression of the reducing power generated in the light reaction.

This picture of the photochemical act suggests that the energy of an electron in P870* should match that of a virtual level in A, if the electron transfer is to occur efficiently. A mismatch of these energy levels could explain such curious effects as the loss of photochemical activity in reaction centers upon dehydration.[17]

C. Dynamics of Chemical Processes after the Primary Photoact

At temperatures from 77°K down to 1.7°K the return from the state $P870^+ \cdot A^-$ to $P \cdot A$ in reaction centers from *Rps. spheroides* follows first order kinetics with a half-time of 30 msec.[24; compare 34] The simplest interpretation is that the electron returns from A^- to $P870^+$ without the involvement of any other molecule; it is a direct back-reaction. At room temperature, the recovery from $P870^+$ to P870 shows first order components with half-times ranging from 100 msec to many minutes. Apparently the electrons can leave A^- and be cycled rapidly or more slowly, through secondary pools, to $P870^+$. Movement of electrons from A^- to the secondary pools is inappreciable at low temperature. If reaction centers or chromatophores are illuminated while being cooled, the system is frozen in a state where electrons are in a secondary pool and $P870^+$ is oxidized.[35] The recovery of P870 at low temperature is then extremely slow. A detailed analysis of the temperature dependence of these secondary electron transfers, involving substances other than A, is lacking.

A block between primary acceptor and secondary pools can also be imposed by ortho-phenanthroline.[36,37] Addition of this substance to reaction centers at room temperature causes the restoration of P870 from $P870^+$ to be rapid after an exciting flash, with a half-time of 100 msec or less, suggesting a direct back-reaction. It cannot yet be said whether this reagent prevents transfer of an electron from A to a secondary acceptor B, or from B to a tertiary one C, or perhaps prevents a relaxation process in A that is prerequisite to further electron transfer.

The direct return of an electron from A^- to $P870^+$, measured with a half-time of 30 msec in reaction centers[24] and 10 to 20 msec in chromatophores of *Rps. spheroides*,[34] represents a waste process in the living cell. Under physiological conditions, there are two events that can stabilize the redox energy against this loss. One is the rapid transfer of an electron from a cytochrome (Cyt) of the c type to $P870^+$, and the other is the transfer from A^- to one or more secondary acceptors B:

$$\text{Cyt} \cdot \text{P870} \cdot \text{A} \cdot \text{B} \xrightarrow{h\nu} \overset{e^-}{\overbrace{\text{Cyt} \cdot \text{P870}^+ \cdot \overset{e^-}{\overbrace{\text{A}^- \cdot \text{B}}}}} \longrightarrow \text{Cyt}^+ \cdot \text{P870} \cdot \text{A} \cdot \text{B}^- \quad (5)$$

The transfer time from Cyt to $P870^+$ can be very short, and varies strongly among species of photosynthetic bacteria; from <2 μsec to several msec at 77°K, and from about 0.3 to 10 μsec at room temperature in cases that have been measured.[38] The large variations could be symptomatic of the exponential dependence of tunneling times on the thickness of an energy barrier.

The transfer time from A$^-$ to B has been inferred in three ways. First, one can measure the onset of reduction of ubiquinone as one component of a light-induced bleaching centered at 275 nm.[39] Ubiquinone is not A; it might not be B either, but electron transfer to B must be at least as fast as transfer to ubiquinone. Ke has reported[40] that the 275 nm absorption change in *Chromatium* chromatophores shows a rise time less than 1 μsec after a flash of exciting light. He attributes the change to ubiquinone reduction, but this identification cannot be considered positive.

A second way to evaluate the transfer of electrons from A$^-$ to B is from the kinetics of the reaction P870$^+$ $\rightarrow$ P870 after a flash. A direct return of electrons from A$^-$ to P870$^+$ would be suggested by a fast (<100 msec) component. Slower kinetic components would indicate that the electron has gone to B (or is in a new, altered state of A) before coming back to P870. This kind of analysis is incomplete but it can be said[36] that in reaction centers at room temperature, without special chemical additions, the transfer of electrons from A$^-$ to B is much faster than that from A$^-$ to P870$^+$. Below about 150°K the transfer from A$^-$ to B becomes negligibly slow. Also there is some indication (unpublished experiments in our laboratory) that, as reaction center preparations become more highly purified, they become depleted of secondary electron acceptors so that a greater proportion of the electrons in A$^-$ come directly back to P870$^+$.

A third way to measure these movements of electrons (actually an extension of the foregoing) is to apply single and double laser flashes (30 nsec duration) and to follow, with a time resolution better than 1 μsec, the resulting transient changes in absorption and fluorescence that reflect the changing states of P870, Cyt, and A. Such measurements were begun by DeVault in Chance's laboratory,[41] and have been extended and refined to a remarkable degree by Parson.[42-44,37] In one study, Parson[43] measured the effects of single and double laser flashes on *Chromatium* chromatophores. He found that after a single flash the P870$^+$ recovers to P870 (by receiving an electron from Cyt) in about 1 μsec. The P870 is then refractory to renewed oxidation by a second flash. This refractoriness is attended by a higher yield of fluorescence from the light-harvesting BChl. It is attributed to the state P870· A$^-$, which cannot trap energy from the light-harvesting apparatus because it cannot react photochemically. The disappearance of the refractory condition, measured both by the capability of P870 to be oxidized and by the return of the fluorescence to a lower yield, then indicates the transfer of electrons from A$^-$ to B. Parson found that 50 per cent recovery from refractoriness happens in about 60 μsec, a time considerably longer than that inferred by Ke[40] for the same material. Parson observed second order kinetics (or perhaps several first order components) for this process. The recovery was faster at lower pH, as if a reaction of the sort

$$A^- + H^+ + B \longrightarrow A + BH \tag{6}$$

were taking place (Parson used the terminology X and Y in place of A and B). Times of the order of 1 to 100 μsec for electron transfer from A^- to B are, of course, more than fast enough to prevent the back reaction (10 to 30 msec) from A^- to $P870^+$.

Rapid electron transfer from A^- to B can become inhibited if the environment is so reducing as to reduce B. Under these conditions, the rapid transfer from Cyt to $P870^+$ is assured, in living cells. Conversely, if the environment is oxidizing, so that the Cyt becomes oxidized and cannot transfer electrons to $P870^+$, rapid functioning of the "A^- to B" mechanism is assured. Thus the two mechanisms for stabilizing primary redox energy are complementary.

Upon leaving A^-, the electrons appear to enter a pool with a redox potential near 0 to -50 mV[45, and compare 32 and 29 with 46] in purple photosynthetic bacteria. This is inferred from the ability of reaction centers and chromatophores to react in a diffusion limited way with external oxidation-reduction indicators. In contrast, illuminated green plant and algal tissues can reduce viologen dyes at rates showing that a potential near -600 mV has been generated and is stable enough to react with the external dye.[47] This shows the reducing power of the primary electron acceptor of green plant Photosystem I. The donor is P700, a molecule (probably Chl a) analogous to P870 and having a midpoint potential near $+430$ mV. The difference between green plant System I and the bacterial system is shown most strikingly by the inability of the latter to reduce dyes at potentials lower than -50 mV. This might be taken as evidence against the importance of higher (more strongly reducing) transient states of A^- in the bacteria.

D. Two Photochemical Systems in Photosynthetic Bacteria?

The foregoing was written in the following postulatory framework; the purple photosynthetic bacteria have one kind of reaction center, with P800, P870, and an unknown acceptor A. This reaction center mediates the photochemical oxidation of a Cyt, typically of the "c" type with midpoint potential about $+350$ mV, but alternatively a "low-potential" cytochrome.[48-50] The reaction center concomitantly forms a reductant that becomes stabilized at a level of about 0 to -50 mV; perhaps reduced ubiquinone. For the most part, the energy represented by this charge separation is conserved as a high-energy substance[51] or thermodynamic state* whose formation is coupled to the return of electrons from the reductant to the oxidized (high-potential) Cyt. The energy thus conserved can be used to form ATP from ADP and orthophosphate[51,52] and also to form stronger reductants from weaker ones,[53] for example, to promote electrons from about zero mV potential to the level of NADH (-320 mV midpoint potential) or reduced ferredoxin (-420 mV). Direct reduction of ferredoxin (and subsequently of NAD^+) by A^- is just thermodynamically possible, given Cramer's values for the midpoint potential of primary reductant in *Chromatium* and in *R. rubrum*, but only if the system is operating in a steady state with most of the A in the reduced form and most of the ferredoxin in the oxidized form.

A different view has gained support recently,[1] that the bacterial photosynthesis involves two distinct patterns of photochemically driven electron flow. One of these (the cyclic system) is the familiar one just described. It uses reaction centers with the combination of P800 and P870 (or P830 and P960 in the case of *Rps. viridis*[54,22]) in conjunction with a high potential (about $+350$ mV) Cyt, that is, "Cyt c_2" or C551 in *R. rubrum* and *Rps. spheroides*, and C555 in *Chromatium*. The other system (noncyclic) is alleged to transfer electrons from substrates (such as H_2S and succinate) through a low potential (about 0 mV), autoxidizable Cyt, photochemically through another kind of reaction center, and on to NAD^+, presumably by way of ferredoxin and NAD^+ reductase.[55] The oxidized "reaction center" BChl for this noncyclic system, analogous to the familiar $P870^+$, is supposed to take electrons directly from the low potential Cyt, C428 in *R. rubrum*[56] and C553 in *Chromatium*.[50] The electron acceptor in this second kind of reaction center should have a potential around -400 mV if it is to mediate the reduction of NAD^+ efficiently.

The first evidence for two kinds of photosystem in the purple bacteria was that the light-induced oxidations of the high- and low-potential cytochromes in *Chromatium* showed different action spectra.[57] A similar observation was reported for *R. rubrum*.[58] In *Chromatium*, the main light-harvesting BChl components are called B810, B850, and B890, after their absorption maxima. Morita showed that B810 is more effective (relative to absorption spectrum) in the oxidation of C553, whereas B890 is predominant in the action spectrum for C555 oxidation.† However, Parson and

---

* The thermodynamic state has been identified, in Mitchell's chemiosmotic hypothesis,[52] as "protonmotive force", as described in the next section.

† Morita also found that the action spectrum for oxidation of cytochromoid cc', or RHP, is different from the other two (greatest effectiveness for B850) and suggested that there might be three distinct photosystems. Others have tended to ignore this complication.

Case[37] have shown that a single laser flash can oxidize a certain amount of
P870 in *Chromatium* cells or chromatophores, and then, depending on the
conditions, this P870$^+$ (apparently of one kind only) can react stoichio-
metrically with either C553 or C555. Aeration and high redox potentials,
which cause chemical oxidation of C553, allow C555 to react with P870$^+$.
At lower redox potentials, as in anaerobic cell suspension, a reaction of
P870$^+$ with C553 is favored. The reaction half-times were measured to be
2 and 0.8 μsec for C555 and C553, respectively. The kinetics of refractoriness after a flash, an indication of electron transfer from primary to secondary acceptors, gave evidence that the same acceptors are involved in the
oxidations of both cytochromes. Parson and Case did not find different
action spectra for the reactions of the two cytochromes. They attributed
the differences reported by Morita to the fact that observations of the
reactions of the different cytochromes require different conditions of redox
poise and background light intensity, and these factors can influence the
efficiencies of energy transfer from the different light-harvesting BChl
components to the reaction centers.

Parson's results tend to weaken the arguments for two or more distinct
photosystems in *Chromatium*, but the case developed by Sybesma[58–60]
for two photosystems in *R. rubrum* appears to be stronger. The absorption
spectrum of *R. rubrum* shows a single "B880" light-harvesting component
and a small band at 800 nm. Part or all of the latter is due to P800, isolated
along with P890* in purified reaction centers.[21] These "familiar" reaction
centers, which show blue-shift of P800 along with bleaching of P890,
mediate the oxidation of the high-potential, nonautoxidizable Cyt c_2 as
part of a cyclic system coupled to phosphorylation. These effects can be
seen in aerobic cell suspensions with background illumination, and in
chromatophores. In whole cells, the action spectrum for this photochemistry shows disproportionately strong effectiveness of the 800 nm
component(s) relative to the major component B880. In anaerobic cell
suspensions exposed only to dim light, the oxidation of Cyt c_2 is slight.
The low-potential C428 is in its reduced form in the dark, and its light-
induced oxidation is conspicuous. The action spectrum for this oxidation
of C428 differs strikingly from that for Cyt c_2 oxidation. The 800 nm component is completely ineffective; the action spectrum shows only a single
band matching B880.

Under the conditions where C428 is oxidized reversibly by light and Cyt

* Those working with *R. rubrum* and *Chromatium* have usually referred to this photo-
active BChl as P890 rather than P870. Actually, its absorption maximum is between
870 and 885 nm, often at 883 nm, with variations among different strains and methods
of preparation.

c_2 is not, the reversible change of absorbance in the near infrared shows only a bleaching near 880 nm, and no sign of a blue-shift of P800. It is as if the P800 is not coupled to the bleachable P890 under these conditions, and any energy absorbed by P800 cannot reach the reaction centers that mediate C428 oxidation.

These results allow at least two interpretations. First, there may be one kind of reaction center, which can react either with Cyt c_2 or with C428. Under the oxidizing conditions that favor a reaction with Cyt c_2, the coupling between P800 and P890 is strong and that between B880 and P890 is somewhat weaker. The more reducing conditions that favor a reaction with C428 cause the coupling between P800 and P890 to become much weaker; one could even speculate that the reduced form of C428 is the agent that weakens the coupling. In washed chromatophores or isolated reaction centers, the coupling between P800 and P890 is strong. Alternatively, there may be two kinds of reaction center, one permanently associated with Cyt c_2 and the other with C428. Only the former kind has P800.

The question remains whether the "two kinds" or P890 in *R. rubrum* have different midpoint potentials and/or are associated with different kinds of electron acceptor. No data have been reported on these matters, nor on quantum efficiencies for the two reaction systems. These data should be easy to obtain and could be most significant.

In *Chromatium*, there is apparently a single P870, of midpoint potential +470 to 490 mV, reacting with either C555 or C553, the latter in less than 1 μsec. Such a rapid, apparently direct reaction between P870$^+$ and the low-potential Cyt seems distressingly wasteful, unless the energy of about 0.5 eV/molecule or 11 kcal/mole can be conserved by a mechanism that has not been discovered.

The question of multiple photochemical systems in photosynthetic bacteria can be thrown into confusion by the occurrence of absorbancy changes due to band shifts of the light-harvesting pigments.[61] It is apparent that these shifts are associated with ion translocation across the membranes of photosynthetic tissues; they are probably caused by electric fields and by changes of membrane configuration. The conditions for the expression of these effects are usually not the same as the conditions for maximum P870 bleaching, giving the impression of two distinct photosystems. One branch of the absorption difference spectrum of a band shift, either an increase or a decrease of absorption, can be interpreted erroneously as the change of a new reaction center pigment. This might be the meaning, for example, of "P905" in *Chromatium*.[62]

In summary, the evidence concerning two systems in bacterial photosynthesis might become consolidated so as to reveal two distinct major functional systems in the sense of Systems I and II of green plants. Alternatively, it might turn out that the differences are more trivial, with basically one kind of photosystem showing various patterns of interaction with cytochromes and other substances.

III. OPTICAL MANIFESTATIONS OF ELECTROCHEMICAL CHANGES ASSOCIATED WITH PHOTOSYNTHETIC MEMBRANES

A. Coupling Mechanisms—Membranes and Phosphorylation

The phenomenology of photosynthetic phosphorylation is for the most part consistent with the following picture[63]: photochemical electron transport is translated into the formation of a high-energy intermediate (a substance or state that can be denoted $\ominus$) which in turn can drive the formation of ATP from ADP and orthophosphate. In healthy materials, the electron transport is linked obligatorily to the formation of $\ominus$. If $\ominus$ is not utilized, as for the formation of ATP, or else dissipated, its accumulation to a saturated level constrains the rate of electron transport. Uncouplers can dissipate $\ominus$, or can weaken the obligatory link between electron flow and the formation of $\ominus$. An uncoupler is usually defined operationally as a substance that inhibits ATP formation and at the same time speeds electron transport. However, it could also be recognized in the absence of ATP formation by its ability to speed electron transport while inhibiting certain effects that have come to be associated[52] with $\ominus$. These effects are the reversible swelling and shrinking of the membrane-bound thylakoids of chloroplasts,[64] the reversible movement (pumping) of H^+ into thylakoids or chromatophores together with the movements of other ions,[65-68] and the band shifts of carotenoids and chlorophylls[69-71] that can be associated with changes of thylakoid or chromatophore membrane potential.[71-74]

A main center of controversy is whether the development of ion (especially H^+) gradients across membranes is an essential link between electron transport and phosphorylation or is only a by-product of some other coupling mechanism. In three major alternative views, $\ominus$ is a substance (chemical coupling hypothesis; see Ref. 51), a thermodynamic state involving ion concentration differences (chemiosmotic hypothesis; see Ref. 52), or a particular conformation of a macromolecule (conformational hypothesis; see Refs. 75, 76). In the chemiosmotic hypothesis, $\ominus$ can be equated to protonmotive force (pmf), the chemical potential that tends to move H^+ through a membrane that separates two regions. One component of this pmf is the concentration potential:

$$\frac{RT}{nF} \ln \left[\frac{(H^+)_i}{(H^+)_o} \right] \tag{7}$$

where the concentrations (more properly, the activities) of H^+ in the inner and outer phases of a membrane-bound region are expressed. The other component of pmf is the electrostatic potential across the membrane. The membrane potential could drive the ATP forming reaction by acting on negatively charged groups involved in the reaction, as well as on H^+. In this hypothesis, pmf is generated by the active translocation (pumping) of H^+ across the photosynthetic membrane.

The possibility of choosing among these mechanisms is being explored mainly by studies of the kinetics and stoichiometry of electron flow, H^+ translocation, and ATP formation in mitochondria, chloroplasts, and chromatophores.[63] There are many complications, such as the reversible binding of H^+ (change of internal buffer capacity) and the effects of pH on redox levels of electron carriers. These matters will not be reviewed further; for the present discussion it is sufficient to note that in chromatophores and chloroplasts the ability to make ATP is usually correlated with the pmf,[77-79] and the pmf is generated by photochemically initiated electron transport (although it can be generated by other means, including the direct method of subjecting materials to an acid-base transition[80]).

B. Ionophorous Substances and Their Effects on Phosphorylation

A variety of artificial and natural membranes have been studied[81-86] with respect to the effects of certain substances on movements and concentration gradients of ions. The natural membranes include those of mitochondria, chloroplast thylakoids, and bacterial cells or chromatophores. These studies, and others involving light and ATP formation,[77-79,87-89 to name a few] have established with reasonable certainty the specific ionophorous actions of some substances at low concentrations.*

In their natural states, with no special agents added, the photosynthetic membranes are relatively impermeable to most ions. Relaxation of an H^+ concentration difference across the chloroplast thylakoid membrane shows a half-time of several seconds.[65] Relaxation of the membrane potential in chromatophores (in the dark following illumination) also takes a few seconds, judging from the kinetics of carotenoid absorption band shifts.[71; see later]

* At higher concentrations the specificity of an ionophore may fail. For example,[89], substituted carbonyl cyanide phenylhydrazones at $10^{-7} M$ concentration make chloroplast thylakoid membranes permeable selectively to H^+, but at $10^{-5} M$ these agents encourage permeation of other univalent cations.

Gramicidin D induces permeability of membranes toward all the common univalent cations. It can therefore prevent the development of H^+ gradients and membrane potentials associated with pumping of any such ions. The substances carbonyl cyanide-m-chlorophenylhydrazone (CCCP) and carbonyl cyanide-p-trifluoromethoxyphenylhydrazone (FCCP) at concentrations about 10^{-7} M (or less) facilitate only the permeation of H^+, and should therefore dissipate pmf generated by H^+ pumping. Dinitrophenol (2,4-DNP) also induces permeability to H^+, at least in mitochondria and chloroplasts.

Valinomycin is selective in allowing permeation of K^+, Rb^+, Cs^+, and apparently also NH_4^+,[71,77] but not H^+ or Na^+. Thus if valinomycin is present, along with sufficient K^+, the passive flow of K^+ should tend to neutralize any membrane potential due to H^+ pumping. The extent of the neutralization will depend on the relative rates of K^+ diffusion and H^+ pumping. In any steady state, the membrane potential must equal the diffusion potential of K^+:

$$E = \frac{RT}{nF} \ln \left[\frac{(K^+)_i}{(K^+)_o} \right] \tag{8}$$

where $(K^+)_i$ and $(K^+)_o$ are the activities of K^+ on the inside and outside of a membrane-enclosed space. Thus a measurement of inner phase and external K^+ concentrations can indicate the membrane potential if valinomycin is present. Note that valinomycin by itself should not eradicate the pmf, because the H^+ concentration gradient is left intact or even enhanced. With valinomycin present, the K^+ inside a thylakoid or chromatophore can facilitate the light-induced inward pumping of H^+ by moving out so as to preserve electrical balance. Normally, this electrical compensation is provided mostly by an influx of Cl^-, which is slower.

Nigericin mediates a stoichiometric exchange of H^+ for K^+ across membranes; for every H^+ that diffuses out, a K^+ must move in or *vice versa*. There is no net movement of charge, and hence no effect on the membrane potential. However, any H^+ gradient across a membrane can be eradicated or diminished while a K^+ gradient builds up. In equilibrium, the ratios of activities are equal:

$$\frac{(H^+)_i}{(H^+)_o} = \frac{(K^+)_i}{(K^+)_o} \tag{9}$$

The ratio for one species of ion thus measures that for the other.

Ammonia acts rather like nigericin, in that it allows exchange of H^+ for another cation, in this case NH_4^+.[90] As NH_3 moves through a membrane,

NH_4^+ is converted to H^+ on one side, and H^+ to NH_4^+ on the other, with no net transport of charge. Ammonium salts, and similarly amines of low molecular weight, would thus be able to dissipate the "H^+ gradient" term of pmf without affecting the membrane potential term.

The properties of these substances are consistent with (and have, indeed, been partly deduced from) their action on light-induced pH changes in suspensions of green plant chloroplasts and bacterial chromatophores. With a few exceptions, the effects on phosphorylation can also be understood, simply by taking pmf as an index of $\ominus$. For example, in the presence of K^+, neither valinomycin (which dissipates membrane potential) nor nigericin (which washes out an H^+ gradient) is a very potent inhibitor of ATP formation in chromatophores of *R. rubrum*. However, the combination of the two substances, which can eradicate both components of the pmf, does produce strong inhibition.[78,79] In similar vein, NH_4Cl inhibits phosphorylation in chloroplast fragments, but only if valinomycin is present so as to allow free movement of NH_4^+ as well as exchange of H^+ for NH_4^+.[77]

The fact that NH_4Cl, unassisted by valinomycin, can uncouple phosphorylation in whole chloroplasts has been excused on the grounds that the "H^+ gradient" term of the pmf is more important than the membrane potential term. This is consistent with the observation that nigericin can uncouple chloroplasts, with the help of K^+ but without valinomycin, whereas valinomycin without nigericin is ineffective.[87] But this explanation only gives more force to a certain exception; neither 2,4-DNP nor FCCP at low concentration uncouples phosphorylation in chloroplasts.[89] These agents should certainly abolish the H^+ gradient, and also the membrane potential if the latter is generated by H^+ pumping. The combination of valinomycin (and K^+) with 2,4-DNP or FCCP causes strong uncoupling in chloroplasts, but this case of synergism lacks an obvious explanation in the foregoing terms. These observations may represent an obstacle to the identification of pmf with $\ominus$, at least in chloroplasts. In chromatophores of *R. rubrum* and *Rps. spheroides*, the identification of pmf with $\ominus$ seems straightforward at present.

With whole cells of *R. rubrum*, light causes an efflux of H^+,[91] in contrast to the influx observed with chromatophores. This is to be expected on topological grounds, because chromatophores seem to result from the pinching off of invaginations of the cytoplasmic membrane.[92] The outer surface of a chromatophore membrane is therefore equivalent to the inner surface of the cytoplasmic membrane of a cell. A consistent picture of ion translocation coupled to photosynthetic electron transport in chromatophores, initiated by a reaction forming $P870^+$ and A^- (see Figure 3, upper part) would then place the primary redox couple across the cell membrane with P870 directed toward the outside and A toward the inside. In chromatophores, this orientation of P870 and A would be reversed.

C. Carotenoid Absorption Band Shifts as Indicators of Membrane Potential

A light-induced shift of the absorption bands of carotenoid pigments, toward the red, occurs in chromatophores of *Rps. spheroides* at 1°K.[34] This observation did not encourage the idea that the shift could be related to the chemistry of ATP formation. However, the magnitude of the shift can be far greater at room temperature than at 1°K, and shows complex kinetics and responses to the environment. A typical time-course of the light-induced shift in *Rps. spheroides* at room temperature is sketched in Figure 3 (lower part). The small component labeled *a* survives at low temperature. The remaining (major) part of the carotenoid shift, and also the more evanescent band shift of light-harvesting BChl[61,71,93] is highly sensitive to treatments that would affect the structure of the tissue, such as dehydration or addition of a detergent.[61,70,71]

Soon, it became clear that the band shifts of pigments in the photosynthetic membranes are intimately connected with the conditions for phosphorylation, in green plant tissues[74] as well as in photosynthetic bacteria.[71-73] In chromatophores of *Rps. spheroides* and *R. rubrum*, the band shifts show a close correlation with the predicted level of $\ominus$,[71,72] and this has turned out to be a direct and simple relation between the band shift and the chromatophore membrane potential.[73] A variety of indirect and direct observations bear this out. For example, the combinations of agents that should dissipate membrane potential, such as valinomycin in the presence of K^+, inhibit the onset and speed the decay of the light-induced band shift.[71] The addition of ATP or pyrophosphate, which should generate $\ominus$ by reversal of the phosphorylating reactions, causes band shifts identical to the light-induced ones.[72] Most strikingly, the band shifts can be induced by treatments designed to create a membrane potential artificially. Thus, in a suspension of *Rps. spheroides* chromatophores exposed to valinomycin, a sudden change in the concentration of K^+ in the medium should generate a diffusion potential across the membrane. This treatment was indeed found to cause a carotenoid band shift,[73] with the amplitude of the shift simply proportional to the membrane potential as computed from Eq. (8). In this experiment, the K^+ within the chromatophores could be released by adding the detergent Triton X-114; this provided an assay of $(K^+)_i$ and therefore afforded a calibration of the relation between band shift amplitude and membrane potential. On that

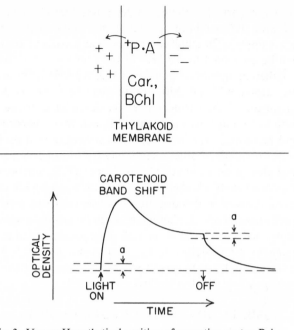

Fig. 3. Upper: Hypothetical position of a reaction center, P.A., in relation to a photosynthetic membrane, so that photochemical electron transport and associated ion movements generate an electric field across the membrane. The left side would be the inside of a thylakoid (closed membrane-bound organelle). Lower: Typical time-course of an absorbancy change showing the band shift of a carotenoid pigment in *Rps. spheroides*; for example the increase in optical density at 490 nm. The peak is passed, and the lower plateau reached, in a few sec. The component labeled *a* is partly a fast, temperature-insensitive component of the carotenoid reaction and partly a change not due to the carotenoid band shift. The band shift can be ascribed to an electric field across the photosynthetic membrane.

basis, the peak of the light-induced shift (Figure 3, lower part) represents a potential of about 420 mV, and the subsequent plateau about 200 mV. Adding a term of 50 mV estimated for the H^+ gradient at an external pH of 7.3,[73] the steady-state pmf in the light is 250 mV, or about 6 kcal per gram-equivalent of H^+ translocated.

These results suggest that the carotenoid band shift is a reliable quantitative index of membrane potential. The phenomenon can be ascribed to an electric field effect on an asymmetric molecule, such that the dipole moments of the excited and ground states are affected differently by the field. The major part of the effect under physiological conditions can be associated with the membrane potential that results from ion pumping (see Figure 3, upper). The small component that persists at low temperature might then be attributed to the field of the primary redox couple, $P870^+$ and A^-. Different species of photosynthetic bacteria differ widely in the size of the carotenoid band shift, and this can be attributed to variations in the alignment of the carotenoid molecules relative to the membrane field as well as to differences in chemical structure of the carotenoids.

It would be most useful to develop this sort of optical probe, so as to measure the potentials of other membranes (mitochondria, nerve, muscle) with reasonable sensitivity ($\Delta OD \gtrsim 10^{-4}$ per mV in practical cases with chromatophores), with excellent time resolution, and with little damage to the tissue. A move in this direction has been made by Emrich et al.,[94] who showed that the dye rhodamine b, when infused into spinach chloroplasts, shows a light-induced band shift with kinetics similar to those of the endogenous band shifts. In the meantime, the carotenoid shift in bacterial chromatophores makes these preparations ideal for the study of ion movements as related to membrane potential.

IV. LUMINESCENCES OF CHLOROPHYLLS *IN VIVO*

A. Luminescence in Relation to the Primary Photochemistry

The chlorophylls and bacteriochlorophylls in photosynthetic tissues emit delayed fluorescence, thermoluminescence, and luminescence induced by a variety of chemical perturbations.[71,95-101] All of these emissions have the same spectrum as the prompt fluorescence, showing that the Chl or BChl has been placed in the lowest singlet excited state. Some of the spectra show that the emission comes from light-harvesting pigments,[102] but functioning reaction centers are also required.[36,103] In photosynthetic bacteria, the mechanism (see later) probably involves first the creation of P870* (singlet excited P870) from a reaction between $P870^+$ and A^- or some other reduced substance. In purified reaction centers, the excitation is quenched so efficiently (in 7×10^{-12} sec)[28] that the emission is undetectably weak. But in chromatophores and cells a quantum of excitation in P870* can become the property of the light-harvesting BChl aggregate, where its lifetime is much greater, about 10^{-9} sec.[104,105] The emission yield is correspondingly much greater, about 2 to 5 per cent[106] rather than 0.04 per cent.[28] Here the normal relationship between light-harvesting BChl and photochemical traps is inverted; the light-harvesting system "traps" quanta from the reaction centers so that they can be detected as luminescence.

A chemiluminescence of chlorophylls *in vitro*, discovered by Goedheer and Vegt,[107] provides a model for the luminescences of BChl in photosynthetic bacteria. The luminescence *in vitro* accompanies the reduction of oxidized BChl,

$$BChl^+ + X^- \longrightarrow BChl^* + X$$
$$BChl^* \longrightarrow BChl + h\nu \tag{10}$$

and has the same spectrum as the prompt fluorescence denoting a transition from the lowest excited singlet state.

Cells or chromatophores of *Rps. viridis* contain P985, analogous to the P870 of *Rps. spheroides* (in purified reaction centers, the band has shifted from 985 to 960 nm and the pigment is called P960). Fleischman has made a variety of experiments with *Rps. viridis*,[99-101] all showing that a necessary condition for luminescence is the re-reduction of oxidized P985. For example, a luminescence can be induced by injecting ferricyanide into a suspension of chromatophores, but only if the suspension has first been illuminated so as to make some photochemical reductant.[99,100] For some time after a flash of light, the chromatophores contain reduced substances $(A^-, B^-, \text{etc.})$ and oxidized cytochrome (Cyt^+). The P985 is in its reduced form, having captured an electron from Cyt. Then when ferricyanide is injected, the newly formed $P985^+$ can react with the stored reductant so as to give a burst of luminescence. The reductant that must react with $P985^+$ to give luminescence need not be formed photochemically; the system can be "charged" by chemical reduction (with reduced methyl viologen or $Na_2S_2O_4$) as well as by light.

In another set of experiments,[101] Fleischman examined the conditions necessary for thermoluminescence in chromatophores from *Rps. viridis*. The basic experiment (compare Ref. 96) was to bring the chromatophores to some low temperature, illuminate them, and then observe the luminescence that accompanied heating. A variety of "low" temperatures were tested. The material was either illuminated or kept dark during the cooling, then illuminated at the low temperature, then held in the dark for a chosen time, and finally heated while luminescence was measured. Fleischman found that "glow curves" are elicited only if $P985^+$ and a reduced substance can interact during heating. The reaction centers must be made to cycle until cytochrome has been oxidized, until then $P985^+$ cannot accumulate. Therefore the necessary preparation for a glow curve requires that two or more photochemical acts take place in a single reaction center.

Analysis of these glow curves showed that the electron trap depth (for the electron that must combine with P^+ to give P^*) was about 0.6 eV.

By providing another observable parameter, these studies of luminescence are helping to unravel the interactions between primary and secondary photoproducts. In green plants, the luminescences will hopefully help to understand the nature of Photosystem II and the chemical pathways leading to O_2 evolution.[96,108]

The delayed fluorescence can be described by a similar model, and because this emission can be observed within milliseconds (or less) after an exciting flash, it can in principle be associated with a reaction between P^+ and A^-.

B. Delayed Fluorescence and Membrane Potential

One of the puzzling aspects of the delayed fluorescence of Chl and BChl *in vivo* has been the profound inhibition exerted by uncouplers of phosphorylation.[108-110] This inhibition was defined more closely by studies in which the intensity of delayed fluorescence was compared with the carotenoid band shift in *Rps. spheroides*.[71] Delayed fluorescence intensity was monitored with a phosphoroscope, operated so that continuous illumination of a chromatophore suspension was approximated by a sequence of flashes spaced about 5 msec apart. Intervals of measurement were sandwiched between these flashes. The same average exciting light intensity could be used to elicit the carotenoid shift. The time-course of delayed fluorescence intensity then showed a resemblance to that of the carotenoid shift (Figure 3, lower). A variety of electron transport inhibitors and uncouplers had exactly parallel effects on the magnitudes and kinetics of the delayed fluorescence and the carotenoid shift. Variations of both phenomena thus reflected a common parameter, at first identified as $\ominus$,[71] but later recognized to be the membrane potential.[73] Variations of the delayed fluorescence might therefore be useful as indicators of changing membrane potential in systems that do not show a carotenoid band shift. This use would have to be made with caution, however, because the delayed fluorescence intensity is certainly modified by other factors having to do with photochemical turnover and fluorescence yield.[108,111]

The luminescence induced by acid-base transition and dependent on prior illumination[97,98,100] can now be understood in either of two ways; through the membrane potential developed by the transition, or through a pH-induced shift in redox levels of primary and secondary electron acceptors.[112]

It remains to be understood how the intensity of delayed fluorescence is modulated by the membrane potential. Fleischman and Crofts made the straightforward suggestion[113] that the membrane field raises the energy of the electron trap A^-, relative to the excited state P^* (Figure 4). Now if the

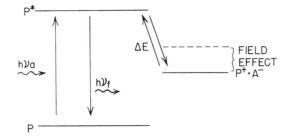

Fig. 4. A sketch of the energy levels of P870 (associated
with an electron acceptor A), of singlet-excited P870, and
of the stabilized state $P^+ \cdot A^-$ in which an electron has
been transferred from P870 to A. The electron is trapped
at a depth ΔE below the singlet excited state. Fleischman
and Crofts have suggested[113] that the depth of this trap is
shifted by the electrostatic field across the photosynthetic
membrane, in order to explain the effect of the field on
delayed fluorescence. Delayed fluorescence would result
when the excited state is regained from the state $P^+ \cdot A^-$,
with activation energy ΔE.

delayed fluorescence depends on the return of an electron from A^- to P^+
so as to give P*, the intensity of delayed fluorescence will involve a Boltzmann factor $\exp(-\Delta E/kT)$. The quantity ΔE is made smaller by the field
across the membrane, and this should increase the intensity of delayed
fluorescence. Collapse of the field, as by a suitable uncoupler, should cause
an exponential decrease of delayed fluorescence intensity by raising ΔE.

An interesting by-product of this suggestion, also anticipated by Fleischman, is related to the observation that NAD^+ reduction in photosynthetic
bacteria depends on the conditions for phosphorylation.[53] Under conditions
of high membrane potential (usually corresponding to a high level of
$\ominus$), the energy of A^- might be raised high enough to drive the reduction
of NAD^+ through ferredoxin. This would compete with the formation
of $\ominus$, which would become depleted. The declining membrane potential
would then bring the level of A^- down to where ferredoxin reduction could
not happen, and the flow of electrons from A^- would be directed again
toward cyclic phosphorylation (formation of $\ominus$). This picture not only
explains the "energy-linked" nature of NAD^+ reduction, it also suggests
a mechanism for the regulated distribution of energy into ATP and NADH.

The foregoing is not the only way to explain how delayed fluorescence
intensity can be related to membrane potential, but it has the heuristic
value of concreteness.

Acknowledgment

I am indebted to Dr. D. E. Fleischman for many illuminating discussions, and for giving me permission to describe some of his most recent ideas in advance of their publication by him.

References

1. G. Hind and J. M. Olson, *Ann. Rev. Plant Physiol.*, **19**, 249 (1968).
2. M. Avron and J. Neumann, *Ann. Rev. Plant Physiol.*, **19**, 137 (1968).
3. N. K. Boardman, *Advances in Enzymol.*, **30**, 1 (1968).
4. P. Joliot, *Photochem. Photobiol.*, **10**, 309 (1969).
5. B. Kok, B. Forbush and M. McGloin, *Photochem. Photobiol.*, in press (1970).
6. T. Yamashita and W. L. Butler, *Plant Physiol.*, **44**, 435 and 1342 (1969).
7. R. L. Heath and G. Hind, *Biochim. Biophys. Acta*, **172**, 290 (1969).
8. G. Döring, G. Renger, J. Vater and H. T. Witt, *Zeitschrift F. Naturforsch.*, **24b**, 1139 (1969).
9. R. Emerson and W. Arnold, *J. Gen. Physiol.*, **15**, 391; and **16**, 191 (1932).
10. C. B. van Niel, *Cold Spring Harbor Symp. Quant. Biol.*, **3**, 138 (1935).
11. L. N. M. Duysens, *Transfer of Excitation Energy in Photosynthesis*, Thesis, State University, Utrecht, The Netherlands, 1952.
12. J. C. Goedheer, *Biochim. Biophys. Acta*, **38**, 389 (1960).
13. L. N. M. Duysens, W. J. Huiskamp, J. J. Vos and J. M. van der Hart, *Biochim. Biophys. Acta*, **19**, 188 (1956).
14. R. K. Clayton, *Photochem. Photobiol.*, **1**, 201 (1962).
15. R. K. Clayton, *Biochim. Biophys. Acta*, **75**, 312 (1963).
16. P. A. Loach, G. M. Androes, A. F. Maksim and M. Calvin, *Photochem. Photobiol.*, **2**, 443 (1963).
17. R. K. Clayton, *Brookhaven Symp. in Biol.*, **19**, 62 (1967).
18. D. W. Reed and R. K. Clayton, *Biochem. Biophys. Research Commun.*, **30**, 471 (1968).
19. R. K. Clayton and R. T. Wang, *Reaction Centers from Rhodopseudomonas spheroides*, in *Methods in Enzymology*, Eds. S. P. Colowick and N. O. Kaplan (Volume "Photosynthesis and Nitrogen Fixation", Ed. A. San Pietro), in press, 1971.
20. D. W. Reed, *J. Biol. Chem.*, **244**, 4936 (1969).
21. G. Gingras and G. Jolchine, in *Progress in Photosynthesis Research*, *I* (Ed. H. Metzner), p. 209. Verlag G. Lichtenstern, Munich, Germany, 1968.
22. J. P. Thornber, J. M. Olson, D. M. Williams and M. L. Clayton, *Biochim. Biophys. Acta*, **172**, 351 (1969).
23. K. Sauer, E. Z. Dratz and L. Coyne, *Proc. Natl. Acad. Sci. U.S.*, **61**, 17 (1968).
24. J. McElroy, G. Feher and D. Mauzerall, *Biochim Biophys. Acta*, **172**, 180 (1969).
25. J. R. Bolton, R. K. Clayton and D. W. Reed, *Photochem. Photobiol.*, **9**, 209 (1969).
26. J. McElroy, G. Feher and D. Mauzerall, *Biophys. J.*, **10**, 204a (1970).
27. R. K. Clayton, *Photochem. Photobiol.*, **5**, 679 (1969).
28. K. L. Zankel, D. W. Reed and R. K. Clayton, *Proc. Natl. Acad. Sci. U.S.*, **61**, 1243 (1969).
29. D. W. Reed, K. L. Zankel and R. K. Clayton, *Proc. Natl. Acad. Sci. U.S.*, **63**, 42 (1969).
30. S. Malkin and B. Kok, *Biochim. Biophys. Acta*, **126**, 413 (1966).

31. B. Kok, S. Malkin, O. Owens and B. Forbush, *Brookhaven Symp. Biol.*, **19**, 446 (1967).
32. W. A. Cramer, *Biochim. Biophys. Acta*, **189**, 54 (1969).
33. S. J. Strickler and R. A. Berg, *J. Chem. Phys.*, **37**, 814 (1962).
34. W. Arnold and R. K. Clayton, *Proc. Natl. Acad. Sci. U.S.*, **46**, 769 (1960).
35. B. C. Mayne and R. K. Clayton, *Proc. 10th Annual Meeting Biophys. Soc.*, WC10 (1966).
36. Unpublished experiments in the laboratories of D. E. Fleischman and R. K. Clayton.
37. W. W. Parson and G. D. Case, *Biochim. Biophys. Acta*, **205**, 232 (1970).
38. T. Kihara and B. Chance, *Biochim. Biophys. Acta*, **189**, 116 (1969).
39. R. K. Clayton, *Biochem. Biophys. Research Commun.*, **9**, 49 (1962).
40. B. Ke, *Biochim. Biophys. Acta*, **172**, 583 (1969).
41. D. deVault and B. Chance, *Biophys. J.*, **6**, 825 (1966).
42. W. W. Parson, *Biochim. Biophys. Acta*, **153**, 248 (1968).
43. W. W. Parson, *Biochim. Biophys. Acta*, **189**, 384 (1969).
44. W. W. Parson, *Biochim. Biophys. Acta*, **189**, 397 (1969).
45. G. L. Nicolson and R. K. Clayton, *Photochem. Photobiol.*, **9**, 395 (1969).
46. P. A. Loach, *Biochem.*, **5**, 592 (1966).
47. B. Kok, in *Currents in Photosynthesis*, Ed. J. B. Thomas and J. C. Goedheer, p. 383. A. Donker, Rotterdam, 1966.
48. J. M. Olson and B. Chance, *Arch. Biochem. Biophys.*, **88**, 26 and 40 (1960).
49. M. A. Cusanovitch and R. G. Bartsch, *Biochim. Biophys. Acta*, **189**, 245 (1969).
50. R. G. Bartsch, *Ann. Rev. Microbiol.*, **22**, 181 (1968).
51. M. E. Pullman and G. Schatz, *Ann. Rev. Biochem.*, **36**, 652 (1967).
52. P. Mitchell, *Biol. Revs.*, **41**, 445 (1966).
53. D. L. Keister and N. J. Yike, *Arch. Biochem. Biophys.*, **121**, 415 (1967).
54. A. S. Holt and R. K. Clayton, *Photochem. Photobiol.*, **4**, 829 (1965).
55. M. C. W. Evans, in *Progress in Photosynthesis Research*, Ed. H. Metzner, Vol. III, p. 1474. Verlag C. Lichtenstern, Munich, Germany, 1968.
56. L. N. M. Duysens, in *Research in Photosynthesis*, Ed. H. Gaffron, p. 164. Interscience, New York, 1957.
57. S. Morita, *Biochim. Biophys. Acta*, **153**, 241 (1968).
58. C. Sybesma and C. F. Fowler, *Proc. Natl. Acad. Sci. U.S.*, **61**, 1343 (1968).
59. C. Sybesma, *Biochim. Biophys. Acta*, **172**, 177 (1969).
60. C. Sybesma and B. Kok, *Biochim. Biophys. Acta*, **180**, 410 (1969).
61. R. K. Clayton, *Proc. Natl. Acad. Sci. U.S.*, **50**, 583 (1963).
62. M. A. Cusanovitch, R. G. Bartsch and M. D. Kamen, *Biochim. Biophys. Acta*, **153**, 397 (1968).
63. C. D. Greville, *Current Topics in Bioenergetics*, **3**, 1 (1969).
64. L. Packer, D. W. Deamer and A. R. Crofts, *Brookhaven Symp. Biol.*, **19**, 281 (1967).
65. G. Hind and A. T. Jagendorf, *J. Biol. Chem.*, **240**, 3195 (1965).
66. L.-V. von Stedingk and H. Baltscheffsky, *Arch. Biochem. Biophys.*, **117**, 400 (1966).
67. L.-V. von Stedingk, *Arch. Biochem. Biophys.*, **120**, 537 (1967).
68. L.-V. von Stedingk, in *Progress in Photosynthesis Research*, Ed. H. Metzner, Vol. III, p. 1410. Verlag C. Lichtenstern, Munich, Germany (1968).
69. L. Smith and J. Ramirez, *Arch. Biochem. Biophys.*, **79**, 233 (1959).
70. R. K. Clayton, *Photochem. Photobiol.*, **1**, 313 (1962).
71. D. E. Fleischman and R. K. Clayton, *Photochem. Photobiol.*, **8**, 287 (1968).

72. M. Baltscheffsky, *Arch. Biochem. Biophys.*, **133**, 46 (1969).
73. J. B. Jackson and A. R. Crofts, *F.E.B.S. Letters*, **4**, 185 (1969).
74. H. M. Emrich, W. Junge and H. T. Witt, *Zeitschr. f. Naturforsch.*, **24b**, 1144 (1969).
75. P. D. Boyer, in *Biological Oxidation*, Ed. T. P. Singer, p. 193. Wiley, New York, 1967.
76. R. A. Dilley, in *Progress in Photosynthesis Research*, Ed. H. Metzner, Vol. III, p. 1354. Verlag C. Lichtenstern, Munich, Germany, 1968.
77. R. E. McCarty, *J. Biol. Chem.* **244**, 4292 (1969).
78. J. B. Jackson, A. R. Crofts and L.-V. von Stedingk, *Eur. J. Biochem.*, **6**, 41 (1968).
79. A. Thore, D. L. Keister, N. Shavit and A. San Pietro, *Biochem.*, **7**, 3499 (1968).
80. A. T. Jagendorf and E. Uribe, *Proc. Natl. Acad. U.S.*, **55**, 170 (1960).
81. J. B. Chappell and A. R. Crofts, *Biochem. J.*, **95**, 393 (1965).
82. J. B. Chappell and K. N. Haarhoff, in *Biochemistry of Mitochondria*, Ed. E. C. Slater, Z. Kanuiga and L. Wojtczak, p. 75. Academic Press, New York, 1967.
83. B. C. Pressman, E. J. Harris, W. S. Jagger and J. H. Johnson, *Proc. Natl. Acad. Sci. U.S.*, **58**, 1949 (1967).
84. P. J. F. Henderson and J. B. Chappell, *Biochem. J.*, **105**, 16P (1967).
85. P. Mueller and D. O. Rudin, *Biochem. Biophys. Research Commun.*, **26**, 398 (1967).
86. L. Packer, *Biochem. Biophys. Research Commun.*, **28**, 1022 (1967).
87. N. Shavit and A. San Pietro, in *Progress in Photosynthesis Research*, Ed. H. Metzner, Vol. III, p. 1392. Verlag C. Lichtenstern, Munich, Germany, 1968.
88. M. Nishimura and B. C. Pressman, *Biochem.*, **8**, 1360 (1969).
89. S. J. D. Karlish, N. Shavit and M. Avron, *Eur. J. Biochem.*, **9**, 291 (1969).
90. A. R. Crofts, *Biochem. Biophys. Research Commun.*, **24**, 127 and 725 (1966).
91. G. E. Edwards and C. R. Bovell, *Biochim. Biophys. Acta*, **172**, 126 (1969).
92. G. Cohen-Bazire and R. Kunisawa, *J. Cell Biol.*, **16**, 401 (1964).
93. W. J. Vredenberg and J. Amesz, *Brookhaven Symp. Biol.*, **19**, 49 (1967).
94. H. M. Emrich, W. Junge and H. T. Witt, *Naturwissenschaften*, **56**, 514 (1969).
95. B. L. Strehler and W. Arnold, *J. Gen. Physiol.*, **34**, 809 (1951).
96. W. Arnold, *Science*, **154**, 1046 (1966).
97. B. C. Mayne and R. K. Clayton, *Proc. Natl. Acad. Sci. U.S.*, **55**, 494 (1966).
98. B. C. Mayne, *Photochem. Photobiol.*, **8**, 107 (1968).
99. D. E. Fleischman and J. A. Cooke, *Biophys. J.*, **9**, A-27 (1969).
100. D. E. Fleischman, in *Progress in Photosynthesis Research*, Ed. H. Metzner, Vol. II, p. 952. Verlag. C. Lichtenstern, Munich, Germany, 1968.
101. D. E. Fleischman, submitted to *Photochem. Photobiol.*, 1970.
102. R. K. Clayton, *J. Gen. Physiol.*, **48**, 633 (1965).
103. R. K. Clayton and W. F. Bertsch, *Biochem. Biophys. Research Commun.*, **18**, 415 (1965).
104. A. B. Rubin and L. K. Osnitskaya, *Mikrobiologia*, **32**, 200 (1963).
105. Govindjee and J. Hammond, verbal communication.
106. R. T. Wang, *Biophys. J.*, **10**, 201a (1970).
107. J. C. Goedheer and G. R. Vegt, *Nature*, **193**, 875 (1962).
108. R. K. Clayton, *Biophys. J.*, **9**, 60 (1969).
109. P. B. Sweetser, C. W. Todd and R. T. Hersh, *Biochim. Biophys. Acta*, **51**, 509 (1961).
110. B. C. Mayne, *Photochem. Photobiol.*, **6**, 189 (1967).
111. J. Lavorel, in *Progress in Photosynthesis Research*, Ed. H. Metzner, Vol. II, p. 883. Verlag C. Lichtenstern, Munich, Germany, 1968.
112. D. E. Fleischman, *Biophys. J.*, **10**, 204a (1970).
113. D. E. Fleischman and A. R. Crofts, verbal communication by D.E.F.

AUTHOR INDEX

Numbers in parentheses are reference numbers and show that an author's work is referred to although his name is not mentioned in the text. Numbers in *italics* indicate the pages on which the full references appear.

SUBJECT INDEX

389

* added paragraph
** added paragraph

* added paragraph